Recent Advances in Nonlinear Analysis

Proceedings of the
International Conference on Nonlinear Analysis

Recent Advances in
Nonlinear Analysis

Hsinchu, Taiwan • 20–25 November 2006

Editors

Michel Chipot
University of Zurich, Switzerland

Chang-Shou Lin
National Taiwan University, Taiwan

Dong-Ho Tsai
National Tsing Hua University, Taiwan

World Scientific

NEW JERSEY • LONDON • SINGAPORE • BEIJING • SHANGHAI • HONG KONG • TAIPEI • CHENNAI

Published by

World Scientific Publishing Co. Pte. Ltd.

5 Toh Tuck Link, Singapore 596224

USA office: 27 Warren Street, Suite 401-402, Hackensack, NJ 07601

UK office: 57 Shelton Street, Covent Garden, London WC2H 9HE

British Library Cataloguing-in-Publication Data
A catalogue record for this book is available from the British Library.

RECENT ADVANCES IN NONLINEAR ANALYSIS
Proceedings of the International Conference on Nonlinear Analysis

ISBN-13 978-981-270-924-0
ISBN-10 981-270-924-X

Printed in Singapore by B & JO Enterprise

PREFACE

This book collects different articles of research on nonlinear analysis. These results were presented in the framework of the International Conference on Nonlinear Analysis which took place in the National Center for Theoretical Sciences at the National Tsing Hua University in Hsinchu, Taiwan in November 2006.

This conference which was a very successful event has covered a large variety of topics in partial differential equations that the reader will discover by looking shortly at the content.

We would like to thank the National Center for Theoretical Sciences for its support in organizing this meeting. Our appreciation goes also to Mrs Schacher who arranged the final version of the whole book and to Ms Zhang and World Scientific for their editing work.

Zürich, November 2007

Michel Chipot
Chang-Shou Lin
Dong-Ho Tsai

CONTENTS

A MATHEMATICAL MODEL FOR BACTERIAL GROWTH DESCRIBED BY A HYSTERESIS OPERATOR

TOYOHIKO AIKI

Department of Mathematics, Faculty of Education, Gifu University, Gifu, 501-1193, Japan

JANA KOPFOVÁ

Mathematical Institute, Silesian University in Opava, 746 01 Opava, Czech Republic

Abstract: We consider a mathematical model for a bacterial growth in a Petri dish. The model consists of partial differential equations and a hysteresis operator describing the relationship between some variables of the equations. We consider the hysteresis relation to be represented by a completed relay operator. We prove the existence of a solution to the system by using the standard approximation method.

1. Introduction

In this paper we consider the mathematical model for a bacterial growth in a Petri dish. The model was proposed by Hoppensteadt-Jäger [2], Hoppensteadt-Jäger-Poppe [3]. Let h be the histidine (amino acid) concentration in a Petri dish, b be the size of the bacterial population, g be the concentration of the growth mediums buffer and v be the metabolic activity of bacteria. In this model the metabolic activity of bacteria v plays a very important and interesting role, and is decided by a hysteresis operator with input functions h and g. We denote by $\Omega \subset \mathbb{R}^N$, $N \geq 2$, the bounded domain with the smooth boundary and put $Q(T) = (0,T) \times \Omega$, $0 < T < \infty$. In [2, 3] they suppose that the growth rate of the bacteria is in proportion to the metabolic activity, that is,

$$\frac{\partial b}{\partial t} = cvb \quad \text{in } Q(T), \tag{1.1}$$

where c is a positive constant. Also, in their model it is supposed that both the histidine and the buffer diffuse and are consumed by the growth of the bacteria. Then, it holds that

$$\frac{\partial h}{\partial t} = D_h \triangle h - \beta v b, \frac{\partial g}{\partial t} = D_g \triangle g - \gamma v b \quad \text{in } Q(T), \tag{1.2}$$

where D_h and D_g are diffusion coefficients of the histidine and the buffer, respectively, and β and γ are positive constants.

The problem was also investigated by Visintin in [7] and Chapter 11 in [8]. Here, we discuss the mathematical description of the metabolic activity along Visintin's idea (Chapters 6 and 11 in [8]). Let $\varphi = \varphi(h, g)$ be a continuous function with $0 \le \varphi \le 1$ on $[0, \infty) \times [0, \infty)$, and $M_0 = \{(h, g) \in \mathbb{R}^2 : \varphi(h, g) = 0\}$, $M = \{(h, g) \in \mathbb{R}^2 : 0 < \varphi(h, g) < 1\}$, $M_1 = \{(h, g) \in \mathbb{R}^2 : \varphi(h, g) = 1\}$ (see Figure 1). Next, we denote by r the corresponding relay operator. For any $u \in C([0, T])$, $0 < T < \infty$, and any $\xi \in \{0, 1\}$, we define the function $w = r(u, \xi) : [0, T] \to \{0, 1\}$ as follows (see Figure 2):

$$w(0) = \begin{cases} 0 & \text{if } u(0) \le 0, \\ \xi & \text{if } 0 < u(0) < 1, \\ 1 & \text{if } u(0) \ge 1; \end{cases}$$

for any $t \in (0, T]$, setting $X_t = \{\tau \in (0, t] : u(\tau) = 0 \text{ or } 1\}$,

$$w(t) = \begin{cases} w(0) & \text{if } X_t = \emptyset, \\ 0 & \text{if } X_t \ne \emptyset \text{ and } u(\max X_t) = 0, \\ 1 & \text{if } X_t \ne \emptyset \text{ and } u(\max X_t) = 1. \end{cases}$$

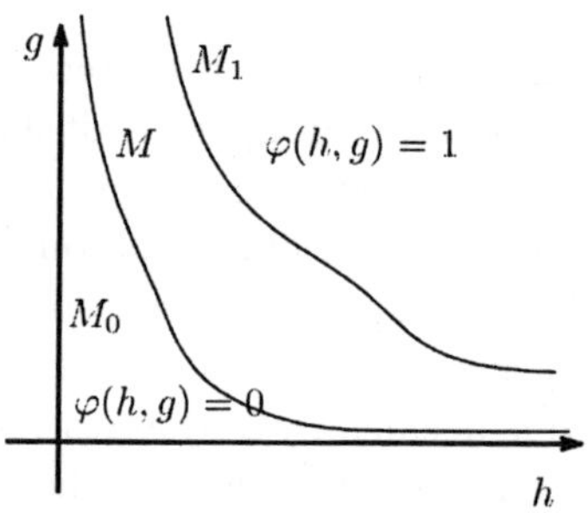

(Figure 1.)

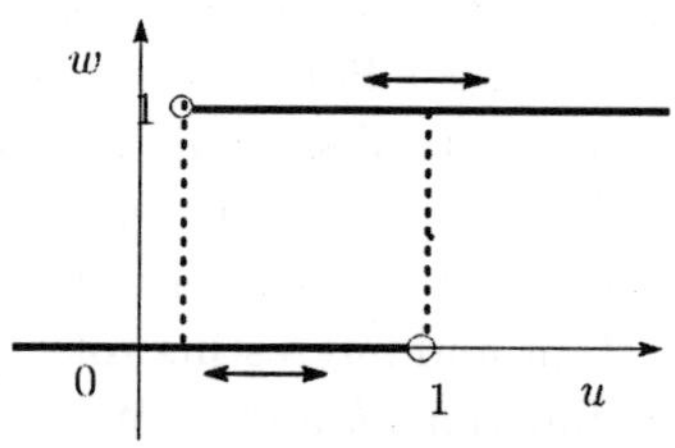

(Figure 2.)

By using these notations it is assumed that

$$v = r(\varphi(h, g), \xi), \tag{1.3}$$

where $\xi \in \{0, 1\}$ is a given number as an initial condition. This means that if both h and g are sufficiently large, that is, $(h, g) \in M_1$, then the bacteria can grow ($v = 1$). Also, if both h and g are sufficiently small, that is, $(h,g) \in M_0$, then the bacteria can not grow ($v = 0$). Otherwise ($(h,g) \in M$), the metabolic activity is decided by the historical data. Hence, we have obtained the system consisting of (1.1), (1.2), (1.3), the homogeneous Neumann boundary condition (1.4) and initial condition (1.5):

$$\frac{\partial h}{\partial \nu} = \frac{\partial g}{\partial \nu} = 0 \quad \text{on } (0,T) \times \partial\Omega, \tag{1.4}$$

where ν is the outward normal vector to the boundary;

$$b(0,x) = b_0(x), h(0,x) = h_0(x), g(0,x) = g_0(x), \tag{1.5}$$

where b_0, h_0 and g_0 are given functions.

The mathematical treatment of the relay operator is difficult, because this operator is closed, we therefore follow the idea of A. Visintin and replace it with a completed relay operator.

considered the system with the completed relay operator instead of the relay operator in [7]. Now, we introduce the completed relay operator, which is denoted by k_ρ. For any $\rho = (\rho_1, \rho_2) \in \mathbb{R}^2$ with $\rho_1 < \rho_2$, $u \in C([0,T])$ and $\xi \in [-1,1]$, $w \in k_\rho(u,\xi)$ if and only if w is measurable in $(0,T)$,

$$\begin{cases} \text{if } u(t) \neq \rho_1, \rho_2, \text{ then } w \text{ is constant in a neighbourhood of } t, \\ \text{if } u(t) = \rho_1, \text{ then } w \text{ is non-increasing in a neighbourhood of } t, \\ \text{if } u(t) = \rho_2, \text{ then } w \text{ is nondecreasing in a neighbourhood of } t, \end{cases} \tag{1.6}$$

$$w(0) \in \begin{cases} \{-1\} & \text{if } u(0) < \rho_1, \\ [-1,\xi] & \text{if } u(0) = \rho_1, \\ \{\xi\} & \text{if } \rho_1 < u(0) < \rho_2, \\ [\xi,1] & \text{if } u(0) = \rho_2, \\ \{1\} & \text{if } u(0) > \rho_2, \end{cases} \tag{1.7}$$

$$w(t) \in \begin{cases} \{-1\} & \text{if } u(0) < \rho_1, \\ [-1,1] & \text{if } \rho_1 \leq u(0) \leq \rho_2, \\ \{1\} & \text{if } u(0) > \rho_2. \end{cases}$$

It is clear that $w \in BV(0,T)$. Thus the multi-valued operator $k_\rho : C([0,T]) \to BV([0,T])$ is well-defined.

Moreover, we assume that the bacteria diffuse:

$$\frac{\partial b}{\partial t} = D_b \Delta b + cvb \text{ in } Q(T), \frac{\partial b}{\partial \nu} = 0 \text{ on } (0,T) \times \partial\Omega, \tag{1.8}$$

where D_b is a positive constant.

Here, we give a brief list of previous works related to the above systems. In [2, 3] numerical results for the system (1.1) $\sim$ (1.5) were obtained. Also, Visintin proved the existence of a solution to the following system in [7]: For $i = 1, 2$

$$\begin{cases} u_{it} - D_i \Delta u_i + c_i s = 0 \text{ in } Q(T), \\ s = \frac{w+1}{2} \text{ on } Q(T), \\ w \in k_\rho(\varphi(u_1, u_2), 2s_0 - 1) \text{ in } Q(T), \\ \frac{\partial u_i}{\partial \nu} = 0 \text{ on } \partial\Omega \times (0, T), \\ u_i(0) = u_{0i}, \end{cases} \quad \begin{cases} b_t = cs \text{ in } Q(T), \\ b(0) = b_0 \text{ in } Q(T), \end{cases}$$

where $\rho_1 = 0, \rho_1 = 1$, c, D_i and c_i are positive constants, $i = 1, 2$, s_0 and b_0 are initial values of s and b, respectively. Although there are not direct relations with bacterial growth, Visintin studied the following problems:

$$\begin{cases} u_t - \Delta u + w = f & \text{in } Q(T), \\ w \in k_\rho(u, w_0) & \text{in } Q(T), \\ u = 0 & \text{on } (0, T) \times \partial\Omega, \\ u(0) = u_0, \end{cases} \quad \begin{cases} u_t - \Delta u + w = f & \text{in } Q(T), \\ w \in r(u, w_0) & \text{in } Q(T), \\ u = 0 & \text{on } (0, T) \times \partial\Omega, \\ u(0) = u_0, \end{cases} \tag{1.9}$$

and

$$\begin{cases} (u + w)_t - \Delta u = g & \text{in } Q(T), \\ w \in k_\rho(u, w_0) & \text{in } Q(T), \\ u = 0 & \text{on } (0, T) \times \partial\Omega, \\ u(0) = u_0, \end{cases} \tag{1.10}$$

where f and g are given functions on $Q(T)$, w_0 is the initial function. For (1.9) the existence theorem was established and for (1.10) the existence and uniqueness were proved in [7, 8].

In the present paper we have slightly modified Visintin's result. Precisely, our main result of this paper is to prove the existence of a solution to the system P := { (1.8), (1.2), (1.11), (1.4), (1.5)}:

$$v = \frac{w+1}{2}, \text{ and } w \in k_\rho(\varphi(h, g), w_0) \text{ on } Q(T). \tag{1.11}$$

In the next section we show the assumptions for data and precise statement of the main result, after we provide the definition and basic properties on the completed relay operator due to [8]. At the end of this paper we prove the existence theorem.

2. Statement of the result

First, we provide a weak and space structured systems of completed relay operator due to [8]. To do so we introduce a function space and the following two continuous functions α_ρ and β_ρ. We denote by $L^2_{w^*}(\Omega; BV(0,T))$ the set of all functions $w : \Omega \to BV(0,T)$ satisfying $\Omega \ni x \to \int_{[0,T]} f dw(x) + \xi \int_0^T w(x)dt$ is measurable on Ω for any $(\xi, f) \in \mathbb{R} \times C_0([0,T])$ and $\|w(x)\|_{BV(0,T)} \in L^2(\Omega)$, where $\int_{[0,T]} f dw(x)$ is the Lebesgue-Stieltjes integral, $C_0([0,T]) = \{f \in C([0,T]); f(0) = f(T) = 0\}$, $\|v\|_{BV(0,T)} = \int_0^T |v|dx + \mathrm{Var}(v)$ and $\mathrm{Var}(v)$ stands the total variation of v on $[0,T]$. Clearly, $L^2_{w^*}(\Omega; BV(0,T))$ is the Banach space with norm $(\int_{[0,T]} \|w(x)\|^2_{BV(0,T)} dx)^{1/2}$. For any $\rho = (\rho_1, \rho_2) \in \mathbb{R}^2$ with $\rho_1 < \rho_2$ we set $\alpha_\rho(\xi) = (\xi - \rho_2)^+ - (\xi - \rho_1)^-, \beta_\rho(\xi) = \xi - \alpha_\rho(\xi)$ for $\xi \in \mathbb{R}$.

Definition 2.1. For any $u \in L^2(\Omega; C([0,T])) \cap L^2_{w^*}(\Omega; BV(0,T))$ and any $\xi \in L^\infty(\Omega; [-1,1])$ $w \in k_\rho(u, \xi)$ if and only if $w \in L^2_{w^*}(\Omega; BV(0,T))$, (1.7) holds a.e. in Ω, and $w \in L^2_{w^*}(\Omega; BV(0,T))$,

$$\int_{Q(T)} w(\alpha_\rho(u) - \eta)dxdt \geq \int_{Q(T)} (|\alpha_\rho(u)| - |\eta|)dxdt \quad \text{for } \eta \in L^1(Q(T)),$$

$$\int_\Omega dx \int_{[0,T]} (\beta_\rho(u) - \beta_\rho(z))dw \geq 0 \text{ for } z \in L^1(\Omega; C([0,T])).$$

Next, for simplicity we can rewrite P as follows:

$$b_t - \kappa \Delta b = cbv \quad \text{on } Q(T), \tag{2.1}$$

$$u_{it} - \kappa_i \Delta u_i + c_i bv = 0 \quad \text{on } Q(T), i = 1,2, \tag{2.2}$$

$$v = \frac{w+1}{2}, w \in k_\rho(\varphi(u_1, u_2), w_0) \quad \text{on } Q(T), \tag{2.3}$$

$$\frac{\partial b}{\partial \nu} = 0, \frac{\partial u_i}{\partial \nu} = 0, i = 1,2, \quad \text{on } (0,T) \times \partial\Omega, \tag{2.4}$$

$$b(0,x) = b_0(x), u_i(0,x) = u_{0i}(x), i = 1,2, \tag{2.5}$$

where κ, c, κ_i, c_i, $i = 1,2$, are positive constants, v_0, b_0, u_{0i}, $i = 1,2$, are initial functions and $w_0 = 2v_0 - 1$.

We define a solution of the system (2.1) $\sim$ (2.5) in the following way.

Definition 2.2. We call that the quadruplet $\{b, u_1, u_2, v\}$ of functions, b, u_1, u_2 and v on $Q(T)$, $T > 0$, is a solution of P on $Q(T)$ if and only if the conditions (C1), (C2), (C3) hold:

(C1) $b \in S(T) := W^{1,2}(0,T; L^2(\Omega)) \cap L^\infty(0,T; H^1(\Omega))$, $u_i \in S(T)$, $i = 1,2$, $v \in L^\infty(Q) \cap L^2_{w^*}(\Omega; BV(0,T))$, and (2.3) holds.

(C2) (2.1), (2.2) and (2.4) hold in the usual weak sense.

(C3) (2.5) holds for a.e. $x \in \Omega$.

In order to give a main result of this paper as Theorem 2.1 we prepare the following notations.

$$f_\rho^*(r) = \begin{cases} 1 & \text{for } r \geq \rho_1, \\ 0 & \text{for } r < \rho_1, \end{cases} \quad f_{\rho*}(r) = \begin{cases} 1 & \text{for } r > \rho_2, \\ 0 & \text{for } r \leq \rho_2. \end{cases}$$

Theorem 2.1. *Let $T > 0$, $\rho = (\rho_1, \rho_2) \in \mathbb{R}^2$, κ, κ_1 and κ_2 be positive numbers, c, c_1 and c_2 be real numbers, and $b_0 \in H^1(\Omega)$, $u_{0i} \in H^1(\Omega)$, $i = 1, 2$, $v_0 \in L^\infty(\Omega)$. Also, assume that $\varphi \in C(\mathbb{R}^2) \cap W^{1,\infty}(\mathbb{R}^2)$, and the compatibility condition for the initial data, $f_{\rho*}(\varphi(u_{01}, u_{02})) \leq w_0 \leq f_\rho^*(\varphi(u_{01}, u_{02}))$ on Ω , where $w_0 = 2v_0 - 1$. Then there exists at least one solution $\{b, u_1, u_2, v\}$ of P on $Q(T)$.*

3. Completed relay operator

In the proof of Theorem 2.1 we shall approximate the completed relay operator k_ρ by play operators. The aim of this section is to introduce the approximation of k_ρ and to show a useful property for the approximation.

First, we show a boundedness of the operator k_ρ.

Lemma 3.1. *(cf. [8]) Let $\rho = (\rho_1, \rho_2)$ with $\rho_1 < \rho_2$ and $\xi \in L^\infty(0, 1)$. If $\varphi \in L^2(\Omega; BV(0, T))$ and $w \in k_\rho(\varphi, \xi)$, then*

$$\Big(\int_\Omega \|w(x)\|^2_{BV(0,T)} dx\Big)^{1/2} \leq \frac{2}{\rho_2 - \rho_1}\Big(\int_\Omega \|\varphi(x)\|^2_{BV(0,T)} dx\Big)^{1/2} + (T+2)|\Omega|^{1/2}.$$

Next, for any $\varepsilon > 0$ we put

$$f_{\rho\varepsilon}^*(r) = \begin{cases} 1 & \text{for } r \geq \rho_1, \\ \frac{1}{\varepsilon}(r - \rho_1) + 1 & \text{for } \rho_1 - 2\varepsilon < r < \rho_1, \\ -1 & \text{for } r \leq \rho_1 - 2\varepsilon, \end{cases}$$

$$f_{\rho\varepsilon*}(r) = \begin{cases} 1 & \text{for } r \geq \rho_2 + 2\varepsilon, \\ \frac{1}{\varepsilon}(r - \rho_2) - 1 & \text{for } \rho_2 < r < \rho_2 + 2\varepsilon, \\ -1 & \text{for } r \leq \rho_2, \end{cases}$$

and define by $I_\varepsilon(\varphi; \cdot) : L^2(\Omega) \to (-\infty, +\infty]$ the indicator function of the interval $[f_{\rho\varepsilon*}(\varphi), f_{\rho\varepsilon}^*(\varphi)]$ for $\varphi \in \mathbb{R}$, that is, for $z \in L^2(\Omega)$

$$I_\varepsilon(\varphi; z) = \int_\Omega \xi dx, \xi(x) \begin{cases} 0 & \text{if } z(x) \in [f_{\rho\varepsilon*}(\varphi), f_{\rho\varepsilon}^*(\varphi)], \\ \infty & \text{otherwise}, \end{cases} \quad \text{for a.e. } x \in \Omega.$$

Here, we consider the following ordinary differential equation,

$$w_t + \partial I_\varepsilon(\varphi; w) \ni 0 \text{ on } [0, T], w(0) = w_0, \tag{3.1}$$

where $\partial I_\varepsilon(\varphi; w)$ is its subdifferential, φ is a continuous function on $[0,T]$ and w_0 is the initial value of w. If $\varphi \in W^{1,2}(0,T;L^2(\Omega))$, then there exists a unique solution $w \in W^{1,2}(0,T;L^2(\Omega))$ of (3.1). Also, it is well-known that in this case the mapping from φ to w corresponds to the play operator. This fact was already pointed out in [8] and used for the analysis for the system including play and stop operators in [8, 6, 4, 1, 5]. Moreover, we can obtain the next lemma concerned with the relationship between the completed operator and the play operator. We note that we can find the assertion of the lemma as a comment in [9] .

Lemma 3.2. *Let $T > 0$, $\rho = (\rho_1, \rho_2) \in \mathbb{R}^2$ with $\rho_1 < \rho_2$, $\varepsilon > 0$, $\rho_\varepsilon = (\rho_1 - \varepsilon, \rho_2 + \varepsilon) \in \mathbb{R}^2$, $\varphi \in W^{1,2}(0,T;L^2(\Omega))$, and $w_0 \in L^\infty(\Omega)$ with $f_{\rho *}(\varphi(0)) \leq w_0 \leq f^*_\rho(\varphi(0))$ a.e. on Ω. Then, $w \in k_{\rho_\varepsilon}(\varphi - \varepsilon w, w_0)$ if and only if $w \in W^{1,2}(0,T;L^2(\Omega))$ is a solution of*

$$w_t(\cdot,x) + \partial I_\varepsilon(\varphi(\cdot,x); w(\cdot,x)) \ni 0 \text{ on } [0,T] \text{ for a.e. } x \in \Omega,$$

$w(0) = w_0$ *a.e. on* Ω.

4. Proof of Theorem 2.1

Throughout this section we use same notations used in the previous sections.

First, we consider the following approximate problem P_ε for $0 < \varepsilon \leq \varepsilon_0 := \frac{\rho_2 - \rho_1}{4}$:

$$b_{\varepsilon t} - \kappa \Delta b_\varepsilon = c b_\varepsilon v_\varepsilon \quad \text{on } Q(T), \tag{4.1}$$

$$u_{i\varepsilon t} - \kappa_i \Delta u_{i\varepsilon} + c_i b_\varepsilon v_\varepsilon = 0 \quad \text{on } Q(T), i = 1,2, \tag{4.2}$$

$$v_\varepsilon = \frac{w_\varepsilon + 1}{2}, w_{\varepsilon t} + \partial I_\varepsilon(\varphi(u_{1\varepsilon}, u_{2\varepsilon}); w_\varepsilon) \ni 0 \text{ on } [0,T], \tag{4.3}$$

$$w_\varepsilon(0) = w_0 \text{ on } \Omega, \tag{4.4}$$

$$\frac{\partial b_\varepsilon}{\partial \nu} = 0, \frac{\partial u_{i\varepsilon}}{\partial \nu} = 0, i = 1,2, \quad \text{on } (0,T) \times \partial\Omega, \tag{4.5}$$

$$b_\varepsilon(0,x) = b_0(x), u_{i\varepsilon}(0,x) = u_{0i}(x), i = 1,2. \tag{4.6}$$

The first lemma is concerned with the well-posedness for P_ε.

Lemma 4.1. *Let $T > 0$, $0 < \varepsilon \leq \varepsilon_0$. If all assumptions of Theorem 2.1 hold, then there exists one and only one solution $\{b_\varepsilon, u_{1\varepsilon}, u_{2\varepsilon}, w_\varepsilon\}$ of P_ε in the following sense: $b_\varepsilon \in S(T)$, $u_{i\varepsilon} \in S(T)$, $i = 1,2$, $w_\varepsilon \in W^{1,2}(0,T;L^2(\Omega))$; (4.1) $\sim$ (4.6) hold in the usual sense.*

This lemma can be proved in the similar way that of Section 5 in [4] so that we omit its proof. Here, we give the proof of Theorem 2.1.

Proof. For $0 < \varepsilon \le \varepsilon_0$ let $\{b_\varepsilon, u_{1\varepsilon}, u_{2\varepsilon}, w_\varepsilon\}$ be a solution of P_ε. Here, for simplicity we put $\varphi_\varepsilon = \varphi(u_{1\varepsilon}, u_{2\varepsilon})$. Then, since $w_\varepsilon \in [f_{\rho\varepsilon *}(\varphi_\varepsilon), f^*_{\rho\varepsilon}(\varphi_\varepsilon)]$ a.e. on $Q(T)$, we have $|w_\varepsilon| \le 1$ a.e. on $Q(T)$ so that $|v_\varepsilon| \le 1$ a.e. on $Q(T)$. Hence, by the standard argument for parabolic equations we observe that $\{b_\varepsilon\}$, $\{u_{i\varepsilon}\}$, $i = 1, 2$, are bounded in $L^\infty(0, T; H^1(\Omega))$ and $W^{1,2}(0, T; L^2(\Omega))$.

Next, Lemmas 3.1 and 3.2 imply $w_\varepsilon \in k_{\rho_\varepsilon}(\varphi_\varepsilon - \varepsilon w_\varepsilon, w_0)$ and

$$
\begin{aligned}
&(\int_\Omega \|w_\varepsilon\|^2_{BV(0,T)} dx)^{1/2} \\
&\le \frac{2}{\rho_2 - \rho_1}(\int_\Omega \|\varphi_\varepsilon - \varepsilon w_\varepsilon\|^2_{BV(0,T)} dx)^{1/2} + (T+2)|\Omega|^{1/2} \\
&\le \frac{2}{\rho_2 - \rho_1}(\int_\Omega \|\varphi_\varepsilon\|^2_{BV(0,T)} dx)^{1/2} + \frac{\varepsilon}{\rho_2 - \rho_1}(\int_\Omega \|w_\varepsilon\|^2_{BV(0,T)} dx)^{1/2} \\
&\quad + (T+2)|\Omega|^{1/2}.
\end{aligned}
$$

On account of the assumption $0 < \varepsilon \le \varepsilon_0$ we have

$$
\begin{aligned}
&\frac{1}{2}(\int_\Omega \|w_\varepsilon\|^2_{BV(0,T)} dx)^{1/2} \\
&\le \frac{2}{\rho_2 - \rho_1}(\int_\Omega \|\varphi_\varepsilon\|^2_{BV(0,T)} dx)^{1/2} + (T+2)|\Omega|^{1/2} \\
&\le \frac{2}{\rho_2 - \rho_1}(\int_\Omega (\int_0^T (|\varphi_\varepsilon| + |\frac{\partial}{\partial t}\varphi_\varepsilon|) dt)^2 dx)^{1/2} + (T+2)|\Omega|^{1/2} \\
&\le \frac{2T^{1/2}}{\rho_2 - \rho_1}(\int_0^T (|\varphi_\varepsilon|^2_{L^2(\Omega)} + |\frac{\partial}{\partial t}\varphi_\varepsilon|^2_{L^2(\Omega)}) dt)^{1/2} + (T+2)|\Omega|^{1/2}.
\end{aligned}
$$

Hence, by the assumption for φ the set $\{w_\varepsilon\}$ is bounded in $L^2(\Omega; BV(0,T))$. Then by applying Proposition 2.3 mentioned in [8] we can take a subsequence $\{\varepsilon_j\} \subset \{\varepsilon\}$ such that $w_{\varepsilon_j} \to w$ weakly* in $L^2_{w^*}(\Omega; BV(0,T))$ and weakly* in $L^\infty(Q(T))$, $b_{\varepsilon_j} \to b$, $u_{i\varepsilon_j} \to u_i$ weakly* in $L^\infty(0, T; H^1(\Omega))$, weakly in $W^{1,2}(0, T; L^2(\Omega))$ and in $C([0,T]; H)$ as $j \to \infty$ for $i = 1, 2$.

Here, we show that $w \in k_\rho(\varphi(u_1, u_2), w_0)$. In fact, it is clear that $w \in L^2_{w^*}(\Omega; BV(0,T))$ and (1.7) holds a.e. on Ω because of the compatibility condition for the initial data. Accordingly, it is sufficient to prove

$$
\int_{Q(T)} w(\alpha_\rho(\varphi_0) - \eta) dxdt \ge \int_{Q(T)} (|\alpha_\rho(\varphi_0)| - |\eta|) dxdt \text{ for } \eta \in L^1(Q(T)), \tag{4.7}
$$

$$\int_\Omega dx \int_{[0,T]} (\beta_\rho(\varphi_0)) - \beta_\rho(z))dw \geq 0 \text{ for } z \in L^1(\Omega; C([0,T])), \tag{4.8}$$

where $\varphi_0 = \varphi(u_1, u_2)$. For $0 < \varepsilon \leq \varepsilon_0$ we set $\rho_\varepsilon = (\rho_1 - \varepsilon, \rho_2 + \varepsilon)$. Then, we see that for each j and $\eta \in L^1(Q(T))$

$$\int_{Q(T)} w_{\varepsilon_j}(\alpha_{\rho_{\varepsilon_j}}(\varphi_{\varepsilon_j} - \varepsilon_j w_{\varepsilon_j}) - \eta)dxdt \geq \int_Q (|\alpha_{\rho_{\varepsilon_j}}(\varphi_{\varepsilon_j} - \varepsilon_j w_{\varepsilon_j})| - |\eta|)dxdt. \tag{4.9}$$

By letting $j \to \infty$ in (4.9) we can get (4.7) since $w_{\varepsilon_j} \to w$ weakly* in $L^\infty(Q(T))$ and $\varphi(u_{1\varepsilon_j}, u_{2\varepsilon_j}) \to \varphi(u_1, u_2)$ in $C([0,T]; L^2(\Omega))$ as $j \to \infty$.

Next, we prove (4.8). Let $z \in L^1(\Omega; C([0,T]))$. Then for each j it holds that $\int_\Omega dx \int_{[0,T]} (\beta_{\rho_{\varepsilon_j}}(\varphi_{\varepsilon_j} - \varepsilon_j w_{\varepsilon_j}) - \beta_{\rho_{\varepsilon_j}}(z))dw_{\varepsilon_j} \geq 0$. Here, for simplicity we put

$$\begin{aligned}
&\int_\Omega dx \left(\int_{[0,T]} (\beta_{\rho_{\varepsilon_j}}(\varphi_{\varepsilon_j} - \varepsilon_j w_{\varepsilon_j}) - \beta_{\rho_{\varepsilon_j}}(z))dw_{\varepsilon_j} \right. \\
&\qquad\qquad \left. - \int_{[0,T]} (\beta_\rho(\varphi_0) - \beta_\rho(z))dw \right) \\
&= \int_\Omega dx \int_{[0,T]} \beta_{\rho_{\varepsilon_j}}(\varphi_{\varepsilon_j} - \varepsilon_j w_{\varepsilon_j})dw_{\varepsilon_j} - \int_\Omega dx \int_{[0,T]} \beta_\rho(\varphi_0)dw_{\varepsilon_j} \\
&\quad + \int_\Omega dx \int_{[0,T]} \beta_\rho(\varphi_0)dw_{\varepsilon_j} - \int_\Omega dx \int_{[0,T]} \beta_\rho(\varphi_0)dw \\
&\quad - \int_\Omega dx \int_{[0,T]} (\beta_{\rho_{\varepsilon_j}}(\varphi_0) - \beta_\rho(\varphi_0))dw_{\varepsilon_j} \\
&\quad -(\int_\Omega dx \int_{[0,T]} \beta_\rho(\varphi_0)dw_{\varepsilon_j} - \int_\Omega dx \int_{[0,T]} \beta_\rho(\varphi_0)dw) \\
&=: \sum_{i=1}^4 I_{ij}.
\end{aligned}$$

By using some properties of Lebesgue-Stieltjes integral we observe that

$$\begin{aligned}
&I_{1j} \\
&\leq \int_\Omega \|w_{\varepsilon_j}\|_{BV(0,T)} |\beta_{\rho_{\varepsilon_j}}(\varphi_{\varepsilon_j} - \varepsilon_j w_{\varepsilon_j}) - \beta_\rho(\varphi_0)|_{C([0,T])} dx \\
&\leq (\int_\Omega \|w_{\varepsilon_j}\|^2_{BV(0,T)} dx)^{1/2} ((\int_\Omega |\varphi_{\varepsilon_j} - \varphi_0|^2_{C([0,T])} dx)^{1/2} \\
&\quad + \varepsilon_j (\int_\Omega |w_{\varepsilon_j}|^2_{C([0,T])} dx)^{1/2} + (\int_\Omega |\beta_{\rho_{\varepsilon_j}}(\varphi_0) - \beta_\rho(\varphi_0)|^2_{C([0,T])} dx)^{1/2}).
\end{aligned}$$

We note that

$$|\varphi_\varepsilon(x) - \varphi_0(x)|^2_{C([0,T])} \le 2\int_0^T |\varphi_\varepsilon(x) - \varphi_0(x)||\varphi_{\varepsilon t}(x) - \varphi_{0t}(x)|dx$$

for a.e. $x \in \Omega$, and for each j

$$\int_\Omega |\varphi_{\varepsilon_j}(x) - \varphi_0(x)|^2_{C([0,T])}dx \le 2|\varphi_{\varepsilon_j t} - \varphi_{0t}|_{L^2(Q(T))}|\varphi_{\varepsilon_j} - \varphi_0|_{L^2(Q(T))}.$$

By elementary calculations we obtain

$$|\beta_{\rho_{\varepsilon_j}}(\varphi_0) - \beta_\rho(\varphi_0)| \le \varepsilon_j \text{ for } j \text{ and } (t,x) \in Q(T).$$

Accordingly, we have $I_{1j} \to 0$ as $j \to \infty$. Also, it is obvious that $I_{ij} \to 0$ as $j \to \infty$ for $i = 2, 3, 4$. Hence, (4.8) holds. Moreover, it is easy to see that (C2) and (C3) hold. □

References

[1] T. Aiki, E. Minchev and T. Okazaki, A prey - predator model with hysteresis effect, SIAM J. Math. Anal., 36 (2005), 2020-2032.

[2] F. C. Hoppensteadt, W. Jäger, Pattern formation by bacteria, Lecture Notes in Biomathematics 38, 68-81, 1980.

[3] F. C. Hoppensteadt, W. Jger, C. Pöppe, A hysteresis model for bacterial growth patterns, Modelling of patterns in space and time (Heidelberg, 1983), 123–134, Lecture Notes in Biomath., 55, Springer, Berlin, 1984.

[4] N. Kenmochi, E. Minchev and T. Okazaki, On a system of nonlinear PDE's with diffusion and hysteresis effects, Adv. Math. Sci. Appl., 14 (2004), 633-664.

[5] E. Minchev, A diffusion-convection prey-predator model with hysteresis, Math. J. Toyama Univ., 27(2004), 51-69.

[6] E. Minchev, T. Okazaki and N. Kenmochi, Ordinary differential systems describing hysteresis effects and numerical simulations, Abstr. Appl. Anal., 7(2002), 563-583.

[7] A. Visintin, Evolution problems with hysteresis in the source term, SIAM J. Math. Anal., 17(1986), 1113–1138.

[8] A. Visintin, *Differential Models of Hysteresis*, Appl. Math. Sci., Vol. 111, Springer-Verlag, Berlin, 1993.

[9] A. Visintin, Quasilinear first-order PDEs with hysteresis, J. Math. Anal. Appl., 312(2005), 401-419.

ON A VECTOR-VALUED SINGULAR PERTURBATION PROBLEM ON THE SPHERE

NELLY ANDRÉ* AND ITAI SHAFRIR†

1. Introduction

Let Γ_1 and Γ_2 be two disjoint smooth, simple closed curves in $\mathbb{R}^2$ of lengths $l(\Gamma_1)$ and $l(\Gamma_2)$, respectively, such that Γ_1 lies inside Γ_2 and the origin 0 lies inside Γ_1. Let $W : \mathbb{R}^2 \to [0, \infty)$ be a smooth function (i.e., at least of class C^4) satisfying

$$W > 0 \text{ on } \mathbb{R}^2 \setminus (\Gamma_1 \cup \Gamma_2) \quad \text{and} \quad W = 0 \text{ on } \Gamma_1 \cup \Gamma_2 \,. \tag{H_1}$$

Since W attains its minimal value zero on $\Gamma_1 \cup \Gamma_2$ we have clearly $W_n = 0$ on Γ_j, $j = 1, 2$, where W_n denotes the derivative in the direction of the exterior normal to Γ_j. We assume then that we are in the generic case, i.e., that

$$W_{nn} > 0 \text{ on } \Gamma_1 \cup \Gamma_2 \,. \tag{H_2}$$

Finally, we add the following coercivity assumption on the behavior of W at infinity: there exist constants $R_0 > 0$ and $C_0 > 0$ such that

$$W(x) \geq C_0 |x| \text{ for } |x| \geq R_0 \,. \tag{H_3}$$

Let G be either a bounded smooth domain in $\mathbb{R}^N$, or a smooth N-dimensional manifold. For each $\varepsilon > 0$ consider the energy functional

$$E_\varepsilon(u) = \int_G |\nabla u|^2 + \frac{W(u)}{\varepsilon^2} \tag{1.1}$$

for $u \in H^1(G, \mathbb{R}^2)$. Let R_c be a positive number such that the circle $S_{R_c} = \{|x| = R_c\}$ *separates* the two curves Γ_1 and Γ_2. We shall assume

*département de mathématiques, université de tours, 37200 tours, france
†department of mathematics, technion – israel institute of technology, 32000 haifa, israel

w.l.o.g. that Γ_1 lies inside S_{R_c} which lies inside Γ_2. The number R_c represents the constraint in the following minimization problem that we shall study:

$$\min\{E_\varepsilon(u) \, : \, u \in H^1(G,\mathbb{R}^2), \fint_G |u| = R_c\}\,, \tag{P_ε}$$

where $\fint_G |u| := \dfrac{1}{\mu(G)}\displaystyle\int_G |u|$. Denoting by u_ε a minimizer in (P_ε), we are interested in the asymptotic behavior of the minimizers $\{u_\varepsilon\}$ and their energies $E_\varepsilon(u_\varepsilon)$, as ε goes to 0.

A first study of this problem was carried out by Sternberg [7]. He proved a Γ-convergence result, which has the following consequences:

(i) If a subsequence $\{u_{\varepsilon_n}\}$ converges to a limit u_0, then u_0 belongs to the set

$$\mathcal{S} := \{u \in BV(G, \Gamma_1 \cup \Gamma_2) \, : \, \fint_G |u| = R_c\}, \tag{1.2}$$

and

$$\mathrm{Per}_G\{u_0 \in \Gamma_1\} = \min\{\mathrm{Per}_G\{u \in \Gamma_1\} \, : \, u \in \mathcal{S}\}\,. \tag{1.3}$$

(ii) The asymptotic expansion for the energy $E_\varepsilon(u_\varepsilon)$, as $\varepsilon \to 0$, is

$$\varepsilon E_\varepsilon(u_\varepsilon) = 2D \min_{u \in \mathcal{S}} \mathrm{Per}_G\{u \in \Gamma_1\} + o(1)\,. \tag{1.4}$$

We refer to the books [5, 1] for the definition of the perimeter and other notions from the theory of BV-functions, that we shall use in the sequel.

The constant D which appears in (1.4) is a certain "distance" between the two curves Γ_1 and Γ_2 which we shall now define. First, for any pair of points $x, y \in \mathbb{R}^2$ we set

$$d_W(x,y) = \inf_{\substack{\gamma \in \mathrm{Lip}([0,1],\mathbb{R}^2),\\ \gamma(0)=0, \gamma(1)=y}} L(\gamma)\,, \tag{1.5}$$

where

$$L(\gamma) = \int_0^1 \left(W(\gamma(t))\right)^{1/2} |\gamma'(t)| \, dt\,. \tag{1.6}$$

Then, we define

$$D := \inf_{\substack{\gamma \in \mathrm{Lip}([0,1],\mathbb{R}^2),\\ \gamma(0)\in\Gamma_1, \gamma(1)\in\Gamma_2}} L(\gamma)\,. \tag{1.7}$$

It was proved by Sternberg (see [7]) that there exists a geodesic $\underline{\gamma}$ realizing the infimum D in (1.7). There may be of course more than one such geodesic; their number may be infinite (as is the case of Γ_1, Γ_2 which are concentric circles and W radially symmetric). We denote the set of all these geodesics by

$$\mathcal{G} = \{\underline{\gamma}^{(i)} : i \in \mathcal{I}\}, \tag{1.8}$$

where $\mathcal{I}$ is some set of indices. For each $i \in \mathcal{I}$ we denote by $\zeta_1^{(i)} = \underline{\gamma}^{(i)}(0) \in \Gamma_1$ and $\zeta_2^{(i)} = \underline{\gamma}^{(i)}(1) \in \Gamma_2$ the endpoints of the geodesic $\underline{\gamma}^{(i)}$, and then set $Z_j = \{\zeta_j^{(i)}\}_{i\in\mathcal{I}}$, for $j = 1, 2$.

The results of Sternberg left some important questions unresolved:
(1) Existence of a converging subsequence is not known, i.e., a compactness result is missing.
(2) Even if we assume convergence of a subsequence towards a limit u_0, which is a map in $\mathcal{S}$ satisfying (1.3), we cannot say *where* on $\Gamma_1 \cup \Gamma_2$, u_0 takes its values.

In [3] we made some progress on this problem in the case where G is a domain in $\mathbb{R}^N$. In particular, we demonstrated the major role played by the geometry of G in determining the asymptotic behavior of $\{u_\varepsilon\}$. In fact, when G is *convex*, and under some additional technical assumptions, the limit u_0 takes only two values, one in Γ_1 and the other one in Γ_2. On the other hand, when G is *nonconvex*, the limit u_0 may be more complicated (i.e., the restriction of u_0 to $\{x \in G : u_0(x) \in \Gamma_j\}$, $j = 1, 2$, is not necessarily identically constant). However, there is still no complete answer to the above questions in general.

In the present article we shall concentrate on the special case where G is the sphere S^2. Thanks to the symmetry properties of the minimizers $\{u_\varepsilon\}$ in this case, we are able to give a quite complete analysis for both the asymptotic behavior of the minimizers and their energies. We believe that the case of the sphere will give some indication on the expected behavior of the minimizers in the case of a convex domain G. First, notice that from the symmetrization method of O. Lopes, see Theorem IV.3 in [6], it follows that for each ε, the minimizer u_ε is axially symmetric with respect to some axis. We shall assume in the sequel, without loss of generality, a common axis of symmetry for $\{u_\varepsilon\}$, i.e.,

$$\text{each } u_\varepsilon \text{ is symmetric w.r.t. the } \boldsymbol{e}_3 \text{ axis.} \tag{1.9}$$

In view of (1.9) we can view each u_ε as a function of a single variable ϕ, i.e.,

$$u_\varepsilon = u_\varepsilon(\phi), \quad \phi \in [-\pi/2, \pi/2]. \tag{1.10}$$

Next we introduce some notation needed in order to state our results. Set

$$m_j = \min_{x \in \Gamma_j} |x| \quad \text{and} \quad M_j = \max_{x \in \Gamma_j} |x| \; (j = 1, 2)\,, \tag{1.11}$$

and

$$\mathcal{M}_j = \{x \in \Gamma_j \,:\, |x| = m_j\}\,,\; j = 1, 2\,.$$

We shall also assume the following:

$$\text{both } \mathcal{M}_1 \text{ and } \mathcal{M}_2 \text{ consist of a finite number of points}\,. \tag{H_4}$$

Consider any $v \in \mathcal{S}$ and let $G_j = \{x \in S^2 \,:\, v(x) \in \Gamma_j\}$, $j = 1, 2$. Denoting by μ the standard measure on the sphere, we have

$$\mu(G_1)m_1 + \mu(G_2)m_2 \le \mu(S^2)R_c = \int_{S^2} |v| \le \mu(G_1)M_1 + \mu(G_2)M_2\,, \tag{1.12}$$

Clearly, (1.12) and (1.11) imply that

$$\frac{m_2 - R_c}{m_2 - m_1} \le \mu(G_1) \le \frac{M_2 - R_c}{M_2 - M_1}\,. \tag{1.13}$$

Recall now the well-known solution to the isoperimetric problem on the sphere. For a given value $t \in (0, 4\pi)$, the domain on S^2 of area surface t with minimal perimeter is a disk on the sphere with surface t. The perimeter of this disk is given by the function

$$I(t) = \sqrt{4\pi t - t^2}\,, \tag{1.14}$$

which is a concave function on $(0, 4\pi)$, symmetric about the middle point 2π. Therefore, from (1.13) we deduce that in order to obtain G_1 with minimal perimeter we must have:

$$\begin{aligned} &\text{either} \quad (i)\, \mu(G_1)/\mu(S^2) = \frac{m_2 - R_c}{m_2 - m_1} \\ &\text{or} \quad (ii)\, \mu(G_1)/\mu(S^2) = \frac{M_2 - R_c}{M_2 - M_1}\,. \end{aligned} \tag{1.15}$$

In order to know which of the two possibilities in (1.15) is preferable we should check which possibility realizes the minimum in $\min\left(\frac{m_2 - R_c}{m_2 - m_1}, 1 - \frac{M_2 - R_c}{M_2 - M_1}\right)$. Without loss of generality, we shall assume in the sequel that possibility (i) holds in (1.15), i.e., that

$$\alpha := \frac{m_2 - R_c}{m_2 - m_1} < 1 - \frac{M_2 - R_c}{M_2 - M_1}\,. \tag{1.16}$$

For α as defined in (1.16), we denote by $\phi_0 \in (0, \pi/2)$ the angle satisfying

$$2\pi\Big(\int_{\phi_0}^{\pi/2} \cos\phi \, d\phi\Big)/4\pi = \alpha\,. \tag{1.17}$$

In other words, the area of the disc on the sphere given, in spherical coordinates, by $\{(\theta, \phi) : \phi \in [\phi_0, \pi/2]\}$, is $4\pi\alpha$. Moreover, thanks to (1.16) we have:

$$(1 - \sin\phi_0)m_1 + (1 + \sin\phi_0)m_2 = 2R_c\,. \tag{1.18}$$

Finally, on each of the curves Γ_j, $j = 1, 2$ we introduce a kind of a distance function to the set $\mathcal{M}_j$ by setting for each $x \in \Gamma_j$,

$$\delta_j(x, \mathcal{M}_j) = \inf\Big\{ \int_0^\infty (|w'|^2 + |w| - m_j)\, dt \,:\, w \in H^1([0,\infty), \Gamma_j),\, w(0) = x \Big\}\,. \tag{1.19}$$

Note that the minimization problem in (1.19) is actually scalar, since w takes its values on a curve. In the next lemma we give some of the properties of the solution to (1.19).

Lemma 1.1. *For any $x \in \Gamma_j$ there exists a minimizer w_j that realizes the minimum in* (1.19). *Moreover, there exists a point $y = y(x) \in \mathcal{M}_j$ such that $\lim_{t\to\infty} w_j(t) = y$ and*

$$\frac{\delta_j(x, \mathcal{M}_j)}{2} = \int_0^\infty |w_j'|^2\, dt = \int_0^\infty (|w_j| - m_j)\, dt\,. \tag{1.20}$$

Proof. The analysis of problem (1.19) is identical to that of scalar problem with a one-well potential. Therefore, all the the assertion of the lemma are known and standard. □

Now we are in position to state our main result.

Theorem 1.1. *Assume that W satisfies hypotheses (H_1)–(H_4),* (1.16) *and that* (1.9) *holds, i.e., each u_ε is of the form* (1.10). *Then, up to replacing each u_ε by $u_\varepsilon \circ \mathcal{R}$, where $\mathcal{R}$ is the reflection w.r.t. the $(\mathbf{e}_1, \mathbf{e}_2)$-plane, we have:*

(i) For a subsequence

$$u_{\varepsilon_n} \to \chi_{(-\pi/2, \phi_0)} x^{(2)} + \chi_{(\phi_0, \pi/2)} x^{(1)}\,, \tag{1.21}$$

uniformly on $(-\frac{\pi}{2} + \delta, \phi_0 - \delta) \cup (\phi_0 + \delta, \frac{\pi}{2} - \delta)$, for every $\delta > 0$, with $x^{(j)} \in \mathcal{M}_j, j = 1, 2$.

(ii) The asymptotic expansion of the energy is given by

$$\frac{E_\varepsilon(u_\varepsilon)}{2\pi} = \cos\phi_0\Big(\frac{2D}{\varepsilon} + \sqrt{2\beta D}\frac{K}{\varepsilon^{1/2}}\Big) + o(\varepsilon^{-\frac{1}{2}}), \tag{1.22}$$

where

$$K = \min\{\delta_1(\zeta_1^{(i)}, \mathcal{M}_1) + \delta_2(\zeta_2^{(i)}, \mathcal{M}_2) \,:\, i \in \mathcal{I}\}\,, \tag{1.23}$$

and $\beta := \frac{\tan\phi_0}{m_2 - m_1}$.

Note that Theorem 1.1 provides us with a *criterion* for identifying the limit in (i). Indeed, the points $x^{(1)}, x^{(2)}$ that realize the limit form the pair of points, one in $\mathcal{M}_1$ and the other one in $\mathcal{M}_2$, which are the closest (in an appropriate sense) to a geodesic in $\mathcal{G}$.

Our results can be generalized without difficulty to the problem on S^N for any $N \geq 3$. Indeed, the axial symmetry of the minimizers is guaranteed by Lopes' method, so that the one dimensional formulation involves the weight function $(\cos\phi)^{N-1}$. Furthermore, although we have not examined in detail such cases, it is very likely that most of the results of this paper may be extended, by the same techniques, to more general potentials W, for example, with zero set consisting of two compact surfaces in $\mathbb{R}^3$, separated by a sphere of radius R_c.

2. A first upper-bound

We shall first introduce some more notation that will be needed in the sequel. Using d_W we define the corresponding distance functions to the curves Γ_1, Γ_2 by

$$\Psi_j(\zeta) = d_W(\zeta, \Gamma_j) := \inf_{x\in\Gamma_j} d_W(\zeta, x)\,,\ j = 1, 2\,. \tag{2.1}$$

We also set

$$\widetilde{\Psi} = \min(\Psi_1, \Psi_2)\,. \tag{2.2}$$

It is well known (c.f. [7, 4]) that for $j = 1, 2$, $\Psi_j \in \mathrm{Lip}(\mathbb{R}^2)$ is a solution of the eikonal-type equation

$$|\nabla\Psi_j(\zeta)|^2 = W(\zeta) \quad \text{a.e. on } \mathbb{R}^2\,. \tag{2.3}$$

It was further shown in [2] that Ψ_j is regular in a neighborhood of Γ_j, i.e.,

$$\exists d_0 > 0 \text{ s.t. } \Psi_j \text{ is of class } C^2 \text{ in } \{x \,:\, \Psi_j(x) < d_0\}\,, \tag{2.4}$$

for $j = 1, 2$. Moreover, we have

$$\Psi_j(x) \sim W(x) \sim \mathrm{dist}^2(x, \Gamma_j) \quad \text{on } \{x \,:\, \Psi_j(x) < d_0\}\,. \tag{2.5}$$

Clearly, $d_W(x_1, x_2) = D$ for every $x_1 \in \Gamma_1$ and $x_2 \in \Gamma_2$. In order to identify the end points of geodesics from $\mathcal{G}$ we shall use yet another distance function between points from Γ_1 and Γ_2. We denote by Ω the domain lying between Γ_1 and Γ_2.

Recall that in [2] it was shown that for each $x_0 \in \Gamma_j$, $j = 1, 2$, there is a curve $\mathcal{G}^{(j)}_{x_0}$ parametrized on $(-\infty, t(x_0)]$ which satisfies the equation

$$\begin{cases} \dot{\mathcal{G}}^{(j)}_{x_0} = \nabla\Psi(\mathcal{G}^{(j)}_{x_0}), \\ \mathcal{G}^{(j)}_{x_0}(-\infty) = x_0 \,, \end{cases} \tag{2.6}$$

such that the union of these curves (over all $x_0 \in \Gamma_j$) covers without intersections $\{\Psi_j \le d_0\}^+$, which is the part of $\{\Psi_j \le d_0\}$ lying between Γ_1 and Γ_2 (see (2.4)). Similarly, the remaining parts $\{\Psi_j \le d_0\}^-$, $j = 1, 2$, can be covered by an analogous family of curves. Using these curves we can now define a projection map $\tilde{s}_j$ from $\{\Psi_j \le d_0\}$ to Γ_j which associates to each $x \in \{\Psi_j \le d_0\}$ the unique point $x_0 = \tilde{s}_j(x) \in \Gamma_j$ for which the curve $\mathcal{G}^{(j)}_{x_0}$ passes by x.

For any small $\delta > 0$ set

$$\Omega_\delta = \{x \in \Omega \, : \, \widetilde{\Psi}(x) > \delta\} \, . \tag{2.7}$$

For any $x_1 \in \Gamma_1$ and $x_2 \in \Gamma_2$ we let $y_1, y_2 \in \partial\Omega_\delta$ be the points determined by $d_W(y_j, x_j) = \delta$, $j = 1, 2$. We then define

$$d_W^{(\delta)}(x_1, x_2) = 2\delta + \inf_{\substack{\gamma \in \mathrm{Lip}([0,1], \bar{\Omega}_\delta), \\ \gamma(0) = y_1, \gamma(1) = y_2}} L(\gamma) \, , \tag{2.8}$$

where $L(\gamma)$ is defined in (1.6). Note that for each $\delta > 0$ we have $d_W^{(\delta)}(x_1, x_2) \ge D$ for every $x_1 \in \Gamma_1$ and $x_2 \in \Gamma_2$ with equality if and only if x_1 and x_2 are the end points of a geodesic in $\mathcal{G}$.

The main result of this section is a simple upper-bound for the energy. It is not optimal, but it gives the exact first term in the energy expansion (of the order $\frac{1}{\varepsilon}$), and the order $\varepsilon^{-1/2}$ of the next term.

Proposition 2.1. *There exists a constant* $C_1 > 0$ *such that*

$$E_\varepsilon(u_\varepsilon) \le 2\pi \cos\phi_0 \big(\frac{2D}{\varepsilon} + \frac{C_1}{\varepsilon^{1/2}} \big) \tag{2.9}$$

Proof. We shall construct a function $v_\varepsilon = v_\varepsilon(\phi)$ which satisfies the constraint in (P_ε), i.e.,

$$\int_{-\pi/2}^{\pi/2} |v_\varepsilon| \cos\phi \, d\phi = 2R_c \, , \tag{2.10}$$

as well as the bound (2.9). Fix two points $x_j \in \mathcal{M}_j$, $j = 1, 2$ and a geodesic $\gamma := \gamma^{(i_0)} \in \mathcal{G}$ (see (1.8)) of length $L = \text{len}(\gamma)$. We shall denote for short by p_1 and p_2 the endpoints $\zeta_1^{(i_0)}$ and $\zeta_2^{(i_0)}$ of γ, respectively, so that choosing the arclength parametrization for γ we have, $\gamma(0) = p_1$ and $\gamma(L) = p_2$. The following function $z(s)$ will be used in a choice of a certain parametrization of the curve γ. It is defined as the solution of the ODE:

$$\frac{dz}{ds} = \sqrt{W(\gamma(z(s)))}\,, \quad z(0) = L/2\,. \tag{2.11}$$

It is easy to see that z is defined on the whole real line and satisfies

$$\lim_{s\to-\infty} z(s) = 0 \text{ and } \lim_{s\to\infty} z(s) = L\,.$$

Furthermore, since

$$\sqrt{W(\gamma(z(s)))} \sim d(\gamma(z(s)), \Gamma_1) \sim z(s) \quad \text{as } s \to -\infty$$

and

$$\sqrt{W(\gamma(z(s)))} \sim d(\gamma(z(s)), \Gamma_2) \sim L - z(s) \quad \text{as } s \to \infty\,,$$

where d stands for the euclidean distance, we have:

$$0 \le z(s) \le C_2 e^{c_1 s} \qquad \text{as } s \to -\infty\,, \tag{2.12}$$

$$0 \le L - z(s) \le C_2 e^{-c_2 s} \qquad \text{as } s \to \infty\,, \tag{2.13}$$

for some positive constants c_1, c_2 and C_1, C_2. From (2.11) we deduce that for any function $v(\phi)$ defined on an interval $[\phi_1, \phi_2]$ by

$$v(\phi) = \gamma(z(c \pm \frac{\phi}{\varepsilon}))\,,$$

we have

$$\begin{aligned}\int_{\phi_1}^{\phi_2} \Big(|v'|^2 + \frac{W(v)}{\varepsilon^2}\Big) \cos\phi \, d\phi &= \frac{2}{\varepsilon^2} \int_{\phi_1}^{\phi_2} W(v) \cos\phi \, d\phi \\ &= \frac{2}{\varepsilon} \int_{\phi_1}^{\phi_2} \sqrt{W(v)} |v'(\phi)| \cos\phi \, d\phi\,.\end{aligned} \tag{2.14}$$

Set $\bar\phi = \phi_0 - \tau\varepsilon^{1/2}$, with $\tau = \tau_\varepsilon$ to be determined later. We define $v_\varepsilon(\phi)$ on $[-\pi/2, \pi/2]$ as follows:

(i) On $[\bar\phi + \varepsilon^{1/2}, \pi/2]$ we set $v_\varepsilon \equiv x^{(1)}$.
(ii) On $[\bar\phi, \bar\phi + \varepsilon^{1/2}]$ we follow the curve Γ_1 in a constant velocity from $p_1\, (= \zeta_1^{(i_0)})$ to $x^{(1)}$ (in the shortest way between the two possibilities).
(iii) On $[\bar\phi - \varepsilon, \bar\phi]$ we let v_ε be the linear function which equals to

$\gamma(z(-\frac{\ln 1/\varepsilon}{c_1}))$ at $\phi = \bar{\phi} - \varepsilon$ and to p_1 at $\phi = \bar{\phi}$.
(iv) On $[\bar{\phi} - \varepsilon - (\frac{1}{c_1} + \frac{1}{c_2})\varepsilon \ln \frac{1}{\varepsilon}, \bar{\phi} - \varepsilon]$ we set

$$v_\varepsilon(\phi) = \gamma\Big(z\big(\frac{\bar{\phi} - \varepsilon - \phi}{\varepsilon} - \frac{\ln 1/\varepsilon}{c_1}\big)\Big).$$

(v) On $[\bar{\phi} - 2\varepsilon - (\frac{1}{c_1} + \frac{1}{c_2})\varepsilon \ln \frac{1}{\varepsilon}, \bar{\phi} - \varepsilon - (\frac{1}{c_1} + \frac{1}{c_2})\varepsilon \ln \frac{1}{\varepsilon}]$ we let v_ε to be the linear function which equals to p_2 $(= \zeta_2^{(i_0)})$ at $\phi = \bar{\phi} - 2\varepsilon - (\frac{1}{c_1} + \frac{1}{c_2})\varepsilon \ln \frac{1}{\varepsilon}$ and to $\gamma(z(\frac{\ln 1/\varepsilon}{c_2}))$ at $\phi = \bar{\phi} - \varepsilon - (\frac{1}{c_1} + \frac{1}{c_2})\varepsilon \ln \frac{1}{\varepsilon}$.
(vi) On $[\bar{\phi} - 2\varepsilon - (\frac{1}{c_1} + \frac{1}{c_2})\varepsilon \ln \frac{1}{\varepsilon} - \varepsilon^{1/2}, \bar{\phi} - 2\varepsilon - (\frac{1}{c_1} + \frac{1}{c_2})\varepsilon \ln \frac{1}{\varepsilon}]$ we follow the curve Γ_2 in a constant velocity from $x^{(2)}$ to p_2 (using the shortest way among the two possibilities).
(vii) On $[-\pi/2, \bar{\phi} - 2\varepsilon - (\frac{1}{c_1} + \frac{1}{c_2})\varepsilon \ln \frac{1}{\varepsilon} - \varepsilon^{1/2}]$ we set $v_\varepsilon \equiv x^{(2)}$.

We shall next compute the contribution to the integral of $|v_\varepsilon|$ from each of the intervals (i)–(vii) and show that a bounded solution for τ is determined by the constraint (2.10). Note first that

$$\begin{aligned}\Big(\int_{\bar{\phi}+\varepsilon^{1/2}}^{\pi/2} \cos\phi\Big)|x^{(1)}| \\ &= \big(1 - \sin(\phi_0 - (\tau - 1)\varepsilon^{1/2})\big)|x^{(1)}| \\ &= \big(1 - \sin\phi_0 + (\tau - 1)\varepsilon^{1/2}\cos\phi_0\big)|x^{(1)}| + O(\varepsilon).\end{aligned} \tag{2.15}$$

Here and in the rest of the proof we denote by $O(f(\varepsilon))$ a function $g(t, \varepsilon)$ satisfying $|g(t, \varepsilon)| \le C(t) f(\varepsilon)$ with $C(t)$ bounded for bounded t. Next, it is easy to verify that

$$\int_{\bar{\phi}}^{\bar{\phi}+\varepsilon^{1/2}} |v_\varepsilon| \cos\phi = \varepsilon^{1/2}\lambda_1(x^{(1)}, p_1)\cos\phi_0 + O(\varepsilon), \tag{2.16}$$

with

$$\lambda_1(x^{(1)}, p_1) = \frac{\int_{J_1(x^{(1)}, p_1)} |z|}{d_{\Gamma_1}(x^{(1)}, p_1)},$$

where $J_1(x^{(1)}, p_1)$ denotes the (shortest) segment of Γ_1 between $x^{(1)}$ and p_1, whose length is denoted then by $d_{\Gamma_1}(x^{(1)}, p_1)$. Clearly

$$\int_{\bar{\phi}-\varepsilon}^{\bar{\phi}} |v_\varepsilon| \cos\phi = O(\varepsilon). \tag{2.17}$$

Similarly, for the intervals in (iv) and (v) we find, respectively,

$$\int_{\bar{\phi}-\varepsilon-(\frac{1}{c_1}+\frac{1}{c_2})\varepsilon \ln \frac{1}{\varepsilon}}^{\bar{\phi}-\varepsilon} |v_\varepsilon| \cos\phi = O(\varepsilon \ln \frac{1}{\varepsilon}), \tag{2.18}$$

and

$$\int_{\bar{\phi}-2\varepsilon-(\frac{1}{c_1}+\frac{1}{c_2})\varepsilon\ln\frac{1}{\varepsilon}}^{\bar{\phi}-\varepsilon-(\frac{1}{c_1}+\frac{1}{c_2})\varepsilon\ln\frac{1}{\varepsilon}} |v_\varepsilon|\cos\phi = O(\varepsilon)\,. \tag{2.19}$$

As in (2.16) we find

$$\int_{\bar{\phi}-2\varepsilon-(\frac{1}{c_1}+\frac{1}{c_2})\varepsilon\ln\frac{1}{\varepsilon}-\varepsilon^{1/2}}^{\bar{\phi}-2\varepsilon-(\frac{1}{c_1}+\frac{1}{c_2})\varepsilon\ln\frac{1}{\varepsilon}} |v_\varepsilon|\cos\phi = \varepsilon^{1/2}\lambda_2(x^{(2)},p_2)\cos\phi_0 + O(\varepsilon)\,, \tag{2.20}$$

where J_2 and λ_2 are defined analogously to J_1 and λ_1. Finally, as in (2.15), we have

$$\begin{aligned}&\Big(\int_{-\pi/2}^{\bar{\phi}-2\varepsilon-(\frac{1}{c_1}+\frac{1}{c_2})\varepsilon\ln\frac{1}{\varepsilon}-\varepsilon^{1/2}} \cos\phi\Big)|x^{(2)}| \\ &\quad= \big(1+\sin\phi_0-(\tau+1)\varepsilon^{1/2}\cos\phi_0\big)|x^{(2)}| + O(\varepsilon\ln\frac{1}{\varepsilon})\,.\end{aligned} \tag{2.21}$$

Summing up the integrals in (2.15)–(2.21) and comparing to (1.18) and (2.10) yields

$$\varepsilon^{1/2}\cos\phi_0\Big(m_1(\tau-1)+\lambda_1(x^{(1)},p_1)+\lambda_2(x^{(2)},p_2)-m_2(\tau+1)\Big) = O(\varepsilon\ln\frac{1}{\varepsilon})\,,$$

from which we can solve for $\tau=\tau_\varepsilon$ that satisfies

$$\tau_\varepsilon = \frac{\lambda_1(x^{(1)},p_1)+\lambda_2(x^{(2)},p_2)-m_1-m_2}{m_2-m_1} + O(\varepsilon^{1/2}\ln\frac{1}{\varepsilon})\,.$$

Clearly τ_ε remains bounded as ε goes to zero.

Next we compute the contribution to the energy $E_\varepsilon(v_\varepsilon)$ from each of the segments (i) to (vii). We denote these contributions by $I_1 - I_7$, respectively. Clearly we have

$$I_1 = I_7 = 0\,, \tag{2.22}$$

and

$$I_2 + I_6 = O(\varepsilon^{-1/2})\,. \tag{2.23}$$

From (2.12)–(2.13) we infer that

$$|p_1-\gamma(z(-\frac{\ln 1/\varepsilon}{c_1}))| = O(\varepsilon) \quad\text{and}\quad |p_2-\gamma(z(\frac{\ln 1/\varepsilon}{c_2}))| = O(\varepsilon)\,,$$

which implies that

$$I_3 + I_5 = O(1)\,. \tag{2.24}$$

Finally, for the contribution I_4 from (iv), we find using (2.14)

$$\frac{I_4}{2\pi} \leq \frac{2D}{\varepsilon} \cos\left(\bar{\phi} - \varepsilon - (\frac{1}{c_1} + \frac{1}{c_2})\varepsilon \ln \frac{1}{\varepsilon}\right) \leq \frac{2D}{\varepsilon}(\cos \phi_0 + C\varepsilon^{1/2}) . \quad (2.25)$$

Combining (2.22)–(2.25) we are led to the desired result

$$E_\varepsilon(v_\varepsilon) \leq 2\pi \cos \phi_0 \left(\frac{2D}{\varepsilon} + \frac{C_1}{\varepsilon^{1/2}}\right) . \qquad \square$$

Note that an immediate consequence of the upper bound (2.9) is that for any $\alpha > 0$ there exists $\beta > 0$ such that

$$\mu(\{x \in S^2 \,:\, W(u_\varepsilon(x)) \geq \alpha\varepsilon^{1/2}\}) \leq \beta\varepsilon^{1/2} . \quad (2.26)$$

3. A key proposition

This section is devoted to the proof of Proposition 3.1 below which establishes a partition of the sphere to two parts, on each of them u_ε is close to either Γ_1 or Γ_2. For any ϕ_1, ϕ_2 satisfying $-\frac{\pi}{2} \leq \phi_1 < \phi_2 \leq \frac{\pi}{2}$ we define the spherical annulus

$$A_{\phi_1,\phi_2} := \{x \in S^2 \,:\, \phi_1 < \phi(x) < \phi_2\} ,$$

where $\phi(x)$ and $\theta(x)$ denote the spherical coordinates of the point x, i.e.,

$$x = (x_1, x_2, x_3) = (\cos\phi(x)\cos\theta(x), \cos\phi(x)\sin\theta(x), \sin\phi(x)) .$$

Proposition 3.1. *For each small $\varepsilon > 0$ there exists $\gamma_1 = \gamma_1(\varepsilon)$ satisfying either:*

$$\begin{aligned} &|\gamma_1 - \phi_0| \leq C_1\varepsilon^{1/2} \\ \textit{and}\ \ &\mu(\{x \in S^2 \,:\, \Psi_1(u_\varepsilon(x)) \leq \varepsilon^{1/2}\}\Delta A_{\gamma_1,\frac{\pi}{2}}) \leq C_2\varepsilon , \end{aligned} \quad (3.1)$$

or

$$\begin{aligned} &|\gamma_1 + \phi_0| \leq C_1\varepsilon^{1/2} \\ \textit{and}\ \ &\mu(\{x \in S^2 \,:\, \Psi_1(u_\varepsilon(x)) \leq \varepsilon^{1/2}\}\Delta A_{-\frac{\pi}{2},-\gamma_1}) \leq C_2\varepsilon , \end{aligned} \quad (3.2)$$

where C_1 and C_2 are positive constants, independent of ε.

Proof. In the sequel we shall denote by c and C different positive constants which do not depend on ε. The upper bound (2.9) combined with (2.5) implies (for small ε) that

$$\int_0^{C\varepsilon^{1/2}} \widetilde{\Psi}(u_\varepsilon(\phi)) \cos\phi \, d\phi \leq C\varepsilon .$$

Therefore, choosing an appropriate C yields the existence of $\gamma_0 = \gamma_0^\varepsilon \in (0, C\varepsilon^{1/2})$ satisfying

$$\widetilde{\Psi}(u_\varepsilon(\gamma_0)) \le \varepsilon^{1/2} . \tag{3.3}$$

We shall assume in the sequel that (3.3) holds with $\widetilde{\Psi}(u_\varepsilon(\gamma_0)) = \Psi_2(u_\varepsilon(\gamma_0))$ (this implies, of course, that $u_\varepsilon(\gamma_0)$ is close to Γ_2).

Next, define, if it exists,

$$\phi_1^+ = \inf\{\phi \ge \gamma_0 \, : \, \Psi_1(u_\varepsilon(\phi)) \le \varepsilon^{1/2}\} ,$$

and then

$$\overline{\phi}_1^+ = \sup\{\phi \le \phi_1^+ \, : \, \Psi_2(u_\varepsilon(\phi)) \le \varepsilon^{1/2}\} .$$

In general, for an even j let

$$\begin{aligned} \phi_j^+ &= \inf\{\phi \ge \phi_{j-1}^+ \, : \, \Psi_2(u_\varepsilon(\phi)) \le \varepsilon^{1/2}\} , \\ \overline{\phi}_j^+ &= \sup\{\phi \le \phi_j^+ \, : \, \Psi_1(u_\varepsilon(\phi)) \le \varepsilon^{1/2}\} , \end{aligned}$$

while for an odd j:

$$\begin{aligned} \phi_j^+ &= \inf\{\phi \ge \phi_{j-1}^+ \, : \, \Psi_1(u_\varepsilon(\phi)) \le \varepsilon^{1/2}\} , \\ \overline{\phi}_j^+ &= \sup\{\phi \le \phi_j^+ \, : \, \Psi_2(u_\varepsilon(\phi)) \le \varepsilon^{1/2}\} . \end{aligned}$$

We stop at $j = k^+ - 1$ if $j = k^+$ is the first index for which either ϕ_j^+ does not exist, or $\phi_j^+ > \frac{\pi}{2} - \varepsilon^{1/2}$. For each $j = 1, \ldots, k^+ - 1$ we have (see (2.14))

$$\begin{aligned} \frac{1}{2\pi} E_\varepsilon(u_\varepsilon, A_{\overline{\phi}_j^+, \phi_j^+}) &\ge \frac{2}{\varepsilon} \cos \phi_j^+ \int_{\overline{\phi}_j^+}^{\phi_j^+} \sqrt{W(u_\varepsilon)} |u_\varepsilon'| \\ &\ge \frac{2}{\varepsilon} (D - 2\varepsilon^{1/2}) \cos \phi_j^+ . \end{aligned} \tag{3.4}$$

In particular, we deduce

$$\frac{1}{2\pi} E_\varepsilon(u_\varepsilon, A_{\overline{\phi}_j^+, \phi_j^+}) \ge \frac{2}{\varepsilon} (D - 2\varepsilon^{1/2}) \sin \varepsilon^{1/2} \ge \frac{C}{\varepsilon^{1/2}} , \tag{3.5}$$

which together with the upper bound (2.9) implies that the process of selection of pairs $(\overline{\phi}_j^+, \phi_j^+)$ must terminate, with the bound $k^+ \le \frac{C}{\varepsilon^{1/2}}$.

Similarly, set, if it exists,

$$\phi_1^- = \sup\{\phi \le \gamma_0 \, : \, \Psi_1(u_\varepsilon(\phi)) \le \varepsilon^{1/2}\} ,$$

and then

$$\overline{\phi_1^-} = \inf\{\phi \geq \phi_1^- \ : \ \Psi_2(u_\varepsilon(\phi)) \leq \varepsilon^{1/2}\}\,.$$

In general, for an even j let

$$\begin{aligned} \phi_j^- &= \sup\{\phi \leq \phi_{j-1}^- \ : \ \Psi_2(u_\varepsilon(\phi)) \leq \varepsilon^{1/2}\}\,, \\ \overline{\phi_j^-} &= \inf\{\phi \geq \phi_j^- \ : \ \Psi_1(u_\varepsilon(\phi)) \leq \varepsilon^{1/2}\}\,, \end{aligned}$$

while for an odd j:

$$\begin{aligned} \phi_j^- &= \sup\{\phi \leq \phi_{j-1}^- \ : \ \Psi_1(u_\varepsilon(\phi)) \leq \varepsilon^{1/2}\}\,, \\ \overline{\phi_j^-} &= \inf\{\phi \geq \phi_j^- \ : \ \Psi_2(u_\varepsilon(\phi)) \leq \varepsilon^{1/2}\}\,. \end{aligned}$$

We stop at $j = k^- - 1$ if $j = k^-$ is the first index for which either ϕ_j^- does not exist, or $\phi_j^- < -\frac{\pi}{2} + \varepsilon^{1/2}$. The same computation as in (3.4)–(3.5) gives

$$\frac{1}{2\pi} E_\varepsilon(u_\varepsilon, A_{\phi_j^-, \overline{\phi_j^-}}) \geq \frac{2}{\varepsilon} \cos \phi_j^- (D - 2\varepsilon^{1/2}) \geq \frac{C}{\varepsilon^{1/2}}\,, \tag{3.6}$$

which implies the bound $k^- \leq \frac{C}{\varepsilon^{1/2}}$. It will be convenient to set also

$$\phi_{k+}^+ = \overline{\phi_{k+}^+} = \frac{\pi}{2} - \varepsilon^{1/2} \quad \text{and} \quad \phi_{k-}^- = \overline{\phi_{k-}^-} = -\frac{\pi}{2} + \varepsilon^{1/2}\,.$$

Since on each $A_{\overline{\phi_i^+}, \phi_i^+}$ and $A_{\phi_i^-, \overline{\phi_i^-}}$ we have $\widetilde{\Psi}(u_\varepsilon) \geq \varepsilon^{1/2}$, it follows from (2.5) and (2.26) that

$$\mu\Big(\bigcup_{i=1}^{k^+-1} A_{\overline{\phi_i^+}, \phi_i^+} \cup \bigcup_{i=1}^{k^--1} A_{\phi_i^-, \overline{\phi_i^-}}\Big) \leq C\varepsilon^{1/2}\,. \tag{3.7}$$

Put

$$V_\varepsilon = \bigcup_{j\geq 1} A_{\phi_{2j-1}^+, \phi_{2j}^+} \cup \bigcup_{j\geq 1} A_{\phi_{2j}^-, \phi_{2j-1}^-}\,. \tag{3.8}$$

In the above only j's which do not go out of range are taken into account, i.e., those satisfying $2j \leq k^+$ in the first union and those satisfying $2j \leq k^-$ in the second. On the one hand we have $\Psi_1(u_\varepsilon) > \varepsilon^{1/2}$ on each $A_{\phi_{2j}^+, \phi_{2j+1}^+}$ and $A_{\phi_{2j+1}^-, \phi_{2j}^-}$, which implies that

$$\{\Psi_1(u_\varepsilon) \leq \varepsilon^{1/2}\} \cap A_{\varepsilon^{1/2} - \frac{\pi}{2}, \frac{\pi}{2} - \varepsilon^{1/2}} \subset V_\varepsilon\,. \tag{3.9}$$

On the other hand, since $\Psi_2(u_\varepsilon) > \varepsilon^{1/2}$ on each $A_{\phi_{2j-1}^+, \phi_{2j}^+}$ and $A_{\phi_{2j}^-, \phi_{2j-1}^-}$, we have

$$V_\varepsilon \subset \{\Psi_1(u_\varepsilon) \leq \varepsilon^{1/2}\} \cup \{\widetilde{\Psi}(u_\varepsilon) \geq \varepsilon^{1/2}\}\,. \tag{3.10}$$

Note that from (2.26) it follows that there exists $\alpha_\varepsilon \in (0,1)$ such that

$$\begin{aligned} &|\mu(\{\Psi_1(u_\varepsilon) \le \varepsilon^{1/2}\}) - 4\pi\alpha_\varepsilon| \le C\varepsilon^{1/2} \\ \text{and}\quad &|\mu(\{\Psi_2(u_\varepsilon) \le \varepsilon^{1/2}\}) - 4\pi(1-\alpha_\varepsilon)| \le C\varepsilon^{1/2}. \end{aligned} \tag{3.11}$$

Note that on the sets $\{\Psi_j(u_\varepsilon) < d_0\}$ we have $|u_\varepsilon - \tilde{s}_j(u_\varepsilon)|^2 \sim W(u_\varepsilon)$ (see (2.5)), hence

$$\begin{aligned} \Big| \int_{\{\Psi_j(u_\varepsilon)\le\varepsilon^{1/2}\}} |u_\varepsilon| - \int_{\{\Psi_j(u_\varepsilon)\le\varepsilon^{1/2}\}} |\tilde{s}_j(u_\varepsilon)| \Big| \\ \le \int_{\{\Psi_j(u_\varepsilon)\le\varepsilon^{1/2}\}} |u_\varepsilon - \tilde{s}_j(u_\varepsilon)| \\ \le C\Big(\int_{\{\Psi_j(u_\varepsilon)\le\varepsilon^{1/2}\}} W(u_\varepsilon)\Big)^{1/2} \le C\varepsilon^{1/2}. \end{aligned} \tag{3.12}$$

By (H_3), (2.5) and (2.26) we also have

$$\begin{aligned} \int_{\{\widetilde{\Psi}(u_\varepsilon)>\varepsilon^{1/2}\}} |u_\varepsilon| \le R_0 \mu(\{\widetilde{\Psi}(u_\varepsilon) > \varepsilon^{1/2}\} \cap \{|u_\varepsilon| \le R_0\}) \\ + \frac{1}{C_0} \int_{\{\widetilde{\Psi}(u_\varepsilon)>\varepsilon^{1/2}\}\cap\{|u_\varepsilon|>R_0\}} W(u_\varepsilon) \le C\varepsilon^{1/2}. \end{aligned} \tag{3.13}$$

Set

$$m_\varepsilon^{(j)} = \frac{1}{\mu(\{\Psi_j(u_\varepsilon) \le \varepsilon^{1/2}\})} \int_{\{\Psi_j(u_\varepsilon)\le\varepsilon^{1/2}\}} |\tilde{s}_j(u_\varepsilon)|, \quad \text{for } j = 1, 2. \tag{3.14}$$

Using (3.12)–(3.14) and the constraint we obtain

$$\begin{aligned} &m_\varepsilon^{(1)} \mu(\{\Psi_1(u_\varepsilon) \le \varepsilon^{1/2}\}) + m_\varepsilon^{(2)} \mu(\{\Psi_2(u_\varepsilon) \le \varepsilon^{1/2}\}) \\ &\quad = \int_{\{\Psi_1(u_\varepsilon)\le\varepsilon^{1/2}\}} |\tilde{s}_1(u_\varepsilon)| + \int_{\{\Psi_2(u_\varepsilon)\le\varepsilon^{1/2}\}} |\tilde{s}_2(u_\varepsilon)| \\ &\quad = \int_{\{\Psi_1(u_\varepsilon)\le\varepsilon^{1/2}\}} |u_\varepsilon| + \int_{\{\Psi_2(u_\varepsilon)\le\varepsilon^{1/2}\}} |u_\varepsilon| + O(\varepsilon^{1/2}) \\ &\quad = 4\pi R_c + O(\varepsilon^{1/2}). \end{aligned} \tag{3.15}$$

Form (3.15) and (3.12) it follows that there exists a number K, which is uniformly bounded, such that

$$m_\varepsilon^{(1)}(\alpha_\varepsilon + K\varepsilon^{1/2}) + m_\varepsilon^{(2)}(1 - \alpha_\varepsilon - K\varepsilon^{1/2}) = R_c. \tag{3.16}$$

Since $m_\varepsilon^{(j)} \in [m_j, M_j]$, $j = 1, 2$, it follows from (3.16), in particular, that α_ε is bounded away from 0 and 1. Therefore, by the definitions of α and I (see (1.14)),

$$I(4\pi(\alpha_\varepsilon + K\varepsilon^{1/2})) \ge I(4\pi\alpha) = 2\pi\cos\phi_0. \tag{3.17}$$

By (3.9)–(3.10),(3.17) and the Lipschitz property of I we conclude that

$$\begin{aligned}2\pi\Big(\sum_{j=1}^{2[\frac{k^+}{2}]}\cos\phi_j^+ + \sum_{j=1}^{2[\frac{k^-}{2}]}\cos\phi_j^-\Big) &= \operatorname{Per} V_\varepsilon \geq I(\mu(V_\varepsilon))\\ &\geq 2\pi\cos\phi_0 - C\varepsilon^{1/2}\,.\end{aligned} \tag{3.18}$$

On the other hand, summing-up the inequalities in (3.4) and (3.6) yields

$$\begin{aligned}&\frac{1}{2\pi}E_\varepsilon\big(u_\varepsilon, \bigcup_{j=1}^{k^+-1} A_{\overline{\phi}_j^+,\phi_j^+} \cup \bigcup_{j=1}^{k^--1} A_{\phi_j^-,\overline{\phi}_j^-}\big)\\ &\geq \frac{2D}{\varepsilon}\big(\sum_{j=1}^{k^+-1}\cos\phi_j^+ + \sum_{j=1}^{k^--1}\cos\phi_j^-\big) - C\varepsilon^{-1/2}\,.\end{aligned} \tag{3.19}$$

Combining (3.19) with the upper-bound (2.9) we obtain

$$\sum_{j=1}^{k^+}\cos\phi_j^+ + \sum_{j=1}^{k^-}\cos\phi_j^- \leq \cos\phi_0 + C\varepsilon^{1/2}\,. \tag{3.20}$$

Note that we used the fact that $\cos\phi_{k^+}^+ = \cos\phi_{k^-}^- = \sin\varepsilon^{1/2} = O(\varepsilon^{1/2})$. Therefore, combining (3.18) with (3.20) we get

$$\Big|\sum_{j=1}^{k^+}\cos\phi_i^+ + \sum_{j=1}^{k^-}\cos\phi_i^- - \cos\phi_0\Big| \leq C\varepsilon^{1/2}\,. \tag{3.21}$$

Our next objective is to show that V_ε consists of "essentially" one annulus.

<u>Claim:</u> We have:

$$\begin{aligned}&\text{either } \mu(V_\varepsilon \Delta A_{\phi_j^+,\phi_{j+1}^+}) = O(\varepsilon),\\ &\text{or } \mu(V_\varepsilon \Delta A_{\phi_{j+1}^-,\phi_j^-}) = O(\varepsilon), \text{ for some } j.\end{aligned} \tag{3.22}$$

<u>Proof of the Claim:</u> If V_ε consists of only one annulus, then either $V_\varepsilon = A_{\phi_1^+,\phi_2^+}$ or $V_\varepsilon = A_{\phi_2^-,\phi_1^-}$ (see (3.8)) and the claim holds. It remains to treat the case where V_ε consists of more than one annulus. Note that since I is concave and $I(0) = 0$, we have

$$I(a+b) \leq I(a) + I(b)\,, \tag{3.23}$$

for all admissible values of a and b. By assumption, we may write V_ε as a *disjoint* union $V_\varepsilon = B_\varepsilon \cup C_\varepsilon$ with $\mu(V_\varepsilon)/2 \leq \mu(B_\varepsilon) < \mu(V_\varepsilon)$. From (1.14) it is clear that there exists $\delta_0 > 0$ such that

$$I(\delta) \geq 2\delta I'(\mu(V_\varepsilon) - \delta)\,, \quad \forall \delta \leq \delta_0\,. \tag{3.24}$$

We distinguish two cases:

(i) $\mu(C_\varepsilon) > \delta_0$.

(ii) $\mu(C_\varepsilon) \leq \delta_0$.

In case (i) we have, for some constant $\eta > 0$, a stronger form of (3.23), namely

$$I(\mu(V_\varepsilon)) + \eta \leq I(\mu(B_\varepsilon)) + I(\mu(C_\varepsilon)). \tag{3.25}$$

From (3.25) and (3.21) it follows that

$$\begin{aligned} I(\mu(V_\varepsilon)) + \eta \leq I(\mu(B_\varepsilon)) + I(\mu(C_\varepsilon)) &\leq \operatorname{Per}(B_\varepsilon) + \operatorname{Per}(C_\varepsilon) = \operatorname{Per}(V_\varepsilon) \\ &\leq 2\pi \cos\phi_0 + C\varepsilon^{1/2}, \end{aligned}$$

which clearly contradicts (3.18), for ε small enough. Therefore, only case (ii) should be considered.

Notice the following simple consequence of the concavity of I:

$$\mu(C_\varepsilon)I'(\mu(B_\varepsilon)) \geq I(\mu(V_\varepsilon)) - I(\mu(B_\varepsilon)). \tag{3.26}$$

By (3)–(3.26) and (3.24) we get

$$\begin{aligned} &2\pi\cos\phi_0 + C\varepsilon^{1/2} \\ &\quad \geq I(\mu(B_\varepsilon)) + I(\mu(C_\varepsilon)) \geq I(\mu(V_\varepsilon)) - \mu(C_\varepsilon)I'(\mu(B_\varepsilon)) + I(\mu(C_\varepsilon)) \\ &\quad \geq I(\mu(V_\varepsilon)) + \frac{1}{2}I(\mu(C_\varepsilon)) \geq 2\pi\cos\phi_0 - C\varepsilon^{1/2} + \frac{1}{2}I(\mu(C_\varepsilon)), \end{aligned}$$

which yields, $I(\mu(C_\varepsilon)) \leq C\varepsilon^{1/2}$, i.e.,

$$\mu(C_\varepsilon) \leq C\varepsilon. \tag{3.27}$$

To conclude the proof of the claim, we shall show that one of the components of V_ε has measure larger than $\mu(V_\varepsilon) - C\varepsilon$. By the above computation it is enough to consider the case where V_ε consists of $r \geq 3$ components (i.e., annuli) that we shall now denote by $A_1, A_2, \ldots, A_r$ with

$$\mu(A_1) \geq \mu(A_2) \geq \cdots \geq \mu(A_r).$$

Furthermore, if $\mu(A_1) \geq \frac{\mu(V_\varepsilon)}{2}$ then the conclusion follows from the argument that led to (3.27). Thus assume that

$$\mu(A_1) < \frac{\mu(V_\varepsilon)}{2}. \tag{3.28}$$

We shall see that this is impossible. Indeed, (3.28) implies the existence of $j_0 \geq 1$ which is the *largest* index for which

$$\mu(A_{j_0+1}) + \mu(A_{j_0+2}) + \cdots + \mu(A_r) \geq \frac{\mu(V_\varepsilon)}{2}.$$

Setting

$$B_\varepsilon = A_{j_0+1} \cup A_{j_0+2} \cup \cdots \cup A_r \quad \text{and} \quad C_\varepsilon = A_1 \cup A_2 \cup \cdots \cup A_{j_0}\,,$$

we obtain from the argument that led to (3.27) that $\mu(C_\varepsilon) \leq C\varepsilon$. But from the definition of j_0 it is easy to see that we must have then $\mu(B_\varepsilon) \leq \frac{3}{4}\mu(V_\varepsilon)$, i.e., $\mu(V_\varepsilon) = \mu(B_\varepsilon) + \mu(C_\varepsilon) \leq \frac{3}{4}\mu(V_\varepsilon) + C\varepsilon$, which is a contradiction (for ε small enough). The proof of the claim is then complete.

Consider now the "fat component" A_1, and assume without loss of generality that it lies in the upper hemisphere, i.e., $A_1 = A_{\phi_j^+,\phi_{j+1}^+}$ for some j. Under this assumption we shall show that (3.1) holds, for $\gamma_1 = \phi_j^+$ (the other possibility leads to (3.2) by an identical argument). By (3.21) we deduce that

$$|\cos\phi_j^+ + \cos\phi_{j+1}^+ - \cos\phi_0| \leq C\varepsilon^{1/2}\,, \tag{3.29}$$

while by (3.18) we have

$$|I(\mu(V_\varepsilon)) - 2\pi\cos\phi_0| \leq C\varepsilon^{1/2}\,. \tag{3.30}$$

Using the Lipschitzity of the (local) inverse of the function I we deduce from (3.30) and (3.22) that

$$\sin\phi_{j+1}^+ - \sin\phi_j^+ = 1 - \sin\phi_0 + O(\varepsilon^{1/2})\,. \tag{3.31}$$

It is easy to conclude from (3.29) and (3.31) that

$$|\phi_j^+ - \phi_0| \leq C\varepsilon^{1/2} \quad \text{and} \quad |\phi_{j+1}^+ - \frac{\pi}{2}| \leq C\varepsilon^{1/2}\,.$$

Finally, we must have $j = 1$, since otherwise we will get a contradiction to (3.21). We can therefore set $\gamma_1 = \phi_1^+$ and (3.1) follows. □

In the sequel we shall assume, without loss of generality, that (3.1) holds. We then set

$$\gamma_2 = \gamma_2(\varepsilon) = \sup\{\phi \leq \gamma_1 \,:\, \Psi_2(u_\varepsilon(\phi)) = \varepsilon^{1/2}\}\,. \tag{3.32}$$

Note that by (3.1) we have

$$E_\varepsilon(u_\varepsilon, A_{\gamma_2,\gamma_1}) \geq \frac{2\cos\gamma_1}{\varepsilon}(D - 2\varepsilon^{1/2}) \geq \frac{2D\cos\phi_0}{\varepsilon} - C\varepsilon^{1/2}\,. \tag{3.33}$$

The next lemma provides pointwise estimates that roughly speaking show that $u_\varepsilon(\phi)$ is "close" to Γ_2 for ϕ below γ_2, while $u_\varepsilon(\phi)$ is "close" to Γ_1 for ϕ above γ_1.

Lemma 3.1. *There exists* $c_3 > 0$ *such that*

$$\Psi_1(u_\varepsilon(\phi))\cos\phi \le c_3\varepsilon^{1/2}\,, \quad \forall\phi \ge \gamma_1\,, \tag{3.34}$$

$$\Psi_2(u_\varepsilon(\phi))\cos\phi \le c_3\varepsilon^{1/2}\,, \quad \forall\phi \le \gamma_2\,. \tag{3.35}$$

Proof. By (2.9) and (3.33), for each $\tilde{\phi} \ge \gamma_1$ we have

$$\begin{aligned}\frac{C}{\varepsilon^{1/2}} &\ge E_\varepsilon(u_\varepsilon, A_{\gamma_1,\tilde{\phi}}) \ge \left|\frac{2}{\varepsilon}\int_{\gamma_1}^{\tilde{\phi}} \cos\phi \frac{d}{d\phi}(\Psi_1(u_\varepsilon(\phi)))\,d\phi\right| \\ &\ge \frac{2}{\varepsilon}\Big(\Psi_1(u_\varepsilon(\tilde{\phi}))\cos\tilde{\phi} - \Psi_1(u_\varepsilon(\gamma_1))\cos\gamma_1 + \int_{\gamma_1}^{\tilde{\phi}} \sin\phi\,\Psi_1(u_\varepsilon(\phi))\,d\phi\Big) \\ &\ge \frac{2}{\varepsilon}\Big(\cos\tilde{\phi}\,\Psi_1(u_\varepsilon(\tilde{\phi})) - \cos\gamma_1\,\varepsilon^{1/2}\Big)\,,\end{aligned}$$

and (3.34) follows.

Next, for each $\tilde{\phi} \le \gamma_2$ we have by the same computation as above

$$\begin{aligned}\frac{C}{\varepsilon^{1/2}} &\ge E_\varepsilon(u_\varepsilon, A_{\tilde{\phi},\gamma_2}) \\ &\ge \left|\frac{2}{\varepsilon}\int_{\tilde{\phi}}^{\gamma_2} \cos\phi\frac{d}{d\phi}(\Psi_2(u_\varepsilon(\phi)))\,d\phi\right| \\ &\ge \frac{2}{\varepsilon}\Big(-\cos\gamma_2\Psi_2(u_\varepsilon(\gamma_2)) + \cos\tilde{\phi}\,\Psi_2(u_\varepsilon(\tilde{\phi})) \\ &\qquad - \int_{\tilde{\phi}}^{\gamma_2}\sin\phi\,\Psi_2(u_\varepsilon(\phi))\,d\phi\Big)\,.\end{aligned} \tag{3.36}$$

Denoting $\tilde{\phi}_+ = \max(\tilde{\phi}, 0)$, we have

$$\begin{aligned}&\int_{\tilde{\phi}}^{\gamma_2}\sin\phi\,\Psi_2(u_\varepsilon(\phi))\,d\phi \\ &\quad= \int_{\tilde{\phi}}^{\tilde{\phi}_+}\sin\phi\,\Psi_2(u_\varepsilon(\phi))\,d\phi + \int_{\tilde{\phi}_+}^{\gamma_2}\sin\phi\,\Psi_2(u_\varepsilon(\phi))\,d\phi \\ &\quad\le \int_{\tilde{\phi}_+}^{\gamma_2}\sin\phi\,\Psi_2(u_\varepsilon(\phi))\,d\phi\,.\end{aligned} \tag{3.37}$$

Since for $\phi \in [0,\gamma_2]$ we have $\sin\phi \le C\cos\phi$ and since by (3.33) and (2.9),

$$\frac{2}{\varepsilon}\int_{\tilde{\phi}_+}^{\gamma_2}\cos\phi\,\Psi_2(u_\varepsilon(\phi))\,d\phi \le \frac{C}{\varepsilon^{1/2}}\,,$$

we obtain from (3.36)–(3.37) that

$$\frac{C}{\varepsilon^{1/2}} \ge \frac{2}{\varepsilon}\cos\tilde{\phi}\,\Psi_2(u_\varepsilon(\tilde{\phi})) - \frac{C}{\varepsilon^{1/2}}\,,$$

and (3.35) follows as well. □

4. Proof of the lower-bound

In this section we prove the lower-bound energy estimate of Theorem 1.1. The next lemma provides a crucial estimate for the distance between γ_1 and γ_2.

Lemma 4.1. *We have* $\gamma_1 - \gamma_2 \sim \varepsilon \ln \frac{1}{\varepsilon}$.

Proof. Put

$$\bar{\phi}_2 = \sup\{\phi \in (\gamma_2, \gamma_1) \, : \, \Psi_2(u_\varepsilon(\phi)) = \frac{d_0}{2}\}, \tag{4.1}$$

$$\bar{\phi}_1 = \inf\{\phi \in (\gamma_2, \gamma_1) \, : \, \Psi_1(u_\varepsilon(\phi)) = \frac{d_0}{2}\}. \tag{4.2}$$

Note that $\widetilde{\Psi}(u_\varepsilon) \geq \frac{d_0}{2}$ on $[\bar{\phi}_2, \bar{\phi}_1]$ (see (2.2)). Therefore, also $W(u_\varepsilon) \geq c > 0$ on $[\bar{\phi}_2, \bar{\phi}_1]$ for some constant c, by (2.5). From the simple upper-bound

$$\frac{1}{\varepsilon^2} \int_{A_{\bar{\phi}_2, \bar{\phi}_1}} W(u_\varepsilon) \leq \frac{C}{\varepsilon}$$

we then easily conclude that

$$\bar{\phi}_1 - \bar{\phi}_2 \leq C\varepsilon . \tag{4.3}$$

It is therefore sufficient to prove the following:

$$\bar{\phi}_2 - \gamma_2 \sim \varepsilon \ln \frac{1}{\varepsilon} \quad \text{and} \quad \gamma_1 - \bar{\phi}_1 \sim \varepsilon \ln \frac{1}{\varepsilon} . \tag{4.4}$$

Clearly it is enough to prove the first estimate in (4.4), as the proof of the second one is identical.

We define the function $\tau_2(\phi)$ by the equation

$$u_\varepsilon(\phi) = \mathcal{G}_{\tilde{s}_2(u_\varepsilon(\phi))}(\tau_2(\phi)) , \quad \phi \in [\gamma_2, \gamma_1] . \tag{4.5}$$

We can write $u'_\varepsilon(\phi)$ as a sum of two orthogonal components, the first, $(u'_\varepsilon)_\nu$ in the direction $\boldsymbol{\nu} := \nabla\psi_2$ and the second, $(u'_\varepsilon)_\sigma$ in the direction of $\nabla\tilde{s}_2$. We therefore have

$$|u'_\varepsilon|^2 = |(u'_\varepsilon)_\nu|^2 + |(u'_\varepsilon)_\sigma|^2 . \tag{4.6}$$

It is easy to verify that the first component is given by

$$(u'_\varepsilon)_\nu = \nabla\Psi_2(u_\varepsilon(\phi))\tau'_2(\phi) . \tag{4.7}$$

Therefore,

$$
\begin{aligned}
\int_{\gamma_2}^{\bar{\phi}_2} & \Big(|u_\varepsilon'|^2 + \frac{W(u_\varepsilon)}{\varepsilon^2}\Big)\cos\phi \\
& \geq \int_{\gamma_2}^{\bar{\phi}_2} \Big(|(u_\varepsilon')_\nu|^2 + \frac{W(u_\varepsilon)}{\varepsilon^2}\Big)\cos\phi \\
& = \int_{\gamma_2}^{\bar{\phi}_2} \Big(|\nabla\Psi_2(u_\varepsilon)|^2(\tau_2')^2 + \frac{W(u_\varepsilon)}{\varepsilon^2}\Big)\cos\phi \\
& = \int_{\gamma_2}^{\bar{\phi}_2} W(u_\varepsilon)\big((\tau_2')^2 + \frac{1}{\varepsilon^2}\big)\cos\phi \\
& = \int_{\gamma_2}^{\bar{\phi}_2} \Big(W(u_\varepsilon)(\tau_2' - \frac{1}{\varepsilon})^2 + \frac{2}{\varepsilon}\frac{d}{d\phi}(\Psi_2(u_\varepsilon))\Big)\cos\phi\,.
\end{aligned}
\tag{4.8}
$$

We also have

$$
\int_{\bar{\phi}_2}^{\gamma_1} \Big(|u_\varepsilon'|^2 + \frac{W(u_\varepsilon)}{\varepsilon^2}\Big)\cos\phi \geq \int_{\bar{\phi}_2}^{\gamma_1} \frac{2}{\varepsilon}\frac{d}{d\phi}(\Psi_2(u_\varepsilon))\cos\phi\,. \tag{4.9}
$$

Note that by (3.1) we have

$$
\begin{aligned}
\int_{\gamma_2}^{\gamma_1} & \frac{d}{d\phi}(\Psi_2(u_\varepsilon))\cos\phi \\
& = (D - \varepsilon^{1/2})\cos\gamma_1 - \varepsilon^{1/2}\cos\gamma_2 \\
& \quad + \int_{\gamma_2}^{\gamma_1} \Psi_2(u_\varepsilon)\sin\phi \geq D\cos\phi_0 - C\varepsilon^{1/2}\,.
\end{aligned}
\tag{4.10}
$$

Combining (4.8)–(4.10) with (3.33) yields

$$
\int_{\gamma_2}^{\bar{\phi}_2} W(u_\varepsilon)(\tau_2' - \frac{1}{\varepsilon})^2 \leq C\varepsilon^{-1/2}\,. \tag{4.11}
$$

Since $W(u_\varepsilon) \geq C\varepsilon^{1/2}$ on $[\gamma_1, \gamma_2]$ we obtain from (4.11) that

$$
\int_{\gamma_2}^{\bar{\phi}_2} (\tau_2' - \frac{1}{\varepsilon})^2 \leq \frac{C}{\varepsilon}\,.
$$

Applying the last estimate together with the Cauchy-Schwarz inequality yields

$$
|\tau_2(\bar{\phi}_2) - \tau_2(\gamma_2) - \frac{\bar{\phi}_2 - \gamma_2}{\varepsilon}| \leq \int_{\gamma_2}^{\bar{\phi}_2} |\tau_2' - \frac{1}{\varepsilon}| \leq C\Big(\frac{\bar{\phi}_2 - \gamma_2}{\varepsilon}\Big)^{1/2}\,. \tag{4.12}
$$

From (2.6) we obtain that

$$
\tau_2(\bar{\phi}_2) = O(1) \quad \text{and} \quad \tau_2(\gamma_2) \sim -\ln\frac{1}{\varepsilon}\,. \tag{4.13}
$$

Plugging (4.13) in (4.12) yields the first estimate in (4.4). □

The next lemma provides an estimate for the distance between γ_2 (or γ_1) and ϕ_0 in terms of the two averages

$$\overline{m}_{1,\varepsilon} = \frac{\int_{\gamma_1}^{\pi/2} |u_\varepsilon| \cos\phi}{1 - \sin\gamma_1} \quad \text{and} \quad \overline{m}_{2,\varepsilon} = \frac{\int_{-\pi/2}^{\gamma_2} |u_\varepsilon| \cos\phi}{1 + \sin\gamma_2} . \tag{4.14}$$

Lemma 4.2. *There exist two constants $\alpha_1, \alpha_2 > 0$ (independent of ε) such that*

$$\cos\gamma_2 - \cos\phi_0 = \alpha_1(\overline{m}_{1,\varepsilon} - m_1) + \alpha_2(\overline{m}_{2,\varepsilon} - m_2) + O(\varepsilon \ln \frac{1}{\varepsilon}) . \tag{4.15}$$

Proof. Because of the constraint and Lemma 4.1 we have

$$\begin{aligned} &\Big(\int_{-\pi/2}^{\phi_0} \cos\phi\Big) m_2 + \Big(\int_{\phi_0}^{\pi/2} \cos\phi\Big) m_1 \\ &\quad = \Big(\int_{-\pi/2}^{\gamma_2} \cos\phi\Big)\overline{m}_{2,\varepsilon} + \Big(\int_{\gamma_2}^{\pi/2} \cos\phi\Big)\overline{m}_{1,\varepsilon} + O(\varepsilon \ln \frac{1}{\varepsilon}) , \end{aligned}$$

which can be rewritten as

$$\begin{aligned} &(\sin\phi_0 - \sin\gamma_2)(m_2 - m_1) \\ &\quad = (\sin\gamma_2 + 1)(\overline{m}_{2,\varepsilon} - m_2) \\ &\qquad + (1 - \sin\gamma_2)(\overline{m}_{1,\varepsilon} - m_1) + O(\varepsilon \ln \frac{1}{\varepsilon}) . \end{aligned} \tag{4.16}$$

Since $\sin\gamma_2 - \sin\phi_0 = O(\varepsilon^{1/2})$ (see (3.1) and Lemma 4.1), it follows from (4.16) that also

$$\overline{m}_{2,\varepsilon} - m_2 = O(\varepsilon^{1/2}) \text{ and } \overline{m}_{1,\varepsilon} - m_1 = O(\varepsilon^{1/2}) . \tag{4.17}$$

Therefore,

$$\begin{aligned} \sin\gamma_2 - \sin\phi_0 = &- \frac{1}{m_2 - m_1}(\sin\phi_0 + 1)(\overline{m}_{2,\varepsilon} - m_2) \\ &- \frac{1}{m_2 - m_1}(1 - \sin\phi_0)(\overline{m}_{1,\varepsilon} - m_1) + O(\varepsilon \ln \frac{1}{\varepsilon}) . \end{aligned} \tag{4.18}$$

Finally, a simple computation using Taylor formula leads from (4.18) to (4.15) with

$$\alpha_1 = \frac{\tan\phi_0(1 - \sin\phi_0)}{m_2 - m_1} \quad \text{and} \quad \alpha_2 = \frac{\tan\phi_0(1 + \sin\phi_0)}{m_2 - m_1} . \tag{4.19}$$

□

Next we shall introduce some more notation. For ϕ's for which $\Psi_j(u_\varepsilon(\phi)) \le d_0$ we define $v_j = v_{j,\varepsilon}(\phi) = \tilde{s}_j(u_\varepsilon(\phi))$, $j = 1, 2$. In particular we set

$$p_1^\varepsilon = v_1(\gamma_1) = \tilde{s}_1(u_\varepsilon(\gamma_1)) \quad \text{and} \quad p_2^\varepsilon = v_2(\gamma_2) = \tilde{s}_2(u_\varepsilon(\gamma_2))\,. \tag{4.20}$$

Recall the distance function $d_W^{(\delta)}$ that was defined in (2.8). We shall next use it for $\delta = \varepsilon^{1/2}$.

Lemma 4.3. *We have*

$$\begin{aligned} &\frac{1}{2\pi} E_\varepsilon(u_\varepsilon) \\ &\ge \int_{-\pi/2+\varepsilon^{1/3}}^{\gamma_2} \Big(|v_2'|^2 + \frac{2\beta d_W^{(\varepsilon^{1/2})}(p_1^\varepsilon, p_2^\varepsilon)}{\varepsilon} (|v_2(\phi)| - m_2) \Big) \cos\phi \\ &+ \int_{\gamma_1}^{\pi/2-\varepsilon^{1/3}} \Big(|v_1'|^2 + \frac{2\beta d_W^{(\varepsilon^{1/2})}(p_1^\varepsilon, p_2^\varepsilon)}{\varepsilon} (|v_1(\phi)| - m_1) \Big) \cos\phi \\ &+ \frac{2 d_W^{(\varepsilon^{1/2})}(p_1^\varepsilon, p_2^\varepsilon) \cos\phi_0}{\varepsilon} - C\varepsilon^{-5/12}\,, \end{aligned} \tag{4.21}$$

where $\beta = \frac{\tan\phi_0}{m_2 - m_1}$.

Proof. Note first that for $\phi \in [-\pi/2 + \varepsilon^{1/3}, \gamma_2]$ we have $\cos\phi \ge C\varepsilon^{1/3}$, so by Lemma 3.1

$$\Psi_2(u_\varepsilon(\phi)) \le C\varepsilon^{1/6}\,, \tag{4.22}$$

hence v_2 is well-defined. Similarly, v_1 is well-defined for $\phi \in [\gamma_1, \pi/2 - \varepsilon^{1/3}]$. By the definition of γ_1 and γ_2, (4.15) and (4.17) we have

$$\begin{aligned} &\frac{1}{2\pi} E_\varepsilon(u_\varepsilon, A_{\gamma_2,\gamma_1}) \ge 2\frac{\cos\gamma_1}{\varepsilon} (d_W^{(\varepsilon^{1/2})}(p_1^\varepsilon, p_2^\varepsilon) - 2\varepsilon^{1/2}) \\ &\ge \frac{2}{\varepsilon}\Big(\cos\phi_0 + \sum_{j=1}^2 \alpha_j(\overline{m}_{j,\varepsilon} - m_j) + O(\varepsilon \ln\frac{1}{\varepsilon})\Big) (d_W^{(\varepsilon^{1/2})}(p_1^\varepsilon, p_2^\varepsilon) - 2\varepsilon^{1/2}) \\ &\ge \frac{2 d_W^{(\varepsilon^{1/2})}(p_1^\varepsilon, p_2^\varepsilon)}{\varepsilon} \big(\cos\phi_0 + \sum_{j=1}^2 \alpha_j(\overline{m}_{j,\varepsilon} - m_j)\big) - \frac{4\cos\phi_0}{\varepsilon^{1/2}} + O(|\ln\varepsilon|)\,. \end{aligned} \tag{4.23}$$

From (4.23) and the upper bound (2.9) we deduce that

$$E_\varepsilon(u_\varepsilon, S^2 \setminus A_{\gamma_1,\gamma_2}) \le \frac{C}{\varepsilon^{1/2}}\,. \tag{4.24}$$

By (4.24)

$$\int_0^{\gamma_2} \Psi_2(u_\varepsilon)\cos\phi \le C\int_0^{\gamma_2} W(u_\varepsilon)\cos\phi \le C\varepsilon^2 E_\varepsilon(u_\varepsilon, S^2\setminus A_{\gamma_1,\gamma_2})$$
$$\le C\varepsilon^2\varepsilon^{-1/2} = C\varepsilon^{3/2},$$

so there exists $\theta_2 \in (0,\gamma_2)$ such that

$$\Psi_2(u_\varepsilon(\theta_2)) \le C\varepsilon^{3/2}. \tag{4.25}$$

From (4.25) we get by the same computation as in (3.36) the lower bound for the *normal* energy (see (4.7)–(4.8)), i.e.,

$$\begin{aligned}
&\frac{1}{2\pi}E_\varepsilon^\nu(u_\varepsilon, A_{\theta_2,\gamma_2})\\
&:= \int_{\theta_2}^{\gamma_2}\left(|(u_\varepsilon')_\nu|^2 + \frac{W(u_\varepsilon)}{\varepsilon^2}\right)\cos\phi\\
&\ge \frac{2}{\varepsilon}\Big(\cos\gamma_2\,\varepsilon^{1/2} - \cos\theta_1\,\Psi_2(u_\varepsilon(\theta_2)) + \int_{\theta_1}^{\gamma_2}\sin\phi\,\Psi_2(u_\varepsilon)\Big)\\
&\ge \frac{2\cos\gamma_2}{\varepsilon^{1/2}} - C\varepsilon^{1/2} \ge \frac{2\cos\phi_0}{\varepsilon^{1/2}} - C\,.
\end{aligned} \tag{4.26}$$

Similarly, there exists $\theta_1 \in (\gamma_1, \gamma_1 + \varepsilon^{1/2})$ such that $\Psi_1(u_\varepsilon(\theta_1)) \le C\varepsilon$. Hence,

$$\begin{aligned}
&\frac{1}{2\pi}E_\varepsilon^\nu(u_\varepsilon, A_{\gamma_1,\theta_1})\\
&\ge \frac{2}{\varepsilon}\int_{\gamma_1}^{\theta_1}(\cos\phi_0 - C\varepsilon^{1/2})\Big|\frac{d}{d\phi}(\Psi_1(u_\varepsilon))\Big|\\
&\ge \frac{2}{\varepsilon}(\cos\phi_0 - C\varepsilon^{1/2})(\varepsilon^{1/2} - C\varepsilon) \ge \frac{2\cos\phi_0}{\varepsilon^{1/2}} - C\,.
\end{aligned} \tag{4.27}$$

Next we take into account also the contribution from the "tangential" energy, i.e., of $(u_\varepsilon')_\sigma$. As in [2] we have

$$(u_\varepsilon')_\sigma = u_\varepsilon'\cdot\frac{\nabla\tilde{s}_2(u_\varepsilon)}{|\nabla\tilde{s}_2(u_\varepsilon)|} = \frac{\frac{d}{d\phi}(\tilde{s}_2(u_\varepsilon))}{|\nabla\tilde{s}_2(u_\varepsilon)|} = \frac{v_2'}{|\nabla\tilde{s}_2(u_\varepsilon)|}, \tag{4.28}$$

at points where $\Psi_2(u_\varepsilon) < d_0$. A simple modification of the argument of the proof of [2] shows that (4.28) implies that

$$|(u_\varepsilon')_\sigma|^2 \ge |v_2'|^2(1 - cd(v_2,\Gamma_2)) \ge |v_2'|^2(1-\kappa\Psi_2^{1/2}(u_\varepsilon)), \tag{4.29}$$

for some constant $\kappa > 0$. Combining (4.26) with (4.29) in conjunction with (4.22) and (4.24) leads to

$$\begin{aligned}\frac{1}{2\pi}&E_\varepsilon(u_\varepsilon, A_{-\pi/2+\varepsilon^{1/3},\gamma_2})\\ &\ge \frac{2\cos\phi_0}{\varepsilon^{1/2}} + \int_{-\pi/2+\varepsilon^{1/3}}^{\gamma_2} \cos\phi(1-\kappa\Psi_2^{1/2}(u_\varepsilon))|v_2'|^2 - C \\ &\ge \frac{2\cos\phi_0}{\varepsilon^{1/2}} + \int_{-\pi/2+\varepsilon^{1/3}}^{\gamma_2} |v_2'|^2\cos\phi - C\varepsilon^{-5/12}.\end{aligned} \tag{4.30}$$

Similarly,

$$\begin{aligned}\frac{1}{2\pi}&E_\varepsilon(u_\varepsilon, A_{\gamma_1,\pi/2-\varepsilon^{1/3}})\\ &\ge \frac{2\cos\phi_0}{\varepsilon^{1/2}} + \int_{\gamma_1}^{\pi/2-\varepsilon^{1/3}} |v_1'|^2\cos\phi - C\varepsilon^{-5/12}.\end{aligned} \tag{4.31}$$

Adding together (4.23) with (4.30) and (4.31) leads to

$$\begin{aligned}\frac{1}{2\pi}E_\varepsilon(u_\varepsilon) \ge{}& \frac{2d_W^{(\varepsilon^{\frac12})}(p_1^\varepsilon,p_2^\varepsilon)\cos\phi_0}{\varepsilon} + \frac{2d_W^{(\varepsilon^{\frac12})}(p_1^\varepsilon,p_2^\varepsilon)}{\varepsilon}\sum_{j=1}^2 \alpha_j(\overline{m}_{j,\varepsilon}-m_j)\\ &+\int_{-\frac{\pi}{2}+\varepsilon^{\frac13}}^{\gamma_2}|v_2'|^2\cos\phi + \int_{\gamma_1}^{\pi/2-\varepsilon^{\frac13}}|v_1'|^2\cos\phi - C\varepsilon^{-\frac{5}{12}}.\end{aligned} \tag{4.32}$$

Next we estimate the second term on the r.h.s. of (4.32). First, write

$$\begin{aligned}\overline{m}_{1,\varepsilon}(1-\sin\gamma_1) &= \int_{\gamma_1}^{\frac{\pi}{2}} |u_\varepsilon|\cos\phi\\ = \int_{\frac{\pi}{2}-\varepsilon^{\frac13}}^{\frac{\pi}{2}}|u_\varepsilon|\cos\phi &+ \int_{\gamma_1}^{\frac{\pi}{2}-\varepsilon^{\frac13}}|v_1|\cos\phi + \int_{\gamma_1}^{\frac{\pi}{2}-\varepsilon^{\frac13}}(|u_\varepsilon|-|v_1|)\cos\phi.\end{aligned} \tag{4.33}$$

Since (H_3) implies that $|u_\varepsilon| \le C(1+W(u_\varepsilon))$ we get, using also (4.24), that

$$\begin{aligned}\int_{\pi/2-\varepsilon^{1/3}}^{\pi/2}|u_\varepsilon|\cos\phi &\le C(1-\cos(\varepsilon^{1/3})) + C\int_{\pi/2-\varepsilon^{1/3}}^{\pi/2} W(u_\varepsilon)\cos\phi\\ &\le C(\varepsilon^{2/3}+\varepsilon^{3/2}).\end{aligned} \tag{4.34}$$

Using (2.5), (4.24) and the Cauchy-Schwarz inequality we obtain

$$\begin{aligned}\Big|\int_{\gamma_1}^{\pi/2-\varepsilon^{1/3}}&(|u_\varepsilon|-|v_1|)\cos\phi\Big|\\ &\le C\int_{\gamma_1}^{\pi/2-\varepsilon^{1/3}}\Psi_1^{1/2}(u_\varepsilon)\cos\phi \le C\varepsilon^{3/4}.\end{aligned} \tag{4.35}$$

From (4.33)–(4.35) we get

$$\overline{m}_{1,\varepsilon}(1-\sin\gamma_1) = \int_{\gamma_1}^{\pi/2-\varepsilon^{1/3}} |v_1|\cos\phi + O(\varepsilon^{2/3}),$$

from which we obtain that

$$\overline{m}_{1,\varepsilon} - m_1 = \frac{1}{1-\sin\gamma_1}\int_{\gamma_1}^{\pi/2-\varepsilon^{1/3}} (|v_1|-m_1)\cos\phi + O(\varepsilon^{2/3}). \qquad (4.36)$$

A similar argument yields

$$\overline{m}_{2,\varepsilon} - m_2 = \frac{1}{1+\sin\gamma_2}\int_{-\pi/2+\varepsilon^{1/3}}^{\gamma_2} (|v_2|-m_2)\cos\phi + O(\varepsilon^{2/3}). \qquad (4.37)$$

Since $|\gamma_2-\phi_0|, |\gamma_1-\phi_0| \leq C\varepsilon^{1/2}$ (see (3.1) and Lemma 4.1), we finally deduce (4.21) from (4.32) and (4.36)–(4.37). □

The following lemma is an immediate consequence of Lemma 4.3.

Lemma 4.4. *There exists a constant C, independent of ε, such that*

$$d_W^{(\varepsilon^{1/2})}(p_1^\varepsilon, p_2^\varepsilon) \leq D + C\varepsilon^{1/2}. \qquad (4.38)$$

Proof. It suffices to combine the upper bound (2.9) with (4.21) (taking into account only the third term on the r.h.s.). □

The next lemma provides a lower bound for the two integrals on the right hand side of (4.21).

Lemma 4.5. *We have*

$$\begin{aligned}\int_{\gamma_1}^{\pi/2-\varepsilon^{1/3}} &\Big(|v_1'|^2 + \frac{2\beta d_W^{(\varepsilon^{1/2})}(p_1^\varepsilon,p_2^\varepsilon)}{\varepsilon}(|v_1(\phi)|-m_1)\Big)\cos\phi \\ &\geq \frac{\cos\phi_0}{\varepsilon^{1/2}}\sqrt{2\beta d_W^{(\varepsilon^{1/2})}(p_1^\varepsilon,p_2^\varepsilon)}\,\delta_1(p_1^\varepsilon,\mathcal{M}_1) + o_\varepsilon(1).\end{aligned} \qquad (4.39)$$

and

$$\begin{aligned}\int_{-\pi/2+\varepsilon^{1/3}}^{\gamma_2} &\Big(|v_2'|^2 + \frac{2\beta d_W^{(\varepsilon^{1/2})}(p_1^\varepsilon,p_2^\varepsilon)}{\varepsilon}(|v_2(\phi)|-m_2)\Big)\cos\phi \\ &\geq \frac{\cos\phi_0}{\varepsilon^{1/2}}\sqrt{2\beta d_W^{(\varepsilon^{1/2})}(p_1^\varepsilon,p_2^\varepsilon)}\,\delta_2(p_2^\varepsilon\mathcal{M}_2) + o_\varepsilon(1).\end{aligned} \qquad (4.40)$$

Proof. Clearly it suffices to prove (4.39). First, note that

$$\begin{aligned}J_1(\varepsilon) &:= \varepsilon^{\frac{1}{2}} \int_{\gamma_1}^{\frac{\pi}{2}-\varepsilon^{\frac{1}{3}}} \Big(|v_1'|^2 + \frac{2\beta d_W^{(\varepsilon^{\frac{1}{2}})}(p_1^\varepsilon, p_2^\varepsilon)}{\varepsilon} (|v_1(\phi)| - m_1) \Big) \cos\phi \\ &\le C\,,\end{aligned} \tag{4.41}$$

by the upper-bound (2.9) and Lemma 4.3. We introduce a new variable t by $\phi = \gamma_1 + t\varepsilon^{1/2}$ and then define a new function by

$$w(t) = v_1(\gamma_1 + t\varepsilon^{1/2})\,, \quad 0 \le t \le \frac{\pi/2 - \varepsilon^{1/3} - \gamma_1}{\varepsilon^{1/2}}\,. \tag{4.42}$$

We have

$$\begin{aligned}J_1(\varepsilon) &\ge \varepsilon^{\frac{1}{2}} \int_{\gamma_1}^{\gamma_1+\varepsilon^{\frac{1}{4}}} \Big(|v_1'|^2 + \frac{2\beta d_W^{(\varepsilon^{\frac{1}{2}})}(p_1^\varepsilon, p_2^\varepsilon)}{\varepsilon} (|v_1(\phi)| - m_1) \Big) \cos\phi\, d\phi \\ &\ge \varepsilon^{\frac{1}{2}}(\cos\gamma_1 - c\varepsilon^{\frac{1}{4}}) \int_{\gamma_1}^{\gamma_1+\varepsilon^{\frac{1}{4}}} \Big(|v_1'|^2 + \frac{2\beta d_W^{(\varepsilon^{\frac{1}{2}})}(p_1^\varepsilon, p_2^\varepsilon)}{\varepsilon} (|v_1(\phi)| - m_1) \Big)\, d\phi \\ &= (\cos\gamma_1 - c\varepsilon^{\frac{1}{4}}) \int_0^{\varepsilon^{-\frac{1}{4}}} \Big(|w'|^2 + 2\beta d_W^{(\varepsilon^{\frac{1}{2}})}(p_1^\varepsilon, p_2^\varepsilon)(|w(t)| - m_1) \Big)\, dt\,. \end{aligned} \tag{4.43}$$

By (4.41) we have

$$\int_0^{\varepsilon^{-1/4}} \Big(|w'|^2 + 2\beta d_W^{(\varepsilon^{1/2})}(p_1^\varepsilon, p_2^\varepsilon)(|w(t)| - m_1) \Big)\, dt \le C\,.$$

Hence, there exists $t_\varepsilon \in (0, \varepsilon^{-1/4}/2)$ such that $y_\varepsilon := w(t_\varepsilon)$ satisfies

$$|y_\varepsilon| - m_1 \le C\varepsilon^{1/4}\,. \tag{4.44}$$

From (4.44) it follows that there exists $z_1^\varepsilon \in \mathcal{M}_1$ such that $|z_1^\varepsilon - y_\varepsilon| = o(1)$. It follows then that there exists a function $\delta(\varepsilon) = o_\varepsilon(1)$ such that

$$|z| - m_1 \le \delta(\varepsilon)\,, \ \forall z \in J_1(y_\varepsilon, z_1^\varepsilon)\,, \tag{4.45}$$

see after (2.16) for the definition of J_1. We shall next define a new function $\widetilde{w}$ on $[0,\infty)$. First, on the interval $[0, t_\varepsilon]$ we set $\widetilde{w} = w$. Then, on the interval $(t_\varepsilon, t_\varepsilon + (\delta(\varepsilon))^{-1/2}]$ we let $\widetilde{w}$ go from $w(t_\varepsilon)$ to z_1^ε along Γ_1 in constant velocity. Finally, we set $\widetilde{w}(t) \equiv z_1^\varepsilon$ on $(t_\varepsilon + (\delta(\varepsilon))^{-1/2}, \infty)$. It is clear from the above construction that $\widetilde{w}$ satisfies

$$\int_{t_\varepsilon}^\infty \Big(|\widetilde{w}'|^2 + 2\beta d_W^{(\varepsilon^{1/2})}(p_1^\varepsilon, p_2^\varepsilon)(|\widetilde{w}(t)| - m_1) \Big)\, dt = o_\varepsilon(1)\,. \tag{4.46}$$

From (4.46) and (1.19) it follows that

$$\int_0^{\varepsilon^{-1/4}} \Big(|w'|^2 + 2\beta d_W^{(\varepsilon^{1/2})}(p_1^\varepsilon, p_2^\varepsilon)(|w(t)| - m_1)\Big)\, dt \\ \geq \sqrt{2\beta d_W^{(\varepsilon^{1/2})}(p_1^\varepsilon, p_2^\varepsilon)}\, \delta_1(p_1^\varepsilon, \mathcal{M}_1) + o_\varepsilon(1)\,. \tag{4.47}$$

Combining (4.47) with (4.43) we are led to (4.39). □

In order to complete the proof of the lower-bound part of (1.22) we need to identify the limit of the points p_1^ε and p_2^ε. This is the purpose of the next lemma.

Lemma 4.6. *Suppose that for a subsequence,* $\lim p_1^{\varepsilon_n} = p_1$ *and* $\lim p_2^{\varepsilon_n} = p_2$. *Then* p_1 *and* p_2 *are the end points of some geodesic from* $\mathcal{G}$.

Proof. Consider a constant unit velocity geodesic γ_n realizing $d_W^{(\varepsilon_n^{1/2})}(p_1^{\varepsilon_n}, p_2^{\varepsilon_n})$, which passes between the points $y_1^{\varepsilon_n} = u_{\varepsilon_n}(\gamma_1)$ and $y_2^{\varepsilon_n} = u_{\varepsilon_n}(\gamma_2)$ (see (4.20)). For each $\delta > 0$ we let $p_{1,n}^\delta$ be the last point on that geodesic where $\Psi_1(p_{1,n}^\delta) = \delta$. Similarly, let $p_{2,n}^\delta$ be the first point on that geodesic where $\Psi_2(p_{2,n}^\delta) = \delta$. Denote by γ_n^δ the part of γ_n between $p_{1,n}^\delta$ and $p_{2,n}^\delta$. Passing to a subsequence if necessary we obtain the convergence of γ_n^δ towards a limiting geodesic with end points p_1^δ and p_2^δ which must satisfy $d_W(p_1^\delta, p_2^\delta) = D - 2\delta$ (because of (4.38)). Repeating this process with a sequence $\delta_m \to 0$ and passing to a diagonal subsequence, we deduce (for a further subsequence) that γ_n converges to a curve $\underline{\tilde{\gamma}}$ joining a point $q_1 \in \Gamma_1$ to a point $q_2 \in \Gamma_1$ with the following property: for each small enough $\delta > 0$ we have $d_W(p_1^\delta, p_2^\delta) = D - 2\delta$, where p_j^δ, $j = 1, 2$, are the points on that geodesic satisfying $\Psi_j(p_1^\delta) = \delta$, respectively. It follows that $\underline{\tilde{\gamma}} \in \mathcal{G}$. The result of the lemma would follow once we show that $q_1 = p_1$ and $q_2 = p_2$.

Looking for a contradiction, assume for example that $q_1 \neq p_1$. Let δ be a small positive number that we shall fix later. We may choose n large enough such that $|p_{1,n}^\delta - p_1^\delta| < \delta$. We now denote by $\tilde{\gamma}_n^\delta$ a re-parametrization on the interval $[0, 1]$ of the part of γ_n between $y_1^{\varepsilon_n}$ and $p_{1,n}^\delta$ such that

$$\sqrt{W(\tilde{\gamma}_n^\delta)}\, |(\tilde{\gamma}_n^\delta)'| \equiv const \quad \text{on } [0,1]\,.$$

Using the same notations as in the proof of Lemma 4.1 we can decompose $(\tilde{\gamma}_n^\delta)'$ as a sum of two orthogonal components: $(\tilde{\gamma}_n^\delta)'_\nu$ in the direction of

$\nabla\Psi_1$, and $(\tilde{\gamma}_n^\delta)'_\sigma$ in the direction of $\nabla\tilde{s}_1$. We have

$$\Big(\int_0^1 \sqrt{W(\tilde{\gamma}_n^\delta)}\,|(\tilde{\gamma}_n^\delta)'|\Big)^2 = \int_0^1 W(\tilde{\gamma}_n^\delta)\left(|(\tilde{\gamma}_n^\delta)'_\nu|^2 + |(\tilde{\gamma}_n^\delta)'_\sigma|^2\right). \tag{4.48}$$

Next, note that

$$\begin{aligned}\int_0^1 W(\tilde{\gamma}_n^\delta)\,|(\tilde{\gamma}_n^\delta)'_\nu|^2 &\geq \Big(\int_0^1 \sqrt{W(\tilde{\gamma}_n^\delta)}\,|(\tilde{\gamma}_n^\delta)'_\nu|\Big)^2\\ &= \Big(\int_0^1 |\nabla\Psi_1(\tilde{\gamma}_n^\delta)|\,|(\tilde{\gamma}_n^\delta)'_\nu|\Big)^2 \geq (\delta - \varepsilon_n^{1/2})^2. \end{aligned}\tag{4.49}$$

Using (2.5) we find

$$\int_0^1 W(\tilde{\gamma}_n^\delta)\,|(\tilde{\gamma}_n^\delta)'_\sigma|^2 \geq c_1\varepsilon_n^{1/2}\Big(\int_0^1 |(\tilde{\gamma}_n^\delta)'_\sigma|\Big)^2 \geq c_2\varepsilon_n^{1/2}, \tag{4.50}$$

for some positive constants c_1 and c_2 which do not depend on ε_n or on δ (for δ small enough). Plugging (4.49)–(4.50) in (4.48) yields (provided δ is chosen small enough),

$$\begin{aligned}\int_0^1 \sqrt{W(\tilde{\gamma}_n^\delta)}\,|(\tilde{\gamma}_n^\delta)'| &\geq \left((\delta - \varepsilon_n^{1/2})^2 + c_2\varepsilon_n^{1/2}\right)^{1/2}\\ &\geq \left(\delta^2 + c_3\varepsilon_n^{1/2}\right)^{1/2} \geq \delta + c_4\frac{\varepsilon_n^{1/2}}{\delta}. \end{aligned}\tag{4.51}$$

From (4.51) we deduce that

$$d_W^{(\varepsilon_n^{1/2})}(p_1^{\varepsilon_n}, p_2^{\varepsilon_n}) \geq D + \varepsilon_n^{1/2} + c_4\frac{\varepsilon_n^{1/2}}{\delta},$$

which contradicts (4.38), provided δ is chosen small enough. □

The next proposition provides the lower-bound estimate needed for assertion (ii) of Theorem 1.1.

Proposition 4.1. *We have*

$$\frac{1}{2\pi}E_\varepsilon(u_\varepsilon) \geq \cos\phi_0\Big(\frac{2D}{\varepsilon} + \sqrt{2\beta D}\frac{K}{\varepsilon^{1/2}}\Big) + o(\varepsilon^{-\frac{1}{2}}). \tag{4.52}$$

Proof. Plugging the estimates (4.39)–(4.40) in (4.21) and using the obvious estimate $d_W^{(\varepsilon^{1/2})}(p_1^\varepsilon, p_2^\varepsilon) \geq D$ yields

$$\begin{aligned}\frac{1}{2\pi}E_\varepsilon(u_\varepsilon) \geq \frac{2D\cos\phi_0}{\varepsilon} &+ \frac{\cos\phi_0}{\varepsilon^{1/2}}\sqrt{2\beta D}\,\Big(\delta_1(p_1^\varepsilon, \mathcal{M}_1)\\ &+ \delta_2(p_2^\varepsilon, \mathcal{M}_2)\Big) + o(\varepsilon^{-1/2}). \end{aligned}\tag{4.53}$$

Therefore, (4.52) is a direct consequence of (4.53), Lemma 4.6 and the definition (1.23) of K. □

5. A refined upper-bound and the convergence result

Next we prove the upper-bound part of the energy expansion (1.22).

Proposition 5.1. *We have*

$$\frac{1}{2\pi}E_\varepsilon(u_\varepsilon) \le \cos\phi_0\Big(\frac{2D}{\varepsilon} + \sqrt{2\beta D}\frac{K}{\varepsilon^{1/2}}\Big) + o(\varepsilon^{-\frac{1}{2}})\,. \tag{5.1}$$

Proof. We are going to refine the construction used in the proof of Proposition 2.1, using the insight we got from the lower-bound estimates we established so far. We first fix $x^{(1)} \in \mathcal{M}_1, x^{(2)} \in \mathcal{M}_2$ and $\underline{\gamma}^{(i_0)} \in \mathcal{G}$ that realizes the minimum for K in (1.23) and denote by $p_1 = \zeta_1^{(i_0)}$ and $p_2 = \zeta_2^{(i_0)}$ the end points of $\underline{\gamma}^{(i_0)}$. For these values of $x^{(1)}, x^{(2)}, p_1$ and p_2 we shall slightly modify the construction of a test function v_ε, that was described in the proof of Proposition 2.1 on 7 intervals, from (i) to (vii). Actually, we shall modify the construction only on the intervals (i)–(ii) and (vi)–(vii). From Lemma 1.1 it follows that there exists a minimizer w_1 that realizes $\delta_1(p_1, \mathcal{M}_1)$ with $\lim_{t\to\infty} w_1(t) = x^{(1)}$. We then define

$$\widetilde{w}_1(t) = w_1(\sqrt{2\beta D}t)\,,\ t \in [0,\infty)\,,$$

so that

$$\int_0^\infty \Big(|\widetilde{w}_1'|^2 + 2\beta D(|\widetilde{w}_1| - m_1)\Big)\,dt = \sqrt{2\beta D}\,\delta_1(p_1, \mathcal{M}_1)\,.$$

Let $\bar{\phi}$ be a number whose distance to ϕ_0 is of the order $O(\varepsilon^{1/2})$, that will be determined later. We first set

$$v_\varepsilon(\phi) = \widetilde{w}_1\Big(\frac{\phi - \bar{\phi}}{\varepsilon^{1/2}}\Big) \quad \text{for } \phi \in \Big[\bar{\phi}, \frac{\phi_0 + \pi/2}{2}\Big]\,. \tag{5.2}$$

Then, on the interval $[\frac{\phi_0+\pi/2}{2}, \pi/2]$ define v_ε in such a way that it follows the curve Γ_1 from the point $v_\varepsilon(\frac{\phi_0+\pi/2}{2})$ to the point $x^{(1)}$ in a constant velocity. From (1.20) and (5.2) it easily follows that

$$\frac{1}{2\pi}E_\varepsilon(v_\varepsilon, A_{\bar{\phi},\pi/2}) = \cos\bar{\phi}\,\sqrt{\beta D/2}\,\delta_1(p_1, \mathcal{M}_1)\varepsilon^{-1/2} + o(\varepsilon^{-1/2})\,. \tag{5.3}$$

The above construction replaces the construction on the intervals (i) and (ii) in the proof of Proposition 2.1. From here we follow exactly that construction on the intervals (iii)–(v). Finally, the construction on the intervals (vi)–(vii) is modified in a similar manner to the above, and yields the analogous estimate to (5.3), namely

$$\frac{1}{2\pi}E_\varepsilon(v_\varepsilon, A_{-\pi/2,\bar\phi-2\varepsilon-(\frac{1}{c_1}+\frac{1}{c_2})\varepsilon\ln\frac{1}{\varepsilon}}) \\ = \cos\bar\phi\,\sqrt{\beta D/2}\,\delta_2(p_2,\mathcal{M}_2)\varepsilon^{-1/2} + o(\varepsilon^{-1/2})\,. \tag{5.4}$$

Combining (5.3) and (5.4) with the estimates from the proof of Proposition 2.1 yields

$$\frac{1}{2\pi}E_\varepsilon(v_\varepsilon) \le \cos\bar\phi\Big(\frac{2D}{\varepsilon} + \sqrt{2\beta D}\,(\delta_1(p_1,\mathcal{M}_1)+\delta_2(p_2,\mathcal{M}_2))\varepsilon^{-1/2}\Big) \\ + o(\varepsilon^{-1/2})\,. \tag{5.5}$$

It remains to fix the value of $\bar\phi$ in such a way that the constraint (2.10) is satisfied. Put

$$\mu_1 = \frac{\int_{\bar\phi}^{\pi/2}|v_\varepsilon|\cos\phi\,d\phi}{1-\sin\bar\phi} \quad\text{and}\quad \mu_2 = \frac{\int_{-\pi/2}^{\bar\phi}|v_\varepsilon|\cos\phi\,d\phi}{1+\sin\bar\phi}\,.$$

By our construction of v_ε and (1.20) we have

$$(\mu_1 - m_1)(1-\sin\bar\phi) = \int_{\bar\phi}^{\pi/2}(|v_\varepsilon| - m_1)\cos\phi \\ = \frac{\delta_1(p_1,\mathcal{M}_1)}{2\sqrt{2\beta D}}(\cos\bar\phi)\varepsilon^{1/2} + o(\varepsilon^{1/2}) \tag{5.6}$$

and

$$(\mu_2 - m_2)(1+\sin\bar\phi) = \int_{-\pi/2}^{\bar\phi}(|v_\varepsilon| - m_2)\cos\phi \\ = \frac{\delta_2(p_2,\mathcal{M}_2)}{2\sqrt{2\beta D}}(\cos\bar\phi)\varepsilon^{1/2} + o(\varepsilon^{1/2})\,. \tag{5.7}$$

We have also a third equation coming from the constraint,

$$\mu_1(1-\sin\bar\phi) + \mu_2(1+\sin\bar\phi) = m_1(1-\sin\phi_0) + m_2(1+\sin\phi_0)\,. \tag{5.8}$$

Writing $\bar\phi = \phi_0 - \tau\varepsilon^{1/2}$ and using first order Taylor expansion yields from (5.6)–(5.8) a perturbed linear system of 3 equations in the 3 unknowns μ_1, μ_2 and τ which has a solution

$$\tau = \sqrt{\frac{\beta}{8D}}(\delta_1(p_1,\mathcal{M}_1) + \delta_2(p_2,\mathcal{M}_2))\cot\phi_0 + o(1)\,. \tag{5.9}$$

For the value of τ given by (5.9) we find by the Taylor expansion for the cosine function that

$$\begin{aligned}\cos\bar{\phi} &= \cos\phi_0 + \tau\varepsilon^{1/2}\sin\phi_0 + o(\varepsilon^{1/2}) \\ &= \cos\phi_0\Big(1 + \sqrt{\frac{\beta}{8D}}(\delta_1+\delta_2)\varepsilon^{1/2}\Big) + o(\varepsilon^{1/2}).\end{aligned} \tag{5.10}$$

Finally, plugging (5.10) in (5.5) gives (5.1). □

Proof of Theorem 1.1 completed. Since the energy estimate (1.22) follows from Proposition 4.1 and Proposition 5.1, it remains to prove the convergence result, i.e., assertion (i). Passing to a subsequence, we may deduce from Lemma 4.6 that $\lim p_1^{\varepsilon_n} = p_1 = \zeta_1^{(i_0)}$ and $\lim p_2^{\varepsilon_n} = p_2 = \zeta_2^{(i_0)}$ for some geodesic $\underline{\gamma}^{(i_0)} \in \mathcal{G}$. Moreover, from (4.53) and (5.1) we deduce that there is a pair of corresponding points, $x^{(j)} \in \mathcal{M}_j$, $j=1,2$, that together with $\underline{\gamma}^{(i_0)}$, realize the minimum K in (1.23). Next we shall show that $u_{\varepsilon_n} \to x^{(1)}$ uniformly on $[\phi_0+\delta, \frac{\pi}{2}-\delta]$ for any $\delta>0$ (the uniform convergence $u_{\varepsilon_n} \to x^{(2)}$ on $[-\frac{\pi}{2}+\delta, \phi_0-\delta]$ is proved in the same manner). A direct consequence of (3.34) is that

$$d(u_\varepsilon, \Gamma_1) \to 0 \quad \text{uniformly on } [\phi_0+\delta, \frac{\pi}{2}-\delta]. \tag{5.11}$$

Therefore, whenever a sequence $\{\phi_n\} \subset [\phi_0+\delta, \frac{\pi}{2}-\delta]$ satisfies $u_{\varepsilon_n}(\phi_n) \to x_1$, then necessarily $x_1 \in \Gamma_1$. Combining (4.21) with the upper-bound (5.1) (or (2.9)) we get

$$\begin{aligned}\int_{\gamma_1}^{\pi/2-\varepsilon^{1/3}} \Big(|v_{1,\varepsilon}'|^2 + \frac{2\beta d_W^{(\varepsilon^{1/2})}(p_1^\varepsilon, p_2^\varepsilon)}{\varepsilon}(|v_{1,\varepsilon}(\phi)| - m_1)\Big)\cos\phi \\ \le C\varepsilon^{1/2},\end{aligned} \tag{5.12}$$

where $v_{1,\varepsilon} = \tilde{s}_1(u_\varepsilon)$. A direct consequence of (5.12) is that for each $\eta>0$ we have

$$\text{meas}\Big(\{\phi \in [\phi_0+\delta, \frac{\pi}{2}-\delta]\} : |v_{1,\varepsilon}(\phi)| - m_1 > \eta\Big) \le C(\eta)\varepsilon^{1/2}. \tag{5.13}$$

Since (5.13) implies that $\lim_{\varepsilon\to 0}|v_{1,\varepsilon}(\phi)| = m_1$ in measure, for a subsequence we have $\lim_{\varepsilon_n\to 0}|v_{1,\varepsilon_n}(\phi)| = m_1$ a.e. on $(\phi_0, \frac{\pi}{2})$. Consider then a point $\tilde{\phi} \in [\phi_0+\delta, \frac{\pi}{2}-\delta]$ such that (possibly for a further subsequence) $\lim_{\varepsilon_n\to 0} u_{\varepsilon_n}(\tilde{\phi}) = \lim_{\varepsilon_n\to 0} v_{1,\varepsilon_n}(\tilde{\phi}) = x_1 \in \mathcal{M}_1$. From the proof of Lemma 4.5 it follows that x_1 must coincide with $x^{(1)}$; otherwise we would get a contradiction to the upper bound (5.1). In the last conclusion we used hypothesis (H_4), which implies that the "distance" according to the expression in (1.19) between two distinct points of Γ_1 is positive. The above argument implies that if $\lim_{\varepsilon_n \to 0} u_{\varepsilon_n}(\phi_n) = \lim_{\varepsilon_n \to 0} v_{1,\varepsilon_n}(\phi_n) = x$, with $\{\phi_n\} \subset [\phi_0+\delta, \frac{\pi}{2}-\delta]$, then $x = x^{(1)}$. The claimed uniform convergence on $[\phi_0 + \delta, \frac{\pi}{2} - \delta]$ follows.

References

[1] L. Ambrosio, N. Fusco and D. Pallara, Functions of Bounded Variation and Free Discontinuity Problems, Oxford Mathematical Monographs. Oxford University Press, New York, 2000.

[2] N. André and I. Shafrir, *On a singular perturbation problem involving a "circular-well" potential*, Trans. Amer. Math. Soc. **359** (2007), 4729–4756.

[3] N. André and I. Shafrir, *On a minimization problem with a mass constraint involving a potential vanishing on two curves*, preprint.

[4] I. Fonseca and L. Tartar, *The gradient theory of phase transitions for systems with two potential wells*, Proc. Roy. Soc. Edinburgh Sect. A **111** (1989), 89–102.

[5] E. Giusti, Minimal Surfaces and Functions of Bounded Variation, Monographs in Mathematics, **80**, Birkhäuser Verlag, Basel, 1984.

[6] O. Lopes, *Radial and nonradial minimizers for some radially symmetric functionals*, Electron. J. Differential Equations **3** (1996) (electronic).

[7] P. Sternberg, *The effect of a singular perturbation on nonconvex variational problems*, Arch. Rational Mech. Anal. **101** (1988), 209–260.

SOME REMARKS ON LIOUVILLE TYPE THEOREMS

H. BREZIS

Laboratoire Jacques-Louis Lions, université Pierre et Marie Curie, 175, rue du Chevaleret, 75013 Paris, France and Rutgers University, Department of Mathematics, Hill Center, Busch Campus, 110, Frelinghuysen Road, Piscataway, NJ 08854, USA, e-mail: brezis@math.rutgers.edu

M. CHIPOT, Y. XIE

Institut für Mathematik, Abt. Angewandte Mathematik, Universität Zürich, Winterthurerstrasse 190, CH–8057 Zürich, Switzerland, e-mail: m.m.chipot@math.unizh.ch, yitian.xie@math.unizh.ch

Abstract: The goal of this note is to present elementary proofs of statements related to the Liouville theorem.

1. Introduction

We denote by $A(x) = (a_{ij}(x))$ a $(k \times k)$-matrix where the functions a_{ij}, $i, j = 1, \ldots, k$ are bounded measurable functions defined on $\mathbb{R}^k$ and which satisfy, for some λ, $\Lambda > 0$, the usual uniform ellipticity condition:

$$\lambda|\xi|^2 \leq (A(x)\xi \cdot \xi) \leq \Lambda|\xi|^2 \quad \text{a.e. } x \in \mathbb{R}^k, \quad \forall \xi \in \mathbb{R}^k. \tag{1.1}$$

We address here the issue of existence of solutions to the equation:

$$-\nabla \cdot (A(x)\nabla u(x)) + a(x)u(x) = 0 \quad \text{in } \mathcal{D}'(\mathbb{R}^k), \tag{1.2}$$

where $a \in L^\infty_{\text{loc}}(\mathbb{R}^k)$ and $a \geq 0$. When $a \not\equiv 0$ and $\nabla \cdot (A\nabla) = \Delta$, the usual Laplace operator, the above equation is the so called stationary Schrödinger equation for which a vast literature is available (see [16], [20]). When $a = 0$ it is well known that every bounded solution to (1.2) has to be constant (see e.g. [5], [11], [12] and also [4], [19] for some nonlinear versions). The case where $a \not\equiv 0$, and $k \geq 3$ is very different and in this case non trivial bounded solutions might exist.

Many of the results in this paper are known in one form or another (see for instance [1], [2], [3], [10], [9], [14], [16], [17]) but we have tried to develop here simple self-contained pde techniques which do not make use of probabilities, semigroups or potential theory as is sometimes the case (see e.g. [2], [3], [8], [17], [18]). One should note that some of our proofs extend also to elliptic systems.

This note is divided as follows. In the next section we introduce an elementary estimate which is used later. In Section 3 we present some Liouville type results, i.e., we show that under some conditions on a, (1.2) does not admit nontrivial bounded solutions. Finally in the last section we give an almost sharp criterion for the existence of nontrivial solutions.

2. A preliminary estimate

Let us denote by Ω a bounded open subset of $\mathbb{R}^k$ with Lipschitz boundary and starshaped with respect to the origin. For any $r \in \mathbb{R}$ we set

$$\Omega_r = r\Omega. \tag{2.1}$$

Let us denote by ϱ a smooth function such that

$$0 \le \varrho \le 1, \quad \varrho = 1 \text{ on } \Omega_{1/2}, \quad \varrho = 0 \text{ outside } \Omega, \tag{2.2}$$

$$|\nabla \varrho| \le c_\varrho, \tag{2.3}$$

where c_ϱ denotes some positive constant.

Lemma 2.1. *Suppose that $u \in H^1_{\mathrm{loc}}(\mathbb{R}^k)$ satisfies (1.2) with $A(x)$ satisfying (1.1). Then there exists a constant C independent of r such that*

$$\begin{aligned} &\int_{\Omega_r} \{|\nabla u|^2 + au^2\}\varrho^2\Big(\frac{x}{r}\Big)\,\mathrm{d}x \\ &\quad \le \frac{C}{r}\Big\{\int_{\Omega_r\setminus\Omega_{r/2}} |\nabla u|^2\varrho^2\Big(\frac{x}{r}\Big)\,\mathrm{d}x\Big\}^{\frac{1}{2}}\Big\{\int_{\Omega_r\setminus\Omega_{r/2}} u^2\,\mathrm{d}x\Big\}^{\frac{1}{2}}, \end{aligned} \tag{2.4}$$

where $|\cdot|$ denotes the usual euclidean norm in $\mathbb{R}^k$.

Proof. By (1.2) we have for every $v \in H^1_0(\Omega_r)$

$$\int_{\Omega_r} A\nabla u \cdot \nabla v + auv\,\mathrm{d}x = 0. \tag{2.5}$$

Taking

$$v = u\varrho^2\Big(\frac{x}{r}\Big) = u\varrho^2 \tag{2.6}$$

yields

$$\int_{\Omega_r} A\nabla u \cdot \nabla\{u\varrho^2\} + au^2\varrho^2\,\mathrm{d}x = 0. \tag{2.7}$$

Since

$$\nabla \varrho^2 = \frac{2\varrho}{r} \nabla \varrho\Big(\frac{x}{r}\Big)$$

we obtain

$$\begin{aligned}\int_{\Omega_r} \{A\nabla u \cdot \nabla u\}\varrho^2 + au^2\varrho^2 \,\mathrm{d}x &= -\frac{1}{r}\int_{\Omega_r \setminus \Omega_{r/2}} A\nabla u \cdot \nabla\varrho \cdot 2\varrho u \,\mathrm{d}x \\ &\leq \frac{C_1}{r}\int_{\Omega_r \setminus \Omega_{r/2}} |\nabla u||u|\varrho \,\mathrm{d}x,\end{aligned}$$

where C_1 is a constant depending on a_{ij} and c_ρ only. Using the ellipticity condition (1.1) it follows easily that

$$\min(1,\lambda)\int_{\Omega_r} \{|\nabla u|^2 + au^2\}\varrho^2 \,\mathrm{d}x \leq \frac{C_1}{r}\int_{\Omega_r \setminus \Omega_{r/2}} |\nabla u||u|\varrho \,\mathrm{d}x.$$

By the Cauchy–Schwarz inequality we have

$$\begin{aligned}&\int_{\Omega_r} \{|\nabla u|^2 + au^2\}\varrho^2 \,\mathrm{d}x \\ &\quad\leq \frac{C_1}{r\min(1,\lambda)}\Big\{\int_{\Omega_r \setminus \Omega_{r/2}} |\nabla u|^2\varrho^2 \,\mathrm{d}x\Big\}^{1/2}\Big\{\int_{\Omega_r \setminus \Omega_{r/2}} u^2 \,\mathrm{d}x\Big\}^{1/2}.\end{aligned}$$

This completes the proof of the lemma. □

3. Some Liouville type results

3.1. *The case where the growth of u is controled*

In this case we have

Theorem 3.1. *Under the asumptions of Lemma 2.1, let u be solution to (1.2) such that for r large,*

$$\frac{1}{r^2}\int_{\Omega_r \setminus \Omega_{r/2}} u^2 \,\mathrm{d}x \leq C' \tag{3.1}$$

where C' is a constant independent of r, then $u =$ constant and if $a \not\equiv 0$ or $k \geq 3$ one has $u = 0$.

Proof. From (2.4) we derive that

$$\int_{\Omega_r} |\nabla u|^2\varrho^2 \,\mathrm{d}x \leq \frac{C}{r}\Big\{\int_{\Omega_r} |\nabla u|^2\varrho^2 \,\mathrm{d}x\Big\}^{1/2}\Big\{\int_{\Omega_r \setminus \Omega_{r/2}} u^2 \,\mathrm{d}x\Big\}^{1/2}$$

and thus

$$\int_{\Omega_{r/2}} |\nabla u|^2 \, \mathrm{d}x \le \int_{\Omega_r} |\nabla u|^2 \varrho^2 \, \mathrm{d}x \le \frac{C}{r^2} \int_{\Omega_r \setminus \Omega_{r/2}} u^2 \, \mathrm{d}x \le CC'.$$

It follows that the nondecreasing function

$$r \mapsto \int_{\Omega_r} |\nabla u|^2 \, \mathrm{d}x$$

is bounded and has a limit when $r \to +\infty$. Going back to (2.4) we have

$$\int_{\Omega_{r/2}} |\nabla u|^2 \, \mathrm{d}x \le \frac{c}{r} \left\{ \int_{\Omega_r \setminus \Omega_{r/2}} |\nabla u|^2 \, \mathrm{d}x \right\}^{\frac{1}{2}} \cdot r$$

for some constant c. This implies

$$\int_{\Omega_{r/2}} |\nabla u|^2 \, \mathrm{d}x \le c \left\{ \int_{\Omega_r} |\nabla u|^2 \, \mathrm{d}x - \int_{\Omega_{r/2}} |\nabla u|^2 \, \mathrm{d}x \right\}^{\frac{1}{2}} \to 0$$

as $r \to +\infty$ and the result follows. □

Remark 3.1. When $k \le 2$ condition (3.1) is satisfied if u is bounded, and in this case the only bounded solution of (1.2) is $u = 0$. **Therefore we will assume throughout the rest of this paper that $k \ge 3$.**

We denote by λ_r the first eigenvalue of the Neumann problem associated to the operator $-\nabla \cdot A\nabla + a$ in $\Omega_r \setminus \Omega_{r/2}$, i.e., we set

$$\begin{aligned} \lambda_r = \operatorname{Inf} \Big\{ \int_{\Omega_r \setminus \Omega_{r/2}} & A\nabla u \cdot \nabla u + a u^2 \, \mathrm{d}x : u \in H^1(\Omega_r \setminus \Omega_{r/2}), \\ & \int_{\Omega_r \setminus \Omega_{r/2}} u^2 \, \mathrm{d}x = 1 \Big\}. \end{aligned} \tag{3.2}$$

One remarks easily that if u is a minimizer of (3.2) so is $|u|$. One can show then that the first eigenvalue is simple. Moreover we have

Theorem 3.2. *Under the assumptions of Lemma 2.1, suppose that for some constants $C_0 > 0$, $\beta < 2$, one has*

$$\lambda_r \ge C_0 / r^\beta \tag{3.3}$$

for r sufficiently large, then the only bounded solution of (1.2) is $u = 0$.

Proof. From the definition of λ_r we have

$$\begin{aligned}\int_{\Omega_r\setminus\Omega_{r/2}} & u^2\,\mathrm{d}x \\ &\le \frac{1}{\lambda_r}\left\{\int_{\Omega_r\setminus\Omega_{r/2}} A\nabla u\cdot\nabla u + au^2\,\mathrm{d}x\right\} \quad \forall\, u\in H^1(\Omega_r\setminus\Omega_{r/2}).\end{aligned} \tag{3.4}$$

Going back to (2.4) we find

$$\begin{aligned}\int_{\Omega_r} & (|\nabla u|^2 + au^2)\varrho^2\Big(\frac{x}{r}\Big)\,\mathrm{d}x \\ &\le \frac{C}{r}\left\{\int_{\Omega_r}(|\nabla u|^2+au^2)\varrho^2\Big(\frac{x}{r}\Big)\,\mathrm{d}x\right\}^{1/2}\left\{\int_{\Omega_r\setminus\Omega_{r/2}} u^2\,\mathrm{d}x\right\}^{1/2},\end{aligned}$$

which leads to

$$\int_{\Omega_r}(|\nabla u|^2+au^2)\varrho^2\Big(\frac{x}{r}\Big)\,\mathrm{d}x \le \frac{C}{r^2}\int_{\Omega_r\setminus\Omega_{r/2}} u^2\,\mathrm{d}x$$

for some constant C independent of r. Using in particular (2.2) we obtain

$$\int_{\Omega_{r/2}} |\nabla u|^2 + au^2\,\mathrm{d}x \le \frac{C}{r^2}\int_{\Omega_r\setminus\Omega_{r/2}} u^2\,\mathrm{d}x, \quad \forall\, r>0. \tag{3.5}$$

From (3.3) and (3.4) we derive that, for some constant C,

$$\begin{aligned}\int_{\Omega_{r/2}} |\nabla u|^2 + au^2\,\mathrm{d}x &\le \frac{C}{r^{2-\beta}}\int_{\Omega_r\setminus\Omega_{r/2}} |\nabla u|^2 + au^2\,\mathrm{d}x \\ &\le \frac{C}{r^{2-\beta}}\int_{\Omega_r} |\nabla u|^2 + au^2\,\mathrm{d}x \quad \forall\, r \geqslant 0.\end{aligned} \tag{3.6}$$

Iterating p-times this formula leads to

$$\int_{\Omega_{r/2}} |\nabla u|^2 + au^2\,\mathrm{d}x \le \frac{C^p}{r^{(2-\beta)p}}\int_{\Omega_{2^{p-1}r}} |\nabla u|^2 + au^2\,\mathrm{d}x.$$

By (3.5) it follows that it holds

$$\int_{\Omega_{r/2}} |\nabla u|^2 + au^2\,\mathrm{d}x \le \frac{C_p}{r^{(2-\beta)p+2}}\int_{\Omega_{2^p r}} u^2\,\mathrm{d}x,$$

for some constant C_p independent of r. If now u is supposed to be bounded by M we get

$$\int_{\Omega_{r/2}} |\nabla u|^2 + au^2\,\mathrm{d}x \le \frac{C_p}{r^{(2-\beta)p+2}}M^2|\Omega_{2^p r}| = \frac{C_p|\Omega|M^2(2^p r)^k}{r^{(2-\beta)p+2}}.$$

($|\Omega_{2^p r}|$ denotes the Lebesgue measure of the set $\Omega_{2^p r}$). Choosing $(2-\beta)p + 2 > k$ the result follows by letting $r\to+\infty$. □

Remark 3.2. Under the assumption of Theorem 3.2 we have obtained in fact that (1.2) can not admit a nontrivial solution with polynomial growth. Of course this result is optimal since $Re(e^z) = e^{x_1}\cos x_2$ is harmonic in $\mathbb{R}^k$ for any $k \geq 2$. One should note that Theorem 3.2 applies also to systems satisfying the Legendre condition when auv is replaced by a nonnegative bilinear form $a(u, v)$ (see [6], [7]).

We now discuss some conditions on a which imply (3.3). We have

Theorem 3.3. *Suppose that for $|x|$ large enough*

$$a(x) \geq \frac{c}{|x|^\beta}, \qquad \beta < 2, \tag{3.7}$$

then (3.3) *holds.*

Proof. We denote by π_r the first eigenfunction corresponding to λ_r, i.e., a minimizer of (3.2). We can assume without loss of generality that

$$\pi_r > 0.$$

We have

$$\int_{\Omega_r \setminus \Omega_{r/2}} A\nabla\pi_r \cdot \nabla v + a\pi_r v \,\mathrm{d}x = \lambda_r \int_{\Omega_r \setminus \Omega_{r/2}} \pi_r v \,\mathrm{d}x \quad \forall v \in H^1(\Omega_r \setminus \Omega_{r/2}).$$

Taking $v = 1$ yields

$$\int_{\Omega_r \setminus \Omega_{r/2}} a(x)\pi_r \,\mathrm{d}x = \lambda_r \int_{\Omega_r \setminus \Omega_{r/2}} \pi_r \,\mathrm{d}x.$$

Using (3.7) we derive, for some constant C',

$$\frac{C'}{r^\beta} \int_{\Omega_r \setminus \Omega_{r/2}} \pi_r \,\mathrm{d}x \leq \lambda_r \int_{\Omega_r \setminus \Omega_{r/2}} \pi_r \,\mathrm{d}x$$

and the result follows. □

We now consider other cases where (3.3) holds, in particular when no decay is imposed to a. We are interested for instance in the case where at infinity a has enough mass locally. We start with the following lemma:

Lemma 3.1. *Let (for instance) $Q = (0, 1)^k$ be the unit cube in $\mathbb{R}^k$. For any $\varepsilon > 0$ and $\mu > 0$ there exists $\delta = \delta(\varepsilon, \mu)$ such that if the function a satisfies*

$$0 \leq a \leq \mu \text{ a.e. } x \in Q, \qquad \int_Q a \,\mathrm{d}x \geq \varepsilon, \tag{3.8}$$

then

$$\delta \int_Q v^2 \,\mathrm{d}x \le \int_Q |\nabla v|^2 + av^2 \,\mathrm{d}x \qquad \forall\, v \in H^1(Q). \tag{3.9}$$

Proof. If not, there exists ε, μ and a sequence of functions a_n, v_n such that a_n satisfies (3.8) and $v_n \in H^1(Q)$ is such that

$$\frac{1}{n} \int_Q v_n^2 \,\mathrm{d}x \ge \int_Q |\nabla v_n|^2 + a_n v_n^2 \,\mathrm{d}x. \tag{3.10}$$

Dividing by $|v_n|_2$ the L^2-norm of v_n we can assume without loss of generality that

$$\int_Q v_n^2 \,\mathrm{d}x = 1. \tag{3.11}$$

By (3.10), (3.11) we have then

$$\int_Q |\nabla v_n|^2 \,\mathrm{d}x \le \frac{1}{n}, \qquad \int_Q v_n^2 \,\mathrm{d}x = 1 \tag{3.12}$$

and v_n is uniformly bounded in $H^1(Q)$. Therefore

$$v_n \to 1 \text{ in } H^1(Q). \tag{3.13}$$

From (3.10) we have

$$\int_Q a_n v_n^2 \,\mathrm{d}x \le \frac{1}{n}. \tag{3.14}$$

Thus

$$\begin{aligned} \varepsilon \le \int_Q a_n \mathrm{d}x &= \int_Q a_n v_n^2 \,\mathrm{d}x + \int_Q a_n(1 - v_n^2) \,\mathrm{d}x \\ &\le \frac{1}{n} + \mu |1 - v_n|_2 |1 + v_n|_2 \to 0 \end{aligned}$$

when $n \to +\infty$. Impossible. This completes the proof of the lemma. □

With the notation of Section 2 we set

$$\Omega = (-1, 1)^k. \tag{3.15}$$

We consider the lattice generated by $Q = (0,1)^k$ – i.e., the cubes

$$Q_i = Q_{z_i} = z_i + Q \quad \forall\, z_i \in \mathbb{Z}^k.$$

Then we have

Theorem 3.4. *Suppose that for n large enough,*

$$\int_{Q_i} a(x)\,\mathrm{d}x \geq \varepsilon \quad \forall\, Q_i \subset \mathbb{R}^k \setminus \Omega_n, \tag{3.16}$$

then

$$\lambda_{2n} \geq \delta \Big/ \Big(\frac{1}{\lambda} \vee 1\Big) \quad \forall\, n \tag{3.17}$$

where δ is defined in Lemma 3.1 and $\vee$ denotes the maximum of two numbers.

Proof. Indeed by Lemma 3.1 after a simple translation from Q_i into Q we have

$$\delta \int_{Q_i} u^2\,\mathrm{d}x \leq \int_{Q_i} |\nabla u|^2 + au^2\,\mathrm{d}x \quad \forall\, Q_i \subset \mathbb{R}^k \setminus \Omega_n \quad \forall\, u \text{ in } H^1(Q_i).$$

This leads clearly to

$$\begin{aligned} \delta \int_{\Omega_{2n}\setminus\Omega_n} & u^2\,\mathrm{d}x \\ & \leq \int_{\Omega_{2n}\setminus\Omega_n} |\nabla u|^2 + au^2\,\mathrm{d}x \\ & \leq \int_{\Omega_{2n}\setminus\Omega_n} \frac{1}{\lambda} A\nabla u \cdot \nabla u + au^2\,\mathrm{d}x \\ & \leq \Big(\frac{1}{\lambda} \vee 1\Big) \int_{\Omega_{2n}\setminus\Omega_n} A\nabla u \cdot \nabla u + au^2\,\mathrm{d}x \quad \forall\, u \in H^1(\Omega_{2n} \setminus \Omega_n). \end{aligned}$$

The result follows then from (3.2). □

Remark 3.3. Combining Theorems 3.2 and 3.4 it follows that (1.2) cannot have a nontrivial bounded solution (or of polynomial growth) when (3.16) holds. This is the case when at infinity

$$a \geq a_0 > 0$$

or more generally

$$a \geq a_p \tag{3.18}$$

where a_p is a periodic function with period Q.

In the case when (3.3) holds with $\beta = 2$ the technique of Theorem 3.2 cannot be applied. However, we will show that the non existence of nontrivial solutions can be established in this case too – i.e., condition (3.3) is not sharp if we impose certain growth condition on $\{a_{ij}(x)\}$. Before turning to this let us prove some general comparison result. For simplicity we will denote also by $\tilde{A}$ the operator $\nabla \cdot A\nabla u = \partial_{x_i}(a_{ij}\partial_{x_j})$.

Proposition 3.1. *Suppose that O is a bounded open subset of $\mathbb{R}^k$. Let a_1, a_2 be two bounded functions satisfying*

$$a_1 \geq a_2 \geq 0 \quad a.e.\ in\ O. \tag{3.19}$$

Let $u_1, u_2 \in H^1(O)$ be such that

$$\begin{cases} -\tilde{A}u_2 + a_2u_2 \geq -\tilde{A}u_1 + a_1u_1 \geq 0 & in\ O, \\ u_2 \geq (u_1 \vee 0) & on\ \partial O, \end{cases} \tag{3.20}$$

then

$$u_2 \geq (u_1 \vee 0) \quad in\ O. \tag{3.21}$$

Proof. The inequality

$$-\tilde{A}u + au \geq 0 \quad \text{in } O$$

means

$$\int_O a_{ij}\partial_{x_j}u\partial_{x_i}v + auv\,\mathrm{d}x \geq 0 \quad \forall v \in H_0^1(O), v \geq 0.$$

Considering $v = u_2^-$ and $-\tilde{A}u_2 + a_1u_2 \geq 0$ leads to $u_2 \geq 0$. Next considering $v = (u_1 - u_2)^+ \in H_0^1(O)$ and (3.20) we obtain

$$\begin{aligned} &\int_O a_{ij}\partial_{x_j}u_1\partial_{x_i}(u_1 - u_2)^+ + a_1u_1(u_1 - u_2)^+\,\mathrm{d}x \\ &\qquad \leq \int_O a_{ij}\partial_{x_j}u_2\partial_{x_i}(u_1 - u_2)^+ + a_2u_2(u_1 - u_2)^+\,\mathrm{d}x. \end{aligned}$$

Hence

$$\int_O a_{ij}\partial_{x_j}(u_1 - u_2)\partial_{x_i}(u_1 - u_2)^+ + (a_1u_1 - a_2u_2)(u_1 - u_2)^+\,\mathrm{d}x \leq 0.$$

Now on $u_1 \geq u_2$ one has $a_1u_1 \geq a_1u_2 \geq a_2u_2$ and it follows that $(u_1 - u_2)^+ = 0$. □

Next we prove

Theorem 3.5. *Assume that there exists R, large enough, such that*

$$a(x) \geq \frac{c_0}{r^2} \quad \forall |x| \geq R > 0$$

where c_0 is a positive constant. In addition to (1.1)*, suppose that* $a_{ij}(x) \in C^1(\mathbb{R}^k \backslash B(0,R))$ *satisfies for some positive* D*:*

$$\partial_{x_i}(a_{ij}(x))x_j \leq D \quad \forall |x| > R.$$

(In the above inequality we make the summation convention of repeated indices). Then the equation

$$-\partial_{x_i}(a_{ij}(x)\partial_{x_j}u) + a(x)u = 0 \tag{3.22}$$

cannot have nontrivial bounded solution.

Proof. Let u_n be the solution to

$$\begin{cases} -\partial_{x_i}(a_{ij}(x)\partial_{x_j}u_n) + a(x)u_n = 0 & \text{in } B(0,n), \\ u_n = 1 & \text{on } \partial B(0,n), \end{cases} \tag{3.23}$$

where $B(0,n)$ denotes the ball of center 0 and radius n. From Proposition 3.1 we obtain that u, solution to (3.23), is such that:

$$-|u|_\infty u_n \leq u \leq |u|_\infty u_n, \tag{3.24}$$

($|u|_\infty$ denotes the L^∞-norm of u). Denote by v_n the function defined as

$$v_n = \begin{cases} c_1 & \text{in } B(0,R) \\ c_2 r^{\beta_1} + c_3 r^{\beta_2} & \text{in } B(0,n)\backslash B(0,R), \end{cases} \tag{3.25}$$

where

$$\beta_1 = -\frac{1}{2}\{(k-2) + \sqrt{(k-2)^2 + 4c'}\} < 0,$$
$$\beta_2 = -\frac{1}{2}\{(k-2) - \sqrt{(k-2)^2 + 4c'}\} > 0$$

and

$$c_1 = c_2 R^{\beta_1} + c_3 R^{\beta_2}$$
$$c_2 = \frac{-\beta_2 R^{\beta_2 - 1}}{n^{\beta_2}\beta_1 R^{\beta_1 - 1} - n^{\beta_1}\beta_2 R^{\beta_2 - 1}}$$
$$c_3 = \frac{\beta_1 R^{\beta_1 - 1}}{n^{\beta_2}\beta_1 R^{\beta_1 - 1} - n^{\beta_1}\beta_2 R^{\beta_2 - 1}}.$$

In the above setting, c' is a positive constant small enough that we will determine later.

We remark that c_2 and c_3 are both positive and that β_1, β_2 are the two roots to the second order equation

$$\beta^2 + (k-2)\beta - c' = 0. \tag{3.26}$$

The choice of $c_i, i=1,2,3$ is such that v_n is a C^1 function and $v_n=1$ on $\partial B(0,n)$.

Now we want to show that v_n, in fact, is a supersolution to (3.23).

It is easy to see that

$$-\partial_{x_i}(a_{ij}(x)\partial_{x_j}v_n)+a(x)v_n\geq 0 \quad \text{in } B(0,R).$$

For any constant β one derives also that

$$\begin{aligned}
&\partial_{x_i}(a_{ij}(x)\partial_{x_j}(r^{\beta}))\\
&=\partial_{x_i}(a_{ij}(x)\beta r^{\beta-2}x_j)\\
&=\partial_{x_i}(a_{ij}(x))\beta r^{\beta-2}x_j+a_{ij}(x)\beta(\beta-2)r^{\beta-4}x_ix_j\\
&\quad +a_{ij}(x)\beta r^{\beta-2}\delta_{ij}.
\end{aligned}$$

Therefore in $B(0,n)\backslash B(0,R)$ this leads to

$$\begin{aligned}
&-\partial_{x_i}(a_{ij}(x)\partial_{x_j}v_n)+a(x)v_n\\
&\geq -\partial_{x_i}(a_{ij}(x))x_j\{c_2\beta_1 r^{\beta_1-2}+c_3\beta_2 r^{\beta_2-2}\}\\
&\quad -a_{ij}(x)x_ix_j\{c_2\beta_1(\beta_1-2)r^{\beta_1-4}+c_3\beta_2(\beta_2-2)r^{\beta_2-4}\}\\
&\quad -a_{ij}(x)\delta_{ij}\{c_2\beta_1 r^{\beta_1-2}+c_3\beta_2 r^{\beta_2-2}\}+\frac{c_0}{r^2}\{c_2r^{\beta_1}+c_3r^{\beta_2}\}\\
&=\left\{-\partial_{x_i}(a_{ij}(x))x_j\{c_2\beta_1 r^{\beta_1-2}+c_3\beta_2 r^{\beta_2-2}\}\right\}\\
&\quad +c_2\left\{-a_{ij}(x)x_ix_j\beta_1(\beta_1-2)r^{\beta_1-4}-a_{ij}(x)\delta_{ij}\beta_1 r^{\beta_1-2}+c_0r^{\beta_1-2}\right\}\\
&\quad +c_3\left\{-a_{ij}(x)x_ix_j\beta_2(\beta_2-2)r^{\beta_2-4}-a_{ij}(x)\delta_{ij}\beta_2 r^{\beta_2-2}+c_0r^{\beta_2-2}\right\}.
\end{aligned}$$

We notice that

$$c_2\beta_1 r^{\beta_1-2}+c_3\beta_2 r^{\beta_2-2}=\frac{\beta_1\beta_2 r^{\beta_2-2}R^{\beta_2-1}\{R^{\beta_1-\beta_2}-r^{\beta_1-\beta_2}\}}{n^{\beta_2}\beta_1R^{\beta_1-1}-n^{\beta_1}\beta_2R^{\beta_2-1}}>0,$$

and thus

$$\begin{aligned}
&-\partial_{x_i}(a_{ij}(x)\partial_{x_j}v_n)+a(x)v_n\\
&\geq \left\{-D\{c_2\beta_1 r^{\beta_1-2}+c_3\beta_2 r^{\beta_2-2}\}\right\}\\
&\quad +c_2\left\{-a_{ij}(x)x_ix_j\beta_1(\beta_1-2)r^{\beta_1-4}-a_{ij}(x)\delta_{ij}\beta_1 r^{\beta_1-2}+c_0r^{\beta_1-2}\right\}\\
&\quad +c_3\left\{-a_{ij}(x)x_ix_j\beta_2(\beta_2-2)r^{\beta_2-4}-a_{ij}(x)\delta_{ij}\beta_2 r^{\beta_2-2}+c_0r^{\beta_2-2}\right\}.
\end{aligned}$$

Taking into account (3.26) –i.e., replacing $\beta_i(\beta_i - 2)$ by $c' - k\beta_i$, yields

$$\begin{aligned}
&-\partial_{x_i}(a_{ij}(x)\partial_{x_j}v_n) + a(x)v_n \\
&\geq c_2 r^{\beta_1 - 2}\left\{[ka_{ij}(x)\frac{x_i x_j}{r^2} - a_{ii}(x) - D]\beta_1 - c' a_{ij}(x)\frac{x_i x_j}{r^2} + c_0\right\} \\
&\quad + c_3 r^{\beta_2 - 2}\left\{[ka_{ij}(x)\frac{x_i x_j}{r^2} - a_{ii}(x) - D]\beta_2 - c' a_{ij}(x)\frac{x_i x_j}{r^2} + c_0\right\} \\
&\geq c_2 r^{\beta_1 - 2}\left\{[ka_{ij}(x)\frac{x_i x_j}{r^2} - a_{ii}(x) - D]\beta_1 - c'\Lambda + c_0\right\} \\
&\quad + c_3 r^{\beta_2 - 2}\left\{[ka_{ij}(x)\frac{x_i x_j}{r^2} - a_{ii}(x) - D]\beta_2 - c'\Lambda + c_0\right\}.
\end{aligned}$$

We can select a D large enough such that the term

$$ka_{ij}(x)\frac{x_i x_j}{r^2} - a_{ii}(x) - D$$

is negative (and bounded). By noticing that $\beta_2 \to 0^+$ when $c' \to 0$ we can then always choose c' small enough such that

$$[ka_{ij}(x)\frac{x_i x_j}{r^2} - a_{ii}(x) - D]\beta_i - c'\Lambda + c_0 > 0.$$

Hence we derive that

$$-\partial_{x_i}(a_{ij}(x)\partial_{x_j}v_n) + a(x)v_n \geq 0,$$

and by Proposition 3.1,

$$u_n \leq v_n.$$

For any bounded subset $\Omega \subset B(0,d)$ in $\mathbb{R}^k$, one has clearly

$$0 \leq v_n \leq \text{Max}\{c_1, c_2 d^{\beta_1} + c_3 d^{\beta_2}\} \to 0 \qquad \text{on } \Omega$$

when $n \to +\infty$ since

$$n^{\beta_2}\beta_1 R^{\beta_1 - 1} - n^{\beta_1}\beta_2 R^{\beta_2 - 1} \to -\infty$$

as $n \to +\infty$. From (3.24) we have also on $B(0,n)$

$$-|u|_\infty v_n \leq u \leq |u|_\infty v_n$$

for any n. Letting $n \to \infty$ leads to that

$$u = 0. \qquad \square$$

Remark 3.4. The above result holds true for an operator in non-divergence form, i.e., under the assumption of Theorem 3.5 the equation

$$-a_{ij}(x)\partial^2_{x_i x_j}u - b_i(x)\partial_{x_i}u + a(x)u = 0$$

with

$$(b, x) \leq D \quad \forall |x| > R$$

cannot have a nontrivial bounded solution.

4. The case of the Laplace operator

In this section we analyze the existence or nonexistence of nontrivial bounded solutions to (1.2) in the case of the Laplacian. Due to the results of the previous section it is clear that existence of nontrivial solutions will impose some kind of decay a. So, let us assume

$$a(x) \in L^\infty_{\mathrm{loc}}(\mathbb{R}^k), a \geq 0, a \not\equiv 0$$

and

$$\int_{|x|>1} a(x)|x|^{-k+2} \, \mathrm{d}x < \infty \tag{4.1}$$

with

$$k \geq 3.$$

Under the above assumptions we can show

Theorem 4.1. *(Grigor'yan [8], see also [2], [3], [15]) Assume ((4.1)). Then there exists a function u such that*

$$0 < u < 1 \quad \text{in } \mathbb{R}^k \tag{4.2}$$

satisfying

$$-\Delta u + au = 0 \quad \text{in } \mathcal{D}'(\mathbb{R}^k). \tag{4.3}$$

Proof. Let u_n be the solution of

$$\begin{cases} -\Delta u_n + au_n = 0 & \text{in } B(0, n), \\ u_n = 1 & \text{on } \partial B(0, n). \end{cases} \tag{4.4}$$

By the maximum principle

$$0 \leq u_n \leq 1 \tag{4.5}$$

and

$$u_{n+1} \leq u_n \quad \text{in } B(0, n). \tag{4.6}$$

Thus $u_n \to u$ which satisfies (4.3). Moreover

$$0 \leq u \leq 1.$$

By the strong maximum principle (and since $a \not\equiv 0$) we have

$$u < 1 \quad \text{in } \mathbb{R}^k.$$

Once more, by the strong maximum principle, it suffices to prove that

$$u \not\equiv 0.$$

Assume, by contradiction, that

$$u \equiv 0. \tag{4.7}$$

Fix a function $\zeta \in C^\infty(\mathbb{R}^k)$, $0 \le \zeta \le 1$ such that

$$\zeta(x) = \begin{cases} 0 & |x| < R \\ 1 & |x| \ge R+1 \end{cases}$$

and R will be determined later. Multiplying (4.4) by $\frac{\zeta(x)}{|x|^{k-2}}$ yields for $n > R+1$, $\nu =$ being the outward unit normal vector:

$$-\int_{|x|=n} \frac{\partial u_n}{\partial \nu} \cdot \frac{1}{n^{k-2}} \, d\sigma - \int_{B(0,n)} u_n \Delta\left(\frac{\zeta}{|x|^{k-2}}\right) dx + \int_{|x|=n} \frac{\partial}{\partial \nu}\left(\frac{\zeta}{|x|^{k-2}}\right) d\sigma + \int_{B(0,n)} a u_n \frac{\zeta}{|x|^{k-2}} \, dx = 0. \tag{4.8}$$

By (4.5)

$$\frac{\partial u_n}{\partial \nu} \ge 0 \quad \text{on } \partial B(0,n) \tag{4.9}$$

and

$$\frac{\partial}{\partial \nu}\left(\frac{\zeta}{|x|^{k-2}}\right) = \frac{2-k}{n^{k-1}} \quad \text{on } \partial B(0,n)$$

so that

$$\int_{|x|=n} \frac{\partial}{\partial \nu}\left(\frac{\zeta}{|x|^{k-2}}\right) d\sigma = (2-k)\sigma_k \tag{4.10}$$

where σ_k denotes the area of S^{k-1}. From (4.8), (4.9) and (4.10) we have

$$-\int_{B(0,n)} u_n \Delta\left(\frac{\zeta}{|x|^{k-2}}\right) dx + \int_{B(0,n)} a u_n \frac{\zeta}{|x|^{k-2}} \, dx \ge (k-2)\sigma_k. \tag{4.11}$$

Notice that $\Delta(\frac{\zeta}{|x|^{k-2}})$ has compact support (in $R < |x| < R+1$) since $\Delta(\frac{1}{|x|^{k-2}}) = c\delta_0$ where δ_0 denotes the Dirac measure at 0. Therefore one has

$$\lim_{n\to\infty} \int_{B(0,n)} u_n \Delta\left(\frac{\zeta}{|x|^{k-2}}\right) dx = 0 \tag{4.12}$$

by assumption (4.7). On the other hand

$$\int_{B(0,n)} au_n \frac{\zeta}{|x|^{k-2}}\,\mathrm{d}x \le \int_{R<|x|} \frac{a(x)}{|x|^{k-2}}\,\mathrm{d}x. \tag{4.13}$$

By (4.11), (4.12) and (4.13) we have

$$\int_{R<|x|} \frac{a(x)}{|x|^{k-2}}\mathrm{d}x \ge (k-2)\sigma_k.$$

Choosing R sufficiently large and using assumption (4.1) yields a contradiction. This completes the proof of $u \not\equiv 0$. □

Remark 4.1. With the same proof and under the assumption (4.1) one can show that (1.2) admits a non trivial solution provided

$$\lim_{R\to\infty} \int_{|x|\ge R} \partial_{x_i}(a_{ij}(x)\partial_{x_j}|x|^{2-k}) < \lambda(k-2)\sigma_k.$$

This is in particular the case when $a_{ij}(x) = \delta_{ij}$ for $|x|$ large.

In the radially symmetric case we can say more.

Theorem 4.2. *Suppose that the solutions to* (1.2) *are radially symmetric, then they do not change sign and are multiple of each other.*

Proof. Let u be a radially symmetric solution to (1.2). Let us first prove that u does not change sign. Changing u into $-u$ we can suppose

$$u(0) \ge 0.$$

We argue by contradiction and assume that u changes sign. If $u(0) > 0$, there exists a r_0 such that

$$u(r_0) = 0.$$

Then

$$\int_{B(0,r_0)} \nabla u \cdot \nabla v + auv\,\mathrm{d}x = 0 \qquad \forall v \in H_0^1(B(0,r_0)). \tag{4.14}$$

Taking $v = u$ we obtain that $u \equiv 0$ in $B(0,r_0)$ and a contradiction. If $u(0) = 0$ then changing u in $-u$ there is a component of the set

$$\{x \big| u(x) > 0\}$$

which is an annulus A. But then we get (4.14) with $B(0,r_0)$ replaced by A and a contradiction as above.

Consider now u, v two solutions to (1.2). If $u \equiv 0$, $u = 0 \cdot v$, or else we have by the first part of the theorem (after changing u into $-u$ if needed):

$$u(0) > 0.$$

Then $w = v - \frac{v(0)}{u(0)} u$ is a solution to (1.2) such that $w(0) = 0$. Since it does not change sign, 0 is a minimum or a maximum and by the maximum principle $w \equiv 0$. This completes the proof of the theorem. □

In the case of radially symmetric solutions we also have:

Proposition 4.1. *Suppose that $a = a(r)$, $r = |x|$ and let u be a bounded positive radially symmetric solution to*

$$-\Delta u + au = 0 \qquad \text{in } \mathcal{D}'(\mathbb{R}^k).$$

We have

$$u(0) > 0,\ u = u(r) \text{ is nondecreasing on } (0, +\infty),\ \lim_{r\to\infty} u(r) = u(\infty) < +\infty.$$

Proof. $u(0) > 0$ results from the previous theorem. In addition we have

$$-u'' - \frac{k-1}{r} u' + au = 0$$

$$\Longrightarrow rau = ru'' + (k-1)u' = \frac{(r^{k-1}u')'}{r^{k-2}} \geq 0. \tag{4.15}$$

Thus $r^{k-1}u'$ is nondecreasing. Since it vanishes at 0 we have $u' \geq 0$ and u is nondecreasing. Hence u has a limit at ∞ since u is bounded. □

As a consequence we have the following property for the solution u that we constructed in the Theorem 4.1.

Theorem 4.3. *Suppose that for $|x| \geq R_0$*

$$a(x) \leq a_0(|x|) \quad \text{with} \int^{+\infty} r a(r)\, \mathrm{d}r < +\infty. \tag{4.16}$$

Then the solution u constructed in Theorem 4.1 verifies

$$\lim_{|x|\to\infty} u(x) = 1.$$

Proof. We introduce

$$\tilde{a} = \begin{cases} |a|_\infty & \text{for } |x| < R_0, \\ a_0(r) & \text{for } |x| \geq R_0. \end{cases}$$

Let $\tilde{u}_n$ be the solution to

$$\begin{cases} -\Delta \tilde{u}_n + \tilde{a}\tilde{u}_n = 0 & \text{in } B(0,n), \\ \tilde{u}_n = 1 & \text{on } \partial B(0,n). \end{cases}$$

By Proposition 3.1 we have

$$0 \leq \tilde{u}_{n+1} \leq \tilde{u}_n \leq u_n \leq 1. \tag{4.17}$$

Of course since $\tilde{a}$ is radially symmetric, so is $\tilde{u}_n$ and it converges to a radially symmetric function $\tilde{u}$ which is a nontrivial solution to (see Theorem 4.1)

$$-\Delta \tilde{u} + \tilde{a}\tilde{u} = 0 \quad \text{in } \mathcal{D}'(\mathbb{R}^k).$$

From Proposition 4.1 we have

$$0 < \lim_{|x|\to\infty} \tilde{u} = \tilde{u}(\infty) \leq 1.$$

Suppose that $\tilde{u}(\infty) < 1$. Consider $\tilde{v}_n$ the solution of

$$\begin{cases} -\Delta \tilde{v}_n + \tilde{a}\tilde{v}_n = 0 & \text{in } B(0,n) \\ \tilde{v}_n = 1 - \tilde{u}(\infty) & \text{on } \partial B(0,n). \end{cases}$$

One has

$$\begin{cases} -\Delta(\tilde{v}_n + \tilde{u}) + \tilde{a}(\tilde{v}_n + \tilde{u}) = 0 & \text{in } B(0,n), \\ \tilde{v}_n + \tilde{u} = 1 + \tilde{u} - \tilde{u}(\infty) \leq 1 & \text{on } \partial B(0,n). \end{cases}$$

Thus, by the maximum principle,

$$\tilde{v}_n + \tilde{u} \leq \tilde{u}_n \quad \text{in } B(0,n).$$

Now clearly $\tilde{v}_n = (1 - \tilde{u}(\infty))\tilde{u}_n$ and thus

$$(1 - \tilde{u}(\infty))\tilde{u}_n + \tilde{u} \leq \tilde{u}_n.$$

Passing to the limit in n we obtain

$$(1 - \tilde{u}(\infty))\tilde{u} + \tilde{u} \leq \tilde{u}$$

which contradicts $\tilde{u}(\infty) < 1$. Thus we have $\tilde{u}(\infty) = 1$.

Now from (4.17) we derive, passing to the limit,

$$\tilde{u} \leq u \leq 1.$$

Since, we already know that $\lim_{|x|\to\infty} \tilde{u}(x) = 1$ the result follows. This completes the proof of the theorem. □

We prove now that condition (4.16) is sharp within the class of radial functions. This was observed in [3] with a different technique (see also [10], [9]). More recently (R. Pinsky [18]) established the sharpness of condition (4.16) in the class of functions a satisfying the additional assumption

$$a(x) \le \frac{C}{(1+|x|)^2}. \tag{4.18}$$

So, let $a(r)$ be a function such that

$$\int^{+\infty} ra(r)\,\mathrm{d}r = +\infty. \tag{4.19}$$

Lemma 4.1. *Under the assumption* (4.19) *there does not exist a bounded nontrivial radially symmetric solution to*

$$-\Delta u + a(r)u = 0 \quad \textit{in } \mathcal{D}'(\mathbb{R}^k). \tag{4.20}$$

Proof. Suppose that (4.20) admits a nontrivial bounded positive solution $u(r)$ (see Theorem 4.3). Integrating the first equality of (4.15) we find

$$\int_0^r sa(s)u(s)\,\mathrm{d}s = \int_0^r su''(s)\,\mathrm{d}s + (k-1)\int_0^r u'(s)\,\mathrm{d}s$$
$$= ru'(r) + (k-2)\{u(r)-u(0)\} = (ru)' + (k-3)u(r) - (k-2)u(0).$$

Integrating again in r yields

$$ru(r) = \int_0^r \left(\int_0^s \xi a(\xi)u(\xi)\,\mathrm{d}\xi\right)\mathrm{d}s - (k-3)\int_0^r u(s)\,\mathrm{d}s + (k-2)u(0)r \tag{4.21}$$

For $s \ge \frac{r}{2}$ we have

$$\int_0^s \xi a(\xi)u(\xi)\,\mathrm{d}\xi \ge u(0)\int_0^{\frac{r}{2}} \xi a(\xi)\,\mathrm{d}\xi.$$

Thus from (4.21) we easily obtain

$$u(r) \ge \frac{1}{r}\int_{\frac{r}{2}}^r \left(\int_0^s \xi a(\xi)u(\xi)\,\mathrm{d}\xi\right)\mathrm{d}s - (k-3)u(\infty) + (k-2)u(0)$$
$$\ge \frac{1}{2}\int_0^{\frac{r}{2}} \xi a(\xi)\,\mathrm{d}\xi \cdot u(0) - (k-3)u(\infty) + (k-2)u(0).$$

By (4.19) the left-hand side of this inequality goes to $+\infty$ with r. This contradicts the boundedness of u and completes the proof of the lemma. □

As a consequence we can now show:

Theorem 4.4. *Suppose that for $|x|$ large*

$$a(x) \geq \bar{a}(r) = \bar{a}(|x|) \tag{4.22}$$

where $\bar{a}$ satisfies (4.19) *then the problem*

$$-\Delta u + au = 0 \quad \textit{in } \mathcal{D}'(\mathbb{R}^k) \tag{4.23}$$

cannot admit nontrivial bounded solutions.

Proof. Suppose that (4.22) holds for $|x| \geq R$. Then define

$$\tilde{a} = \begin{cases} 0 & \text{when } |x| \leq R, \\ \bar{a}(r) & \text{when } |x| > R. \end{cases}$$

$\tilde{a}$ is a radially symmetric function satisfying (4.19). Let u be a bounded solution to (4.23). Let u_n, v_n be the solution of

$$-\Delta u_n + \tilde{a} u_n = 0 \quad \text{in } B(0,n), u_n = |u|_\infty \quad \text{on } \partial B(0,n), \tag{4.24}$$

$$-\Delta v_n + a v_n = 0 \quad \text{in } B(0,n), v_n = |u|_\infty \quad \text{on } \partial B(0,n), \tag{4.25}$$

where $|u|_\infty$ denotes the L^∞-norm of u. It follows from Proposition 3.1 and the maximum principle that

$$u < v_n \leq u_n, \quad 0 \leq u_{n+1} \leq u_n \leq |u|_\infty. \tag{4.26}$$

Changing u into $-u$ one if needed, can assume that the set

$$\{u > 0\} = \{x \in \mathbb{R}^k \mid u(x) > 0\}$$

has a positive measure. Now, clearly, by the uniqueness of the solution to (4.24), u_n is radially symmetric. By (4.26) u_n converges to u_∞ solution of

$$-\Delta u_\infty + \tilde{a} u_\infty = 0 \quad \text{in } \mathbb{R}^k$$

and u_∞ is radially symmetric. By Lemma 4.1 this implies that $u_\infty = 0$. Hence from (4.26) we get

$$u \leq 0$$

which contradicts the fact that $\{u > 0\}$ is of positive measure. □

Remark 4.2. Theorem 4.4 applies for instance when

$$a(x) = \frac{C_0}{|x|^2}$$

for a constant C_0 and $|x|$ large enough. Let λ_r be given by (3.2). Then for r large enough we have

$$\frac{c}{r^2} \leq \lambda_r \leq \frac{C}{r^2} \tag{4.27}$$

for some constants c, C. In other words the technique of Theorem 3.2 cannot work in this case. To show (4.27), recall the definition (3.2) and use the constant function

$$u = 1/|\Omega_r \setminus \Omega_{r/2}|^{1/2} \in H^1(\Omega_r \setminus \Omega_{r/2})$$

($|\cdot|$ is the Lebesgue measure); we obtain

$$\lambda_r \leq \frac{1}{|\Omega_r \setminus \Omega_{r/2}|} \int_{\Omega_r \setminus \Omega_{r/2}} a(x)\,\mathrm{d}x = \frac{C_0}{|\Omega_r \setminus \Omega_{r/2}|} \int_{\Omega_r \setminus \Omega_{r/2}} \frac{\mathrm{d}x}{|x|^2} \leq \frac{C}{r^2}$$

for r large enough. To obtain the left-hand side inequality of (4.27) we remark (see Theorem 3.3) that for r large enough

$$\begin{aligned}
\lambda_r &= \int_{\Omega_r \setminus \Omega_{r/2}} a\pi_r\,\mathrm{d}x \Big/ \int_{\Omega_r \setminus \Omega_{r/2}} \pi_r\,\mathrm{d}x \\
&= \int_{\Omega_r \setminus \Omega_{r/2}} C_0\pi_r/|x|^2\,\mathrm{d}x \Big/ \int_{\Omega_r \setminus \Omega_{r/2}} \pi_r\,\mathrm{d}x \\
&\geq \frac{c}{r^2} \int_{\Omega_r \setminus \Omega_{r/2}} \pi_r\,\mathrm{d}x \Big/ \int_{\Omega_r \setminus \Omega_{r/2}} \pi_r\,\mathrm{d}x = \frac{c}{r^2},
\end{aligned}$$

with $c = C_0$ for $\Omega_r = B(0, r)$. This completes the proof of (4.27).

We conclude this note with the following result.

Theorem 4.5.

Suppose that (4.3) *admits a bounded solution, then it admits a positive solution.*

If

$$\int_{|x|>1} a(x)|x|^{-k+2}\,\mathrm{d}x = \infty \tag{4.28}$$

then (4.3) *cannot admit nontrivial bounded solution such that*

$$0 < c \leq u. \tag{4.29}$$

Proof. We first prove the existence of a positive solution. If $u < 0$, $-u$ is a positive solution. So, we can assume that u changes sign. Then introduce u_n solution of

$$-\Delta u_n + au_n = 0 \quad \text{in } B(0,n), u_n = |u|_\infty \quad \text{on } \partial B(0,n). \tag{4.30}$$

One has

$$0 < u_{n+1} \leq u_n \leq |u|_\infty \tag{4.31}$$

and u_n converges to some fonction u_∞ for instance in $L^1_{\text{loc}}(\mathbb{R}^k)$. Then u_∞ is a solution of (4.3). Moreover by the maximum principle one has $u \leq u_n$ on $B(0,n)$ and thus $u \leq u_\infty$. u_∞ cannot vanish identically and is the positive solution we are looking for.

Suppose now that u is a nonnegative bounded solution to

$$-\Delta u = -au.$$

Set $U(r) = \int_{\partial B_1} u(r\sigma)\, d\sigma$ where B_1 denotes the unit ball of $\mathbb{R}^k$. Then

$$-(r^{k-1}U')' = -r^{k-1} \int_{\partial B_1} a(r\sigma)u(r\sigma)\, d\sigma \tag{4.32}$$

hence

$$-r^{k-1}U' = -\int_0^r s^{k-1} \int_{\partial B_1} a(r\sigma)u(s\sigma)\, d\sigma ds \tag{4.33}$$

and $U(r) = \int_{\partial B_1} u(r\sigma)\, d\sigma$ is nondecreasing. Moreover U is a solution of the second order differential equation (4.32). A particular solution is given (see (4.33)) by

$$U = \int_0^r \frac{1}{s^{k-1}} \int_0^s t^{k-1} \int_{\partial B_1} a(t\sigma)u(t\sigma)\, d\sigma dt ds.$$

The solution of the homogeneous equation is given by

$$\frac{A}{r^{k-2}} + B.$$

Thus we have

$$U(r) = \frac{A}{r^{k-2}} + B + \int_0^r \frac{1}{s^{k-1}} \int_0^s t^{k-1} \int_{\partial B_1} a(t\sigma)u(t\sigma)\, d\sigma dt ds. \tag{4.34}$$

Since u is bounded, so is U and necessarily $A = 0, B \geq 0$. From (4.34) we derive

$$U(r) = B + \int_0^r \frac{1}{s^{k-1}} \int_0^s t^{k-1} \int_{\partial B_1} a(t\sigma)u(t\sigma)\, d\sigma dt ds. \tag{4.35}$$

Integrating by parts we get

$$U(r) = B - \frac{1}{(k-2)r^{k-2}} \int_0^r t^{k-1} \int_{\partial B_1} a(t\sigma)u(t\sigma)\,\mathrm{d}\sigma\mathrm{d}t$$

$$+ \int_0^r \frac{1}{(k-2)t^{k-2}} t^{k-1} \int_{\partial B_1} a(t\sigma)u(t\sigma)\,\mathrm{d}\sigma\mathrm{d}t$$

$$= B + \frac{1}{k-2} \int_0^r \int_{\partial B_1} ta(t\sigma)(1 - \frac{t^{k-2}}{r^{k-2}})u(t\sigma)\,\mathrm{d}\sigma\mathrm{d}t. \tag{4.36}$$

When

$$\int_{|x|>1} \frac{a(x)}{|x|^{k-2}}\,\mathrm{d}x = \int_1^{+\infty} \int_{\partial B_1} ta(t\sigma)\,\mathrm{d}\sigma\mathrm{d}t = +\infty,$$

then the equation (4.3) cannot have a solution such that

$$0 < c \leq u \leq C.$$

Indeed from (4.36) we would get

$$U(r) \geq B + \frac{1}{k-2} \int_1^{\frac{r}{2}} \int_{\partial B_1} ta(t\sigma)\,\mathrm{d}\sigma\mathrm{d}t(1 - \frac{1}{2^{k-2}})c \to +\infty$$

which contradicts the fact that u and U are bounded. □

Remark 4.3. Using this result one recovers easily the Lemma 4.1.

Acknowledgement: The second and third authors have been supported by the Swiss National Science foundation under the contracts #20-105155, #20-113287 #20-111543 and #20-117614. The second author is very grateful to the department of mathematics of Rutgers University for the kind of hospitality during the preparation of the manuscript. We would like also to thank Y. Pinchover and R. Pinsky for useful discussions.

References

[1] S. Agmon: On positive solutions of elliptic equations with periodic coefficients in $\mathbb{R}^N$, spectral results and extensions to elliptic operators on Riemannian manifolds. In: Differential equations, North-Holland Math. Stud. 92, North-Holland, Amsterdam, (1984).

[2] C.J.K. Batty: Asymptotic stability of Schrödinger semigroups: path integral methods. Math. Ann. 292, 457–492, (1992).

[3] W. Arendt, C.J.K. Batty, P. Bénilan: Asymptotic stability of Schrödinger semigroups on $L^1(\mathbb{R}^N)$. Math. Z. 209, 511–518, (1992).

[4] Berestycki, I. Capuzzo-Dolcetta, L. Nirenberg: Superlinear indefinite elliptic problems and nonlinear Liouville theorems. Topol. Methods Nonlinear Anal. 4, 59–78, (1994).

[5] L.C. Evans: *Partial differential equations.* Graduate Studies in Mathematics, 19, A.M.S., (1998)

[6] M. Giaquinta: *Multiple integrals in the calculus of variations and nonlinear elliptic systems.* Annals of math studies 105, Princeton University Press, (1983).

[7] E. Giusti: *Direct Methods in the Calculus of variations.* World Scientific (2003).

[8] A. Grigor'yan: Bounded solutions of the Schrödinger equation on noncompact Riemannian manifolds. J. Sov. Math. 51, 2340–2349, (1990).

[9] A. Grigor'yan: Analytic and geometric background of recurrence and non-explosion of the brownian motion on Riemannian manifolds. Bull. AMS, 36, 135–2492, (1999).

[10] A. Grigor'yan, W. Hansen: A Liouville property for Schrödinger operators. Math. Ann. 312, 659–716, (1998).

[11] M. Meier: Liouville theorem for nonlinear elliptic equations and systems. Manuscripta Mathematica, 29, 207–228, (1979).

[12] J. Moser: On Harnack's theorem for elliptic differential equations. Comm. Pure Appl.Math., 14, 577–591, (1961).

[13] L. Moschini: New Liouville theorems for linear second order degenerate elliptic equations in divergence form. Ann. Inst. H. Poincaré Anal. Non Linéaire 22, 11–23, (2005).

[14] Y. Pinchover: On the equivalence of Green functions of second order elliptic equations in $\mathbb{R}^n$. Diff. and Integral Equations 5, 481–493, (1992).

[15] Y. Pinchover: Maximum and anti-maximum principles and eigenfunctions estimates via perturbation theory of positive solutions of elliptic equations. Math. Ann. 314, 555–590, (1999).

[16] Y. Pinchover: Topics in the theory of positive solutions of second-order elliptic and parabolic partial differential equations. Preprint.

[17] R. G. Pinsky: *Positive harmonic functions and diffusion: An integrated analytic and probabilistic approach.* Cambridge studies in advanced mathematics, 45, (1995).

[18] R. G. Pinsky: A probabilistic approach to a Liouville-type problem for Schrödinger operators. Preprint, (2006)

[19] M. Rigoli, A. Setti: A Liouville theorem for a class of superlinear elliptic equations on cones. Nonlinear Differential Equations Appl. 9, 15–36, (2002).

[20] B. Simon: Schrödinger semigroups. Bull. Am. Math. Soc. 7, 447-526, (1982).

FREE BOUNDARY PROBLEMS OF THE NONLINEAR HEAT EQUATIONS COUPLED WITH THE NAVIER-STOKES EQUATIONS

TAKESI FUKAO*

General Education, Gifu National College of Technology, 2236-2 Kamimakuwa, Motosu-shi, Gifu, 501-0495 Japan

Abstract: In this paper, we consider the existence problem for two free boundary problems of nonlinear heat equations coupled with the Navier-Stokes equations. The first one is called the Stefan problem and the second one is called the phase field equations. In both cases we determine the convective vector by the weak solution of the Navier-Stokes equations in the unknown sub-domain. Firstly, we survey an existence result of the weak solution which satisfies the variational formulation of the Stefan problem, by approximation of the interface. Secondly, in the phase field equations, by comparing with the above result we define an artificial interface that is the level carve of the order parameter. In 2-dimensional case, it seems that applying the L^p-theory of parabolic type we can obtain the weak solution for the phase field equations coupled with the Navier-Stokes equations.

1. Solid-liquid phase transitions

We consider a material which has two physical phases, liquid and solid. We are interested in the dynamics of their interface, taking account of a convective flow of the material in the liquid. Our idea is to describe this phenomenon by the second order partial differential equations, namely the heat equation with convection governed by the Navier-Stokes equations in the unknown liquid region. Our purpose is to show the existence of some solutions for the initial and boundary value problem. The phase field equations is one of the famous model which describes the solid-liquid phase transition phenomena. This model has the strong relationship between the Stefan problem which is also well-known as the effective model. About the Stefan problem we have some result [10] from the stand point of the practical situation, in which the convective vector is treated by the weak

*supported by a grant-in-aid for encouragement of young scientists (b) (no.18740095), jsps.

solutions of the Navier-Stokes equations in the liquid region. But we do not know that the solution of the Stefan problem is smooth. So it is not easy to guarantee the liquid region as the exact open set. On the other hand, in the case of the prototype phase filed equations, it seems that we can gain the regularity of the solutions by applying the L^p-theory of parabolic type.

We use the following notations: $H := L^2(\Omega)$, $Y := L^4(\Omega)$, $V := H^1(\Omega)(:= W^{1,2}(\Omega))$, $X := W^{1,4}(\Omega)$ with the usual norms, and Y^*, V^* and X^* are the dual spaces of Y, V and X; we denote by $\langle \cdot, \cdot \rangle_{V^*,V}$ the duality pair between V^* and V. Especially H is a Hilbert space with standard inner product $(\cdot,\cdot)_H$ and we have the following relations:$X \subset V \hookrightarrow Y \subset H \subset Y^* \hookrightarrow V^* \subset X^*$, where $\hookrightarrow$ means that the imbedding is compact. Moreover we use the following notations for vector valued function spaces: $\boldsymbol{\mathcal{D}}_\sigma(\Omega) := \{\boldsymbol{z} \in \boldsymbol{C}_0^\infty(\Omega); \mathrm{div}\boldsymbol{z} = 0 \text{ in } \Omega\}$, $\boldsymbol{H} := \boldsymbol{L}^2_\sigma(\Omega)$, $\boldsymbol{Y} := \boldsymbol{L}^4_\sigma(\Omega)$, $\boldsymbol{V} := \boldsymbol{H}^1_\sigma(\Omega)$, $\boldsymbol{X} := \boldsymbol{W}^{1,4}_\sigma(\Omega)$, where for $N = 2$ or 3, $\boldsymbol{L}^2_\sigma(\Omega)$, $\boldsymbol{L}^4_\sigma(\Omega)$, $\boldsymbol{H}^1_\sigma(\Omega)$ and $\boldsymbol{W}^{1,4}_\sigma(\Omega)$ are closures of $\boldsymbol{\mathcal{D}}_\sigma(\Omega)$ in spaces $\boldsymbol{L}^2(\Omega) := H^N$, $\boldsymbol{L}^4(\Omega) := Y^N$, $\boldsymbol{H}^1(\Omega) := V^N$ and $\boldsymbol{W}^{1,4}(\Omega) := X^N$, respectively. They are equipped with the usual product norms, and $\boldsymbol{Y}^*, \boldsymbol{V}^*$ and $\boldsymbol{X}^*$ are the dual spaces of $\boldsymbol{Y}, \boldsymbol{V}$ and $\boldsymbol{X}$; we denote by $\langle \cdot, \cdot \rangle_{\boldsymbol{V}^*,\boldsymbol{V}}$ the duality pair between $\boldsymbol{V}^*$ and $\boldsymbol{V}$. We see that $\boldsymbol{H}$ is a Hilbert space with the usual inner product $(\cdot,\cdot)_{\boldsymbol{H}}$ and the following relations hold: $\boldsymbol{X} \subset \boldsymbol{V} \hookrightarrow \boldsymbol{Y} \subset \boldsymbol{H} \subset \boldsymbol{Y}^* \hookrightarrow \boldsymbol{V}^* \subset \boldsymbol{X}^*$.

2. Enthalpy formulation of the Stefan problem

For $0 < T < +\infty$ and $N = 2$ or 3, $\Omega \subset \mathbb{R}^N$ be a bounded domain with smooth boundary $\Gamma := \partial\Omega$ such that Ω is occupied by a material having two phases of liquid and solid states; $\Omega_\ell(t)$ (resp. $\Omega_s(t)$) is the liquid (resp. solid) region at time t, and these regions are separated by an unknown interface $S(t)$, namely $\Omega := \Omega_\ell(t) \cup S(t) \cup \Omega_s(t)$ for each $t \in [0,T]$. Now we denote by $\theta := \theta(t,x)$ the temperature field, by $\boldsymbol{v} := \boldsymbol{v}(t,x)$ the velocity of the flow of material and by $p_\ell := p_\ell(t,x)$ the pressure field. In the above setting we consider the following problem (SP)

$$\frac{\partial u}{\partial t} + \boldsymbol{v} \cdot \nabla u - \Delta\beta(u) = f \quad \text{in } Q := (0,T) \times \Omega, \tag{2.1}$$

$$\frac{\partial \boldsymbol{v}}{\partial t} + (\boldsymbol{v} \cdot \nabla)\boldsymbol{v} - \Delta \boldsymbol{v} = \beta(u)\boldsymbol{g} - \nabla p_\ell \quad \text{in } Q_\ell(u), \tag{2.2}$$

$$\mathrm{div}\boldsymbol{v} = 0 \quad \text{in } Q_\ell(u), \tag{2.3}$$

$$\boldsymbol{v} = 0 \quad \text{in } Q_s(u) \cup S(u), \tag{2.4}$$

$$\frac{\partial \beta(u)}{\partial \boldsymbol{n}} + n_0 \beta(u) = h, \quad \boldsymbol{v} = 0 \quad \text{on } \Sigma := (0,T) \times \Gamma, \tag{2.5}$$

$$u(0) = u_0, \quad \boldsymbol{v}(0) = \boldsymbol{v}_0 \quad \text{in } \Omega, \tag{2.6}$$

where f, h, $\boldsymbol{g}$, u_0 and $\boldsymbol{v}_0$ are given functions. $\boldsymbol{n}$ is the unit vector outward normal to Γ. n_0 is a positive constant. The above problem (SP) stands for the weak variational formulation of the classical Stefan problem, introducing a new parameter u, which is called the enthalpy, and a function $\beta : \mathbb{R} \to \mathbb{R}$ defined by

$$u := \begin{cases} \theta & \text{if } \theta < 0, \\ r \in [0, L] & \text{if } \theta = 0, \\ \theta + L & \text{if } \theta > 0, \end{cases} \qquad \beta(r) := \begin{cases} k_s r & \text{if } r < 0, \\ 0 & \text{if } 0 \le r \le L, \\ k_\ell (r - L) & \text{if } r > L, \end{cases}$$

where k_s, k_ℓ and L are positive constants. It is most important that how to define the unknown domains $Q_\ell(u)$, $Q_s(u)$ and $S(u)$. For $i = s, \ell$ we define

$$Q_i(u) := \bigcup_{t \in (0,T)} \{t\} \times \Omega_i(t), \quad S(u) := \bigcup_{t \in (0,T)} \{t\} \times S(t),$$

where $\Omega_\ell(t)$, $\Omega_s(t)$ and $S(t)$ are

$$\Omega_\ell(t) := \left\{ x \in \Omega; u(t,x) > \frac{L}{2} \right\}, \quad \Omega_s(t) := \left\{ x \in \Omega; u(t,x) < \frac{L}{2} \right\},$$

$$S(t) := \left\{ x \in \Omega; u(t,x) = \frac{L}{2} \right\}.$$

It is easy to derive the following variational identity from (2.1), (2.5) and (2.6):

$$-\int_0^T \langle \eta', u \rangle_{V^*,V} dt + \int_0^T (\nabla \beta(u) - u\boldsymbol{v}, \nabla \eta)_{\boldsymbol{L}^2(\Omega)} dt + n_0 \int_0^T (\beta(u), \eta)_{L^2(\Gamma)} dt$$

$$= \int_0^T (f, \eta)_H dt + \int_0^T (h, \eta)_{L^2(\Gamma)} dt + (u_0, \eta(0))_H \quad \text{for all } \eta \in W, \tag{2.7}$$

where $W := \{\eta \in H^1(Q); \eta(T) = 0\}$.

In the paper of Rodrigues [20], the well-posedness for the problem (2.7) was treated, when the convective vector $\boldsymbol{v}$ is prescribed. Also the papers of [11] gave the result for this problem in the case when the domain Ω depends on time t. Recently Casella and Giangi [3] was treated the penalized problem for (SP).

Next we consider the Navier-Stokes equations in the liquid region. We employ the standard framework for the Navier-Stokes equations. Accordingly, we define the bilinear functional $a(\cdot,\cdot) : \boldsymbol{V} \times \boldsymbol{V} \to \mathbb{R}$ and trilinear functional $b(\cdot,\cdot,\cdot) : \boldsymbol{V} \times \boldsymbol{V} \times \boldsymbol{V} \to \mathbb{R}$ by

$$a(\boldsymbol{u},\boldsymbol{w}) := \sum_{i,j=1}^{N} \int_\Omega \frac{\partial u_j}{\partial x_i}\frac{\partial w_j}{\partial x_i} dx, \quad b(\boldsymbol{u},\boldsymbol{v},\boldsymbol{w}) := \sum_{i,j=1}^{N} \int_\Omega u_i \frac{\partial v_j}{\partial x_i} w_j dx,$$

for all $\boldsymbol{u}, \boldsymbol{v}$, and $\boldsymbol{w} \in \boldsymbol{V}$. A weak variational formulation of Navier-Stokes equations is described as follows:

$$\begin{aligned} &-\int_0^T (\boldsymbol{\eta}', \boldsymbol{v})_{\boldsymbol{H}} dt + \int_0^T a(\boldsymbol{v},\boldsymbol{\eta}) dt + \int_0^T b(\boldsymbol{v},\boldsymbol{v},\boldsymbol{\eta}) dt \\ &= \int_0^T (\beta(u)\boldsymbol{g}, \boldsymbol{\eta})_{\boldsymbol{H}} dt + (\boldsymbol{v}_0, \boldsymbol{\eta}(0))_{\boldsymbol{H}} \quad \text{for all } \boldsymbol{\eta} \in \boldsymbol{W}(u), \end{aligned} \tag{2.8}$$

where

$$\boldsymbol{W}(u) := \left\{ \boldsymbol{\eta} \in L^4(0,T;\boldsymbol{X}); \boldsymbol{\eta}' \in L^2(0,T;\boldsymbol{H}), \operatorname{supp} \boldsymbol{\eta} \subset Q_\ell(u), \boldsymbol{\eta}(T) = 0 \right\}.$$

From the mathematical point of view it is extremely difficult to handle this weak variational formulation of our problem (SP), especially (2.8) because of the lack of regularity of the enthalpy u. In the paper [10], in order to avoid this difficulty we replace the liquid region $\Omega_\ell(t)$ and solid region $\Omega_s(t)$ by their approximations

$$\Omega_{\ell,\varepsilon}(t) := \left\{ x \in \Omega; (\rho_\varepsilon * u)(t,x) > \frac{L}{2} \right\}, \tag{2.9}$$

and

$$\Omega_{s,\varepsilon}(t) := \left\{ x \in \Omega; (\rho_\varepsilon * u)(t,x) < \frac{L}{2} \right\}, \tag{2.10}$$

respectively, where $\rho_\varepsilon := \rho_\varepsilon(x)$ is the usual ε-mollifier in $\mathbb{R}^3$ with respect to the space variable x, and the class $\boldsymbol{W}(u)$ of test functions by

$$\boldsymbol{W}_\varepsilon(u) := \left\{ \boldsymbol{\eta} \in L^4(0,T;\boldsymbol{X}); \boldsymbol{\eta}' \in L^2(0,T;\boldsymbol{H}), \operatorname{supp} \boldsymbol{\eta} \subset Q_{\ell,\varepsilon}(u), \boldsymbol{\eta}(T) = 0 \right\},$$

where for $i = \ell, s$

$$Q_{i,\varepsilon}(u) := \bigcup_{t \in (0,T)} \{t\} \times \Omega_{i,\varepsilon}(t).$$

Now we introduce a class of weak solutions, which is a reasonable approximation to the original one, as follows:

Definition 2.1. For a fixed $\varepsilon > 0$, the pair $\{u, \boldsymbol{v}\}$ of functions $u \in L^\infty(Q)$ and $\boldsymbol{v} \in \boldsymbol{L}^2(Q)$ is called a weak solution of $(\mathrm{SP})_\varepsilon$, if (D1)-(D3) are satisfied:

(D1) $\beta(u) \in L^2(0,T;V) \cap L^\infty(Q)$ and $u \in C_w(0,T;H)$;
(D2) $\boldsymbol{v} \in L^\infty(0,T;\boldsymbol{H}) \cap L^2(0,T;\boldsymbol{V})$, $\boldsymbol{v}$ is weakly continuous from $[0,T]$ into $\boldsymbol{H}$, and $\boldsymbol{v} = 0$ a.e. on $Q_{s,\varepsilon}(u)$;
(D3) u and $\boldsymbol{v}$ satisfy (2.7) and (2.8) with $\boldsymbol{W}(u)$ replaced by $\boldsymbol{W}_\varepsilon(u)$.

Theorem 2.1. [10] *Assume $f \in L^\infty(Q)$, $\boldsymbol{g} \in \boldsymbol{W}^{1,\infty}(\Omega)$, $h \in L^\infty(\Sigma)$, $u_0 \in L^\infty(\Omega)$ and $\boldsymbol{v}_0 \in \boldsymbol{H}$. Then, for each small positive number ε, there exist at least one solution $\{u_\varepsilon, \boldsymbol{v}_\varepsilon\}$ of $(SP)_\varepsilon$ such that the following uniform estimates in $\varepsilon \in (0,1]$:*

$$|u_\varepsilon|_{L^\infty(Q)} + |\nabla\beta(u_\varepsilon)|_{L^2(0,T;\boldsymbol{L}^2(\Omega))} \le R_0 \tag{2.11}$$

and

$$|\boldsymbol{v}_\varepsilon|_{L^\infty(0,T;\boldsymbol{H})} + |\boldsymbol{v}_\varepsilon|_{L^2(0,T;\boldsymbol{V})} \le R_1, \tag{2.12}$$

where R_0 and R_1 are positive constants which depend on $|u_0|_{L^\infty(\Omega)}$, $|f|_{L^\infty(Q)}$, and $|h|_{L^\infty(\Sigma)}$ independent of $\varepsilon \in (0,1]$.

As to the limit of solutions $\{u_\varepsilon, \boldsymbol{v}_\varepsilon\}$ we have the following result.

Theorem 2.2. [10] *Under the same assumption of Proposition 2.1, let $\{u_\varepsilon, \boldsymbol{v}_\varepsilon\}_{\varepsilon>0}$ be the family of approximate solutions constructed by Proposition 2.1. Then there exists a sequence $\{\varepsilon_n\}$ of positive numbers with $\varepsilon_n \to 0$ (as $n \to +\infty$) such that $u_n := u_{\varepsilon_n} \to u$ weakly in $L^2(Q)$ and $\boldsymbol{v}_n := \boldsymbol{v}_{\varepsilon_n} \to \boldsymbol{v}$ weakly in $L^2(0,T;\boldsymbol{V})$. Moreover u and $\boldsymbol{v}$ satisfy (2.7).*

Remark 2.1. As to the vector function $\boldsymbol{v}$ obtained in theorem 2.2 as a limit of $\{\boldsymbol{v}_n\}$, it is not clear whether $\boldsymbol{v}$ is a solution of the variational form (2.8) of the Navier-Stokes equations, because the class $\boldsymbol{W}(u)$ of test functions for the limit u of $\{u_n\}$ is not able to be defined without the regularity of u.

3. Phase field equations with convection

The system of the phase field equations is also well-known that describes the solid-liquid phase transition phenomena. Mathematically, the Boussinesq system, namely the system of a heat equation and the Naiver-Stokes equations has been studied in many papers, Foias, Manley, and Temam [7], Morimoto [18], Inoue and Ôtani [14], Diaz and Galiano [5], Lorca and Boldrini [17], Fukao and Kubo [12] and so on. In the paper of Planas and Boldrini [19] the existence problem for the free boundary problem of the phase field type was treated, where the domain was separated to three regions; pure solid, pure liquid, and mushy region. And they consider the Navier-Stokes equations in the not solid regions. The Carman-Kozeny penalty term has an important role from the physical view point. In this paper, the different interpretation of the domain are considered. We separate the region by an artificial interface. Such a model is well-known as the enthalpy formulation of the Stefan problem. However, from the Remark 2.1, there is the difficulty of the free boundary problem because of the lack of regularity of the weak solution of the Stefan problem. Namely, if we admit an artificial interface that is defined by the order parameter, then we can treat the Navier-Stokes equations without the approximation of the interface.

Hereafter let $N = 2$. We consider the following system (PF):

$$\frac{\partial \theta}{\partial t} + \boldsymbol{v} \cdot \nabla \theta + \frac{\partial \chi}{\partial t} + \boldsymbol{v} \cdot \nabla \chi - \Delta \theta = f \quad \text{in } Q, \tag{3.1}$$

$$\frac{\partial \chi}{\partial t} + \boldsymbol{v} \cdot \nabla \chi - \Delta \chi + \chi^3 - \chi = \theta \quad \text{in } Q, \tag{3.2}$$

$$\frac{\partial \boldsymbol{v}}{\partial t} + (\boldsymbol{v} \cdot \nabla)\boldsymbol{v} - \Delta \boldsymbol{v} = \theta \boldsymbol{g} - \nabla p_\ell \quad \text{in } Q_\ell(\chi), \tag{3.3}$$

$$\text{div}\boldsymbol{v} = 0 \quad \text{in } Q_\ell(\chi), \tag{3.4}$$

$$\boldsymbol{v} = 0 \quad \text{in } Q_s(\chi) \cup S(\chi), \tag{3.5}$$

$$\frac{\partial \theta}{\partial \boldsymbol{n}} = 0, \quad \frac{\partial \chi}{\partial \boldsymbol{n}} = 0, \quad \boldsymbol{v} = 0 \quad \text{on } \Sigma := (0,T) \times \Gamma, \tag{3.6}$$

$$\theta(0) = \theta_0, \quad \chi(0) = \chi_0, \quad \boldsymbol{v}(0) = \boldsymbol{v}_0 \quad \text{in } \Omega, \tag{3.7}$$

where θ_0 and χ_0 are given functions. It is most important that how to define the unknown domains $Q_\ell(\chi)$, $Q_s(\chi)$ and $S(\chi)$. For $i = s, \ell$ we define

$$Q_i(\chi) := \bigcup_{t \in (0,T)} \{t\} \times \Omega_i(t), \quad S(\chi) := \bigcup_{t \in (0,T)} \{t\} \times S(t),$$

where $\Omega_\ell(t)$ and $\Omega_s(t)$ are determined by the artificial interface between solid and liquid as follows:

$$\Omega_\ell(t) := \{x \in \Omega; \chi(t,x) > 0\}, \quad \Omega_s(t) := \{x \in \Omega; \chi(t,x) < 0\},$$

$$S(t) := \{x \in \Omega; \chi(t,x) = 0\}.$$

We use the notation:

$$W_p := W_p^{2,1}(Q) := \left\{ u \in L^p(Q); \frac{\partial u}{\partial t}, \frac{\partial u}{\partial x_i}, \frac{\partial^2 u}{\partial x_i \partial x_j} \in L^p(Q) \right\}.$$

Then $W_3 \hookrightarrow C(\overline{Q})$, see Ladyženskaja, Solonnikov, and Ural'ceva [[16], p.80]. Put $u^+ := \max\{0, u\}$, $u^- := \max\{0, -u\}$. Under these settings, we define our solution.

Definition 3.1. The triplet $\{\theta, \chi, \boldsymbol{v}\} \in L^2(Q) \times C(\overline{Q}) \times \boldsymbol{L}^2(Q)$ is called a solution of our system (PF) if there exist sequences of functions $\{\theta_\varepsilon, \chi_\varepsilon\} \subset L^\infty(0,T;H) \cap L^2(0,T;V) \times W_3$ and $\{\boldsymbol{v}_\varepsilon\} \subset L^\infty(0,T;\boldsymbol{H}) \cap L^2(0,T;\boldsymbol{V})$ such that they are approximate solutions for (PF) in the following variational sense of (3.1)

$$-\int_0^T \langle \eta', \theta_\varepsilon + \chi_\varepsilon \rangle_{V^*,V} dt - \int_0^T ((\theta_\varepsilon + \chi_\varepsilon)\boldsymbol{v}_\varepsilon, \nabla\eta)_{\boldsymbol{L}^2(\Omega)} dt + \int_0^T (\nabla\theta_\varepsilon, \nabla\eta)_{\boldsymbol{L}^2(\Omega)} dt$$

$$= \int_0^T (f, \eta)_H dt + (\theta_0 + \chi_0, \eta(0))_H \quad \text{for all } \eta \in W, \tag{3.8}$$

where $W := \{\eta \in L^2(0,T;V); \eta' \in L^2(0,T;V^*), \eta(T) = 0\}$, and of (3.3), (3.4)

$$-\int_0^T \langle \boldsymbol{\eta}', \boldsymbol{v}_\varepsilon \rangle_{\boldsymbol{V}^*,\boldsymbol{V}} dt + \int_0^T a(\boldsymbol{v}_\varepsilon, \boldsymbol{\eta}) dt + \int_0^T b(\boldsymbol{v}_\varepsilon, \boldsymbol{v}_\varepsilon, \boldsymbol{\eta}) dt + \frac{1}{\varepsilon} \int_0^T (\chi_\varepsilon^- \boldsymbol{v}_\varepsilon, \boldsymbol{\eta})_{\boldsymbol{H}} dt$$

$$= \int_0^T (\theta_\varepsilon \boldsymbol{g}, \boldsymbol{\eta})_{\boldsymbol{H}} dt + (\boldsymbol{v}_0, \boldsymbol{\eta}(0))_{\boldsymbol{H}} \quad \text{for all } \boldsymbol{\eta} \in \boldsymbol{W}, \tag{3.9}$$

where $\boldsymbol{W} := \{\boldsymbol{\eta} \in L^2(0,T;\boldsymbol{V}); \boldsymbol{\eta}' \in L^2(0,T;\boldsymbol{V}^*), \boldsymbol{\eta}(T) = 0\}$, and then

$$\theta_\varepsilon \to \theta \quad \text{weakly in } L^2(0,T;V), \tag{3.10}$$

$$\theta_\varepsilon \to \theta \quad \text{in } L^2(0,T;H), \tag{3.11}$$

$$\chi_\varepsilon \to \chi \quad \text{weakly in } L^2(0,T;V), \text{ and uniformly in } Q, \tag{3.12}$$

$$\boldsymbol{v}_\varepsilon \to \boldsymbol{v} \quad \text{weakly in } L^2(0,T;\boldsymbol{V}), \tag{3.13}$$

$$\boldsymbol{v}_\varepsilon \to \boldsymbol{v} \quad \text{in } \boldsymbol{L}^2_{loc}(Q_\ell(\chi)) \quad \text{as } \varepsilon \to 0. \tag{3.14}$$

As well, there exist positive constants M_1, M_2, M_3, M_4, and M_5 independent of ε such that

$$|\theta_\varepsilon|_{L^\infty(0,T;H)} + |\theta_\varepsilon|_{L^2(0,T;V)} \le M_1 := M_1(|f|_{L^2(Q)}, |\theta_0|_H, |\chi_0|_H), \quad (3.15)$$

$$|\chi_\varepsilon|_{L^2(0,T;V)} \le M_1, \quad (3.16)$$

$$|\chi_\varepsilon|_{L^\infty(Q)} \le M_2 := M_2(M_1, |\chi_0|_{L^\infty(\Omega)}, |\Omega|), \quad (3.17)$$

$$|\boldsymbol{v}_\varepsilon|_{L^\infty(0,T;\boldsymbol{H})} + |\boldsymbol{v}_\varepsilon|_{L^2(0,T;\boldsymbol{V})} \le M_3 := M_3(M_1, |\boldsymbol{v}_0|_{\boldsymbol{H}}), \quad (3.18)$$

$$\int_0^T \int_\Omega \chi_\varepsilon^- |\boldsymbol{v}_\varepsilon|^2 dxdt \le \varepsilon M_3, \quad (3.19)$$

$$|\chi_\varepsilon|_{W_3} \le M_4 := M_4(M_2, M_3, |\chi_0|_{W^{4/3,3}(\Omega)}). \quad (3.20)$$

$$|\theta'_\varepsilon|_{L^2(0,T;V^*)} \le M_5 := M_5(M_1, M_3, M_4, |f|_{L^2(Q)}). \quad (3.21)$$

To obtain a strong estimate with respect to W_3 we completely imitate the result of Planas and Boldrini by L^p-theory of parabolic type, see Planas and Boldrini [[19], Theorem 2], essential due to Hoffman and Jang [13]. By virtue of Aubin's compactness theorem, see Simon [21], with estimates (3.15), (3.21), and the imbedding $W_3 \hookrightarrow C(\overline{Q})$ with the estimate (3.20), there exist some subsequence for $\varepsilon > 0$, which we also denote by $\{\theta_\varepsilon, \chi_\varepsilon\}$ for simplicity, and functions $\theta \in L^\infty(0,T;H) \cap L^2(0,T;V)$ and $\chi \in W_3$ such that the convergences (3.10), (3.11), and (3.12) hold. Here for any $s_1, s_2 \in [0,T]$ and any $\omega \subset \Omega$ with $(s_1, s_2) \times \omega \subset Q_\ell(\chi)$, applying the method of the compact cylinder essentially due to Fujita and Sauer [8], we have the following estimate:

Lemma 3.1. *For any sufficiently small $\varepsilon > 0$ there exists a positive constant M_6 independent of ε such that*

$$|\boldsymbol{v}'_\varepsilon|_{L^2(s_1,s_2;\boldsymbol{H}^1_\sigma(\omega)^*)} \le M_6 := M_6(M_1, M_3, |\boldsymbol{g}|_{\boldsymbol{W}^{1,\infty}(\Omega)}). \quad (3.22)$$

Proof. From the equation (3.9) taking the test function $\boldsymbol{\eta} \in \boldsymbol{W}$ with $\operatorname{supp} \boldsymbol{\eta} \subset (s_1, s_2) \times \omega \subset Q_\ell(\chi)$. Then from the uniform convergence (3.12) we see that there exists a small $\varepsilon^* > 0$ such that $\operatorname{supp} \boldsymbol{\eta} \subset Q_\ell(\chi_\varepsilon)$ for all ε with $0 < \varepsilon < \varepsilon^*$, namely $\chi_\varepsilon^- \boldsymbol{\eta} = 0$ on ω . So we have the following estimate

$$\int_{s_1}^{s_2} \langle \boldsymbol{v}'_\varepsilon, \boldsymbol{\eta} \rangle_{\boldsymbol{H}^1_\sigma(\omega)^*, \boldsymbol{H}^1_\sigma(\omega)} dt$$
$$\le M_3 |\boldsymbol{\eta}|_{L^2(s_1,s_2;\boldsymbol{V})} + M_3^4 |\boldsymbol{\eta}|_{L^2(s_1,s_2;\boldsymbol{V})} + 2M_1 |\boldsymbol{g}|_{\boldsymbol{W}^{1,\infty}(\Omega)} |\boldsymbol{\eta}|_{L^2(s_1,s_2;\boldsymbol{H})}.$$

Thus $M_6 := M_3 + M_3^4 + 2M_1|\boldsymbol{g}|_{\boldsymbol{W}^{1,\infty}(\Omega)}$. □

The estimate (3.18) implies that there exists some subsequence for $\varepsilon > 0$, which we also denote by $\{\boldsymbol{v}_\varepsilon\}$ for simplicity, and a function $\boldsymbol{v}$ such that the convergence (3.13) holds. Define the set $\boldsymbol{V}_\sigma(\omega)$ is the the closure of $\{\boldsymbol{u} \in \boldsymbol{C}^\infty(\omega); \mathrm{div}\boldsymbol{u} = 0\}$ with respect to $\boldsymbol{V}$. Then the relation $\boldsymbol{V} \subset \boldsymbol{V}_\sigma(\omega) \hookrightarrow \boldsymbol{L}^2_\sigma(\omega) \hookrightarrow \boldsymbol{H}^1_\sigma(\omega)^*$ holds. Now we have $|\boldsymbol{v}_\varepsilon|_{L^2(s_1,s_2;\boldsymbol{V}_\sigma(\omega))} \le M_2$ and (3.22). So Aubin's compactness theorem implies that $\boldsymbol{v}_\varepsilon \to \boldsymbol{v}$ in $L^2(s_1, s_2; \boldsymbol{L}^2_\sigma(\omega))$ as $\varepsilon \to 0$. Noting that this is valid for every relatively compact and open cylindrical subdomain of the form $(s_1, s_2) \times \omega$ in $Q_\ell(\chi)$, we conclude (3.14), since any compact subset of $Q_\ell(\chi)$ can be covered by a finite number of subdomains of the form $(s_1, s_2) \times \omega$. Finally, the estimate (3.19) implies that $\chi^-|\boldsymbol{v}|^2 = 0$ a.e. in Q, namely $\boldsymbol{v} = 0$ a.e. on $Q_s(\chi)$. Thus we see that θ, χ, and $\boldsymbol{v}$ satisfies the Definition 3.1. So we can see that if we assume $f \in L^2(Q)$, $\boldsymbol{g} \in \boldsymbol{W}^{1,\infty}(\Omega)$, $\theta_0 \in H$, $\chi_0 \in W^{4/3,3}(\Omega)$ with $\partial\chi_0/\partial\boldsymbol{n} = 0$ on Γ, and $\boldsymbol{v}_0 \in \boldsymbol{H}$, see Planas and Boldrini [[19], Theorem 2]. Then, there exists at least one solution $\{\theta, \chi, \boldsymbol{v}\}$ for our system (PF).

References

[1] J. P. Aubin, Un théorème de compacité. C. R. Acad. Sci. Paris, **256**(1963), 5042–5044.

[2] G. Caginalp, An analysis of a phase field model of a free boundary, Arch. Rat. Mech. Anal., **92**(1986), 205-245.

[3] E. Casella and M. Giangi, An analytical and numerical study of the Stefan problem with convection by means of an enthalpy method, Math. Methods Appl. Sci., **24**(2001), 623–639.

[4] A. Damlamian, N. Kenmochi, and N. Sato, Phase field equations with constraints, pp.391–404 in *Nonlinear Mathematical Problems in Industry*, GAKUTO Internat. Ser. Math. Sci. Appl., **Vol.2**, Gakkōtosho, Tokyo, 1993.

[5] J. I. Diaz and G. Galiano, Existence and uniqueness of solutions of the Boussinesq system with nonlinear thermal diffusion, Topol. Methods Nonlinear Anal., **11**(1998), 59–82.

[6] G. J. Fix, Phase field methods for free boundary problems, pp 580–589 in *Free Boundary Problems: Theory and Applications*, Pitman Rese. Notes Math. Ser., **Vol.79**, Longman, London, 1983.

[7] C. Foias, O. Manley and R. Temam, Attractors for the Benard problem: existence and physical bounds on their fractal dimension, Nonlinear Anal. **11**(1987), 939–967.

[8] H. Fujita and N. Sauer, On existence of weak solutions of the Navier-Stokes equations in regions with moving boundaries, J. Fac. Sci., Univ. Tokyo., Sec. IA. Math., **17**(1970), 403–420.

[9] T. Fukao, Free boundary problem for phase field equations and Navier-Stokes equations without the Carman-Kozeny penalty, preprint.

[10] T. Fukao and N. Kenmochi, Stefan problems with convection governed by Navier-Stokes equations, Adv. Math. Sci. Appl., **15**(2005), 29–48.

[11] T. Fukao, N. Kenmochi and I. Pawłow, Stefan problems in non-cylindrical domains arising in Czochralski process of crystal growth, Control Cybernet., **32**(2003), 201–221.

[12] T. Fukao and M. Kubo, Nonlinear degenerate parabolic equations for a thermohydraulic model. to appear in Discrete and Continuous Dynamical Systems, Supplement Volume 2007.

[13] K.-H. Hoffman and L. Jiang, Optimal control of a phase filed model for solidification, Numer. Funct. Anal. Optim. **13**(1992), 11–27.

[14] H. Inoue and M. Ôtani, Strong solutions of initial boundary value problems for heat convection equations in noncylindrical domains, Nonlinear Anal., **24**(1995), 1061–1090.

[15] N. Kenmochi, Solvability of nonlinear evolution equations with time-dependent constraints and applications, Bull. Fac. Edu., Chiba Univ., **30**(1981), 1–87.

[16] O. A. Ladyženskaja, V. A. Solonnikov and N. N. Ural'ceva, *Linear and quasi-linear equations of parabolic type*, Translations of Mathematical Monographs, **Vol.23**, Amer. Math. Soc., 1968.

[17] S. A. Lorca and J. L. Boldrini, The initial value problem for a generalized Boussinesq model, Nonlinear Anal., **36**(1999), 457–480.

[18] H. Morimoto, Nonstationary Boussinesq equations. J. Fac. Sci., Univ. Tokyo., Sec. IA. Math., **39**(1992), 61–75.

[19] G. Planas and J. L. Boldrini, A bidimensional phase-field model with convection for change phase of an alloy, J. Math. Anal. Appl. **303**(2005), 669–687,

[20] J. F. Rodrigues, Variational methods in the Stefan problem, pp.147–212 in *Phase Transitions and Hysteresis*, Lecture Notes Math., **Vol.1584**, Springer-Verlag, 1994.

[21] J. Simon, Compact sets in the spaces $L^p(0,T;B)$, Ann. Mate. Pura. Appl., **146** (1987), 65–96.

[22] R. Temam, *Navier-Stokes equations, Theory and numerical analysis*, 3rd edition, Studies in Mathematics and its Applications, **Vol.2**, North Holland Amsterdam, New York-Oxford, 1984.

[23] A. Visintin, *Models of phase transitions*, Progr. Nonlinear Differential Equations Appl., Birkhäuser, Boston, 1996.

BLOW-UP AT SPACE INFINITY FOR NONLINEAR HEAT EQUATIONS

YOSHIKAZU GIGA, YUKIHIRO SEKI AND NORIAKI UMEDA
Graduate School of Mathematical Sciences, University of Tokyo, 3-8-1 Komaba, Meguro-ku Tokyo 153-8914, Japan

1. Introduction

This is a survey paper on blow-up phenemena at space infinity for nonlinear heat equations. We are interested in a blow-up problem of the Cauchy problem for nonlinear heat equations

$$\begin{cases} u_t = \Delta u + f(u), & x \in \mathbf{R}^N,\ t > 0, \\ u(x,0) = u_0(x), & x \in \mathbf{R}^N. \end{cases} \tag{1.1}$$

Here the nonlinear term f and initial data u_0 are assumed to satisfy the following conditions;

$$\begin{aligned} &f \text{ is a locally Lipschitz continuous function in } [0,\infty) \text{ and} \\ &\text{fulfills } f(\xi) > 0 \text{ for } \xi > 0 \text{ and } \int_1^\infty \frac{d\xi}{f(\xi)} < \infty; \end{aligned} \tag{$A1$}$$

$$u_0 \text{ is a nonnegative bounded continuous function in } \mathbf{R}^N. \tag{$A2$}$$

The last condition of $(A1)$ forces f to grow superlinearly at infinity.

A nonlinear evolution equation may have a unique local-in-time solution in a suitable function space and it can be extend as a solution together with evolution of time so long as it belongs to the function space. However, in general, the Cauchy problem is not solvable globally in time; a solution may blow up in finite time. That is, there may exist a finite time $T < \infty$ such that the solution ceases to exist in the function space at the time T. This phenomenon is called *blow-up* in finite time and we call such a time T *blow-up time.* The Cauchy problem (1.1) has a unique local-in-time solution $u = u(\cdot, t)$ in $L^\infty(\mathbf{R}^N)$ for any nonnegative initial data $u_0 \in L^\infty(\mathbf{R}^N)$. However, it may blow up in finite time. For instance, if the initial data does not decrease at space infinity, the solution of (1.1) does blow up in finite time. We are interested in the blow-up times of solutions and detailed behavior of solutions at the blow-up times. In particular, we discuss solutions which blow up at space infinity as we will state later.

2. Various Concepts

Let u be a blow-up solution to the problem (1.1) with initial data u_0. The notation $\|\cdot\|_\infty$ stands for usual sup norm in $\mathbf{R}^N$. By a blow-up time, we mean

$$t_b(u_0) := \sup\{\tau > 0\ ;\ u(\cdot,t) \text{ is bounded in } \mathbf{R}^N (\text{ i.e. } \|u(t)\|_\infty < \infty) \\ \text{for } 0 < t < \tau\}.$$

If $t_b(u_0) < \infty$, the solution cannot live in $L^\infty(\mathbf{R}^N)$ beyond the time $t_b(u_0)$. Therefore it does hold that $\|u(t)\|_\infty \to \infty$ as $t \nearrow t_b(u_0)$. A point in $\mathbf{R}^N$ where u is not locally bounded is called a blow-up point. The blow-up set of u is defined as the set of all blow-up points. Many researchers have tried to characterize blow-up sets. There is a huge literature on this topic. Notice that *a blow-up set may possibly be empty even if a finite time blow-up occurs.* We are just going to discuss such phenomena. We define blow-up at space infinity as follows:

Definition 2.1. A solution u of (1.1) with blow-up time $t_b(u_0) < \infty$ is said to *blow up at space infinity* if there exists a sequence $\{(x_n, t_n)\} \subset \mathbf{R}^N \times (0, t_b(u_0))$ such that

$$|x_n| \to \infty,\ t_n \nearrow t_b(u_0) \quad \textit{and} \quad u(x_n, t_n) \to \infty.$$

We define $T_M = t_b(M)$ with $M = \|u_0\|_\infty$, which coincides with the blow-up time of the solution $v_M(t)$ of the corresponding problem for ordinally differential equation;

$$\begin{cases} v' = f(v), & t > 0, \\ v(0) = M. \end{cases} \tag{2.1}$$

It is easily seen that a unique solution to (2.1) exists and it is expressed as

$$v_M(t) = G^{-1}(T_M - t) \qquad \text{with} \qquad G(v) = \int_v^\infty \frac{d\xi}{f(\xi)}, \tag{2.2}$$

where G^{-1} is the inverse function of G. Clearly, $T_M = G(M)$. For example, if $f(u) = u^p$, $(p > 1)$ then $v_M(t) = \kappa(T_M - t)^{-1/(p-1)}$ with $\kappa = (p-1)^{-(p-1)}$ and $T_M = (p-1)^{-1}M^{-p+1}$ and if $f(u) = e^u$, then $v_M(t) = -\log(T_M - t)$ with $T_M = e^{-M}$. By the comparison principle, we have $\|u(\cdot,t)\|_\infty \le v_M(t)$

for any solution u to the problem (1.1). Thus, in general, $t_b(u_0) \geq T_M$.

Definition 2.2. Let u_0 belong to $L^\infty(\mathbf{R}^N)$ and set $M = \|u_0\|_\infty$. A solution u to the problem (1.1) with initial data u_0 is said to blow up at *minimal blow-up time* or *the least (possible) blow-up time* provided that $t_b(u_0) = T_M$.

By definition and the strong maximum principle ([12]), one is able to show that a solution with minimal blow-up time necessarily blows up at space infinity:

Theorem 2.3. *Assume (A1) and (A2). Suppose that a solution u of (1.1) with initial data u_0 blows up at minimal blow-up time. Then*

$$\|u(\cdot,t)\|_\infty = \lim_{R\to\infty} \sup_{|x|\geq R} u(x,t) = v_M(t) \quad \text{in } [0,T_M).$$

Hence, in particular, the initial data u_0 should satisfy

$$\lim_{R\to\infty} \sup_{|x|\geq R} u_0(x) = M. \tag{2.3}$$

Moreover, the solution u blows up at space infinity. Namely, there exists a sequence $\{(x_n,t_n)\} \subset \mathbf{R}^N \times (0,T_M)$ such that

$$|x_n| \to \infty,\ t_n \nearrow T_M \quad \text{and} \quad u(x_n,t_n) \to \infty \text{ as } n \to \infty.$$

Corollary 2.4. *Assume the same hypotheses with Theorem 2.3. Then there exists a "direction" $\psi \in S^{N-1}$, where S^{N-1} is $(N-1)$-dimensional unit sphere, such that*

$$\lim_{k\to\infty} \frac{x_{n(k)}}{|x_{n(k)}|} = \psi$$

for some subsequence $\{(x_{n(k)},t_{n(k)})\} \subset \{(x_n,t_n)\}$.

The proofs of these results are due to [14]. We give a sketch of the proofs in section 4.

Remark 2.5. Condition (2.3) is not sufficient to raise a blow-up at space infinity. We will introduce some necessary and sufficient conditions on initial data for a solution to blow up at space infinity at minimal blow-up time.

Definition 2.6. Let u be a blow-up solution to the problem (1.1). A "direction" $\psi \in S^{N-1}$ is said to be a *blow-up direction* of u if there exists a sequence $\{(x_n, t_n)\} \subset \mathbf{R}^N \times (0, t_b(u_0))$ such that

$$|x_n| \to \infty, \ \frac{x_n}{|x_n|} \to \psi, \ t_n \nearrow t_b(u_0) \ \text{ and } \ u(x_n, t_n) \to \infty \quad \text{as } n \to \infty.$$

If the solution has at least one blow-up direction, we say that *directional blow-up* arises. Corollary 2.4 asserts that if the solution of (1.1) blows up at minimal blow-up time, then directional blow-up does occur. If a direction $\eta \in S^{N-1}$ is not a blow-up direction, we call it *non-blow-up direction.* Our interests are characterizing blow-up set and blow-up directions by behavior of initial data.

Blow-up problems for nonlinear parabolic equations have been studied for a long time. However, many researchers consider the Cauchy problem only with initial data which decay at space infinity or the Dirichlet problem in a bounded domain. Blow-up at space infinity does not occur for such problems. Only a few researchers studied blow-up at space infinity as far as we know. Let us recall a few results concerning blow-up at space infinity.

The first result in this topic is due to Lacey [10]. He considered the Cauchy-Dirichlet problem in one-dimensional half line;

$$\begin{cases} u_t = \Delta u + f(u), & \text{in } (0,\infty) \times (0,T), \\ u(0,t) = 1, & t \in (0,T), \\ u(x,0) = u_0(x), & \text{in } (0,\infty) \end{cases} \tag{2.4}$$

and proved that the solution blows up only at space infinity.

Giga and Umeda [6] considered the Cauchy problem for equation (1.1) with $f(u) = u^p$ $(p > 1)$ in higher dimension. Assuming $\lim_{|x|\to\infty} u_0(x) = \|u_0\|_\infty$ and $u_0 \not\equiv \|u_0\|_\infty$, they proved that the solution blows up at minimal blow-up time and blow-up occurs only at space infinity. They ([7, 8]) dealt with the Cauchy problem (1.1) with general nonlinear term f satisfying condition (GU). They weakened condition on initial data, which leads to a refinement sufficient condition for blow-up at space infinity and discussed directional blow-up. They characterized blow-up directions according to the behavior of the mean value of u_0 on ball. Although they also deal with sign-changing solutions, we discuss only nonnegative solutions for simplicity.

Seki, Suzuki and Umeda [14] not only generalized the results of [7] to degenerate quasilinear parabolic equations $u_t = \Delta\phi(u) + f(u)$ but also gave necessary and sufficient conditions for a solution to blow up at minimal blow-up time and conditions for a direction to be a blow-up direction. The equation is a generalization of porus medium equation with reaction term; $u_t = \Delta u^m + f(u)$ with $m \geq 1$. Moreover, the nonlinear term f can be taken from very wide class of functions. For example, when $m = 1$, $f(u) = (1+u)\{\log(1+u)\}^\beta$ with $\beta > 2$ is allowed in [14]. One of the author obtained the same results for a quasilinear parabolic equation which is a generalization of fast diffusion equation $u_t = \Delta u^m + f(u)$ with $0 < m < 1$ in [13], although the assumption of f is a little stronger than [14].

This note is organized as follows. We introduce typical results in the next section and show main ideas of their proofs in §4. In the final section we show, as applications of their results, some examples of directional blow-up, such as a solution which has a single blow-up direction and a solution whose set of blow-up directions coincides with arbitrary given closed set in S^{N-1}. We also prove the solution does not blow up at minimal blow-up time if the initial data is an almost periodic function.

3. Typical results

We set $\rho(x) = (\int_{\mathbf{R}^N} e^{-|y|}dy)^{-1}e^{-|x|}$. The mean value function of u_0 with weight ρ is defined by

$$A_\rho(x; u_0) = \int_{\mathbf{R}^N} \rho(y-x)u_0(y)dy,$$

which plays significant roles. For a direction $\psi \in \mathbf{R}^N$, we consider some conditions on u_0 for ψ to be a blow-up direction.

There exists a sequence $\{x_n\} \subset \boldsymbol{R}^N$ such that

$$|x_n| \to \infty, \ \frac{x_n}{|x_n|} \to \psi \ \text{and} \ A_\rho(x_n; u_0) \to \|u_0\|_\infty. \qquad (A3_\psi)$$

Definition 3.1. Let u_0 be a bounded continuous function in $\mathbf{R}^N$ and set $M := \|u_0\|_\infty$. A direction $\psi \in S^{N-1}$ is said to be a *direction of mean convergence* of u_0 (to M) if condition $(A3_\psi)$ is satisfied for ψ.

Actually, condition $(A3_\psi)$ can be converted to the equivalent conditions not invoking the weight ρ. Namely, there are some equivalent conditions for ψ to be a direction of mean convergence of u_0:

There exists a sequence $\{x_n\} \subset \boldsymbol{R}^N$ such that

$$|x_n| \to \infty, \ \frac{x_n}{|x_n|} \to \psi \ \text{and} \ u_0(x + x_n) \to \|u_0\|_\infty \ \text{a.e. in } \mathbf{R}^N \text{ as } n \to \infty. \qquad (A4_\psi)$$

There exists a sequence $\{x_n\} \subset \mathbf{R}^N$ *such that, for each* $R > 0$,

$$|x_n| \to \infty, \ \frac{x_n}{|x_n|} \to \psi \ \text{and} \ \frac{1}{|B_R|} \int_{B_R(x_n)} u_0(x)dx \to \|u_0\|_\infty \ \text{as} \ n \to \infty. \tag{$A5_\psi$}$$

There exists a sequence $\{x_n\} \subset \mathbf{R}^N$ *and a sequence* $\{R_n\}$ ($R_n > 1$, $R_n \to \infty$ *as* $n \to \infty$) *such that*

$$|x_n| \to \infty, \ \frac{x_n}{|x_n|} \to \psi \ \text{and} \ \inf_{r \in [1, R_n]} \frac{1}{|B_r|} \int_{B_r(x_n)} u_0(x)dx \to \|u_0\|_\infty \ \text{as} \ n \to \infty. \tag{$A6_{\psi,\{R_n\}}$}$$

Here and hereafter, $B_R(a)$ denotes a N-dimensional open ball with radius $R > 0$ centered at $a \in \mathbf{R}^N$ and $B_R = B_R(0)$. We refer to [14] for the proof of the equivalence.

Theorem 3.2. *Assume* ($A1$) *and* ($A2$). *Let* $M = \|u_0\|_\infty$ *and let* $\psi \in S^{N-1}$ *be a direction of mean convergence of* u_0. *Then* u *blows up at minimal blow-up time and* ψ *is a blow-up direction. Moreover, for each* $R > 0$,

$$\lim_{n \to \infty} \sup_{x \in B_R(x_n)} |u(x,t) - v_M(t)| = 0. \tag{3.1}$$

The convergence is uniform on every compact subset of interval $(0, T_M)$.

If we assume a certain growth condition on nonlinear term f, we are able to show that the blow-up set is empty unless the initial data is constant, and completely characterize blow-up directions (at minimal blow-up time) by behavior of initial data. Moreover, we are able to obtain a necessary and sufficient condition on initial data for a solution to blow up at minimal blow-up time. The following condition is equivalent to the condition (B) in [8]:

$$\begin{gathered} \text{There exist constants } \ \xi_0 > 0 \ \text{ and } p > 1 \ \text{ such that} \\ \frac{f(\xi)}{\xi^p} \ \text{ is nondecreasing for } \xi \geq \xi_0. \end{gathered} \tag{GU}$$

This condition means that f grows faster more than polynomial growth.

So far, the next condition is the weakest assumption to show these results, although it looks more complicated.

$$
\begin{aligned}
&\text{There exist } \Phi \in C^2(0,\infty),\ c>0 \text{ and } \eta_1 \ge 0 \text{ such that}\\
&\quad \Phi(\eta) > 0,\ \ \Phi'(\eta) \ge 0 \text{ and } \Phi''(\eta) \ge 0 \text{ for } \eta \ge \eta_1;\\
&\quad \int_1^\infty \frac{d\eta}{\Phi(\eta)} < \infty;\\
&\quad f'(\eta)\Phi(\eta) - f(\eta)\Phi'(\eta) \ge c\Phi(\eta)\Phi'(\eta) \quad \text{for } \eta \ge \eta_1.
\end{aligned} \tag{FM}
$$

This kind of condition was originally introduced by Friedman and McLeod [5] to show that, what is called, a single-point blow-up does occur for bell shaped radially decreasing initial data u_0 and it has been re-formulated into weaker version by Fujita and Chen [4] and Chen [1]. One can also find a quasilinear version of this condition in Mochizuki and Suzuki [11].

Theorem 3.3. *Assume* $(A1)$, $(A2)$ *and* (GU) *(or* (FM)*). Let* u *be a solution of* (1.1) *with initial data* u_0 *which has minimal blow-up time. Then the following hold:*

(i) *If the initial data* u_0 *is not a constant, then the solution* u *blows up only at space infinity, that is, its blow-up set is empty.*

(ii) *A direction is a blow-up direction of* u *if and only if it is a direction of mean convergence of* u_0.

Combining Corollary 2.4 and Theorem 3.3, we obtain necessary and sufficient conditions for a solution of (1.1) blows up at minimal blow-up time.

Theorem 3.4. *Assume* $(A1)$, $(A2)$ *and* (GU) *(or* (FM)*). Suppose that* $u_0 \not\equiv 0$. *Then a solution* u *of* (1.1) *blows up at minimal blow-up time if and only if the initial data* u_0 *has at least one direction of mean convergence.*

Corollary 3.5. *Assume the same hypotheses with Theorem 3.4. Let* u *be a solution of* (1.1)*.Then* u *blows up at minimal blow-up time if and only if one of the following two conditions for initial data* u_0 *holds:*

$$
\begin{aligned}
&\textit{There exists a sequence } \{x_n\} \subset \mathbf{R}^N \textit{ such that}\\
&\quad |x_n| \to \infty \ \textit{ and } \ u_0(x+x_n) \to \|u_0\|_\infty \text{ a.e. } \textit{in } \mathbf{R}^N \text{ as } n \to \infty;
\end{aligned} \tag{3.2}
$$

$$
\sup_{x \in \mathbf{R}^N} A_\rho(x; u_0) = \|u_0\|_\infty. \tag{3.3}
$$

Remark 3.6. Giga and Umeda [7] first showed two sufficient conditions on initial data for $\psi \in S^{N-1}$ to be a blow-up direction or non-blow-up direction, respectively and proved that every direction satisfies each of the conditions by a supplementary argument. On the other hand, Seki, Suzuki and Umeda [14] established the formulation of Theorem 3.3(ii) via entirely different approach, adopting a regularizing argument (see Lemma 4.1.2). As above mentioned, their assumption on initial data is equivalent to that of Giga and Umeda [7]. Thus, one is also able to prove Theorem 3.3(ii) by Giga and Umeda's approach if one uses the regularizing argument.

4. Ideas of proofs

The proof of Theorem 2.3. The assertion is clear if $u_0(x) \equiv M = \|u_0\|_\infty$. Thus we may assume that $u_0(x) \not\equiv M$. Let u be a solution of (1.1) with minimal blow-up time T_M. Contrary to the conclusion, assume that there were $t_1 \in [0, T_M)$ for which

$$\lim_{R\to\infty} \sup_{|x|\geq R} u(x, t_1) < v_M(t_1). \tag{4.1}$$

If $t_1 \neq 0$, then we see $L := \sup_{x\in\mathbf{R}^N} u(x, t_1) < v_M(t_1)$ in view of (4.1) and the strong maximum principle ([12]). A comparison argument gives $u(x,t) \leq v_L(t - t_1)$ in $[t_1, t_1 + T_L)$, where v_L is the solution of (2.2) with initial data L replacing M and $T_L = t_b(L)$. Then we see $t_b(u_0) \geq t_1 + T_L > T_M$, which contradicts the assumption that u blows up at minimal blow-up time T_M, that is, $t_b(u_0) = T_M$.

As for the case $t_1 = 0$, we take a radially symmetric, radially nonincreasing and continuous function $w_0(x)$ in $\mathbf{R}^N$ satisfying $u_0(x) \leq w_0(x) \leq M$ in $\mathbf{R}^N$. Then the solution w of (1.1) with initial data w_0 is radially symmetric and radially nonincreasing, and blows up at minimal blow-up time T_M. Then we reach a contradiction in the same way as the case $t_1 > 0$. □

The proof of Theorem 3.2. Let u (or v) be a supersolution (or subsolution) of (1.1) in $\mathbf{R}^N \times (0, T_M)$ satisfying $u, v \leq M$ in $\mathbf{R}^N \times [0, T)$ for some $M > 0$. Denote by K the Lipschitz constant of function f on $[0, M]$. Then there is an increasing function $C_{K,M}(t)$ such that for $0 < t < T$,

$$\int_{\mathbf{R}^N} [v(x,t) - u(x,t)]_+ \rho(x) dx \leq C_{K,M}(t) \int_{\mathbf{R}^N} [v(x,0) - u(x,0)]_+ \rho(x) dx$$

([14]). For a sequence $\{x_n\}$ satisfying $(A3_\psi)$, we set $u_n(x,t) := u(x + x_n, t)$

and substitute $u = u_n$ together with $v = v_M$. Then we have

$$\int_{\mathbf{R}^N} \{v_M(t) - u_n(x,t)\}\rho(x)dx \le C_{K,M}(t)\{M - A_\rho(x_n; u_0)\} \to 0 \quad \text{as } n \to \infty.$$

On the other hand, for each $\epsilon \in (0, T_M)$, we see that $u(\cdot, \epsilon) \in BC^1(\mathbf{R}^N)$, that is, $u(\cdot, \epsilon)$ is bounded and continuous in $\mathbf{R}^N$ up to the first derivative. Thus,

$$\{u_n\} \text{ is uniformly bounded and equicontinuous in } B_R \times [\epsilon, T_M - \epsilon]$$

for every $R > 0$. Therefore, we are able to extract a subsequence $\{u_{n'}\}$ such that $\{u_{n'}\}$ converges locally uniformly to some continuous function w by virtue of Ascoli-Arzela theorem. Consequently, we obtain

$$\int_{B_R} |v_M(t) - w(x,t)|\rho(x)dx = 0, \quad t \in (0, T_M).$$

Namely, $w = v_M$. Since the limit is independent of the choice of a subsequence, we conclude $u_n \to v_M$. Thus, (3.1) holds and ψ is a blow-up direction. □

Remark 4.0.1. Theorem 3.2 can be proved even for degenerate quasilinear parabolic equations of the form $u_t = \Delta\phi(u) + f(u)$ ([14]). For that case, we employ the results on modulus of continuity due to DiBenedetto (Lemma 5.2 of [3]) in order to get equicontinuity of $\{u_n\}$.

We present two different procedures to show Theorem 3.3. It is Theorem 3.3(i) that is a basic result in both approaches. The first one is a semilinear version of [14], which comes from the method originally introduced by Friedman and McLeod [5] to show that a single-point blow-up does occur for bell shaped radially decreasing initial data. This is valid for a very wide class of nonlinear term f, such as $f(u) = (u+1)\{\log(u+1)\}^\beta$ with $\beta > 2$. The second one relies on a certain non-blow-up criterion around a given point. It was established by Giga and Kohn (Theorem 2.1 of [9]) for differential inequality $u_t - \Delta u \le K(1+|u|)^p$, $(K > 0,\ p > 1)$. Our criterion is its direct extension for equations of (1.1) with general nonlinear term f satisfying a certain growth condition. Although it does not allow so wide class of nonlinear term as in the first one, this approach has superiority that it is applicable even to general semilinear parabolic equation whose linear part has variable coefficients and to vector valued equations.

4.1. *The proof of Theorem 3.3 (Part 1)*

By the strong maximum principle, we have $u(x,t) < v_M(t)$ in $\mathbf{R}^N \times (0, T_M)$. Take $t_1 \in (0, T_M)$ and fix it. Let R be a positive number and let $w_0 \in C^2(B_R)$ be a radially symmetric and radially nondecreasing function satisfying $u(x,t_1) \le w_0(x) \le v_M(t_1)$, $w_0(x) = v_M(t_1)$ for $|x| = R$ and $\Delta w_0 + f(w_0) \ge 0$ in B_R. Denote by w the solution to the problem

$$\begin{cases} w_t = \Delta w + f(w), & \text{in } B_R \times (t_1, T_M), \\ w = v_M(t), & \text{on } \partial B_R \times (t_1, T_M), \\ w(x,t_1) = w_0(x), & \text{in } B_R. \end{cases} \tag{4.2}$$

Lemma 4.1.1. *Assume (A1), (A2) and condition (FM). Then for any compact subset $\Omega \subset B_R$,*

$$\sup_{(x,t)\in\Omega\times(\tau,T_M)} w(x,t) < \infty.$$

Proof. We only use well-known technique which has been used in the study of blow-up sets since it was developed by [5] to show a single-point blow-up phenomenon. See Proposition 2.6 of [14] for detail. □

If $A_\rho(0; u(t_1)) \le L$ for some $L < M$, there is a point $z \in \mathbf{R}^N$ such that $u(z,t_1) \le L'$ for some $L' \in (L, M)$. We consider the solution w to the problem (4.2) as a supersolution of (1.1). In order to construct a function $w_0(x)$ as used in (4.2) independently of z, we make use of equicontinuity of the solution u or a result on modulus of continuity (Lemma 5.2 of [3]). As a result, for any $L'' \in (L', M)$ there exists $r_0 > 0$ not depending on z such that $u(x,t_1) \le L''$, for $|x - z| < r_0$. Using Lemma 4.1.1 with sufficiently large $R > 0$, we are able to show the following lemma:

Lemma 4.1.2. *Assume (A1), (A2) and condition (FM). If $A_\rho(0; u_0) \le L$ for some $L < M := \|u_0\|_\infty$, then there exists a constant $C_{M,L}$ such that*

$$u(0,t) \le C_{M,L} \quad \textit{for } t \in (0, T_M).$$

The proof of Theorem 3.3. The statement of Theorem 3.3(i) is immediate consequence of Lemma 4.1.1. Indeed, we are able to construct, locally in $\mathbf{R}^N$, a supersolultion of (1.1) having no blow-up points, taking $R > 0$ large enough. We shall prove (ii) of Theorem 3.3. We have only to show that if u_0 does not satisfy $(A3_\psi)$, then ψ cannot be a blow-up direction. Assume that $(A3_\psi)$ does not hold for some $\psi \in S^{N-1}$. Then there exists an open neighborhood D of ψ such that

$$\sup_{x/|x|\in D} A_\rho(x;u_0) \le L < M.$$

For $z \in \mathbf{R}^N$ such that $z/|z| \in D$, set $u_z(x,t) := u(x+z,t)$. Application of Lemma 4.1.2 to u_z implies

$$u_z(0,t) = u(z,t) \le C_{M,L} \quad \text{for } t \in (0,T_M).$$

Since $C_{M,L}$ is independent of z, we see, with the aid of Lemma 4.1.2, that ψ is a non-blow-up direction. □

4.2. *The proof of Theorem 3.3 (Part 2)*

We first establish a criterion to see whether a point in $\mathbf{R}^N$ is a non-blow-up point. Its prototype is found in Giga and Kohn (Theorem 2.1 of [9]).

Lemma 4.2.1. *Assume* (*A*1), (*A*2) *and condition* (*GU*). *Let p be the constant appearing in the condition* (*GU*) *and let a be a point in $\mathbf{R}^N$. Then there exists a constant $\delta_0 \in (0,1]$ having the following property:*
(i) *Suppose that $1 < p \le 3$. If for some $\delta \in (0,\delta_0)$, $r > 0$ and $\tau \in (0,T_M)$,*

$$u(x,t) \le \delta v_M(t), \qquad \textit{for } (x,t) \in B_r(a) \times (\tau, T_M), \tag{4.3}$$

and u solves the equation (1.1) *in $B_r(a)\times(0,T_M)$, then u is locally bounded around the point a at $t = T_M$.*
(ii) *Suppose that $p > 3$. Then the statement of* (i) *holds true with $\delta_0 = 1$.*

Remark 4.2.2. This result is stated in Theorem 3.7 of [7]. However, there is a flaw in their proof. Indeed, we have to divide the argument into three cases according to the value of p, where p is the constant appearing in the condition (*GU*). Unfortunately, the argument in Theorem 3.7 of [7] works only for the case $1 < p < 3$. However, the proof is completed by reduction to the case $p \ge 3$. We take the opportunity to correct the proof.

The proof of Lemma 4.2.1. Translating the coordinates in space variables and by scaling, we may assume, without loss of generality, that $a = 0$ and $r = 1$. In the proof we denote by C a generic positive constant possibly changing from line to line. Consider a cutoff function $\phi \in C_0^\infty(B_1)$ satisfying $0 \le \phi \le 1$ and $\phi \equiv 1$ on $B_{1/2}$. The function $w(t) = w(x,t) := \phi(x)u(x,t)$ satisfies equation

$$w_t = \Delta w - g(x,t) + \phi(x) f(u), \qquad \text{in } B_1 \times (0,T_M),$$

where $g(x,t) = 2\nabla \cdot (u\nabla\phi) - u\Delta\phi$. Then the representation formula of solutions gives us

$$w(t) = e^{(t-\tau)\Delta}w(\tau) + \int_\tau^t e^{(t-s)\Delta}(-g(s) + \phi f(u(s)))ds,$$

where $\{e^{t\Delta}\}_{t\geq 0}$ is the heat semigroup in B_1 with Dirichlet boundary condition. Using the L^∞-L^∞ estimates, we have

$$\|e^{(t-s)\Delta}g(s)\|_{L^\infty(B_1)} \leq C(t-s)^{-1/2}\|u(s)\|_{L^\infty(B_1)}.$$

Thus we obtain

$$\begin{aligned}\|w(t)\|_{L^\infty(B_1)} \leq & \|w(\tau)\|_{L^\infty(B_1)} + \int_\tau^t C(t-s)^{-1/2}\|u(s)\|_{L^\infty(B_1)}ds \\ & + \int_\tau^t \|w(s)\|_{L^\infty(B_1)}\left\|\frac{f(u(s))}{u(s)}\right\|_{L^\infty(B_1)}ds. \qquad (4.4)\end{aligned}$$

(In [7], estimating the integrand of the last term of (4.4) as (4.5) below, they use the Gronwall type inequality (Lemma 2.3 of [9]). However, since the integrand of the middle term of (4.4) has a singularity, the assumption of the lemma does not satisfied.)

By condition (GU) and (2.2), it is easily seen that $v_M(t) \leq C(T_M - t)^{-1/(p-1)}$. There are three cases to consider according to the value of p. The first case is $p > 3$. The integrand of the second term is bounded by $C(t-s)^{-1/2}(T_M - s)^{-1/(p-1)}$, so that its integral with respect to s over interval $[\tau, t]$ is bounded by a constant when $p > 3$. Since $\xi \mapsto f(\xi)/\xi$ is nondecreasing by condition (GU), it follows from assumption (4.3) that

$$\int_\tau^t \left\|\frac{f(u(s))}{u(s)}\right\|_{L^\infty(B_1)} ds \leq \delta^{p-1}\int_\tau^t \frac{f(v_M(s))}{v_M(s)}ds. \qquad (4.5)$$

Therefore usual Gronwall's inequality implies

$$\|w(t)\|_{L^\infty(B_1)} \leq C\exp\left(\int_\tau^t \left\|\frac{f(u(s))}{u(s)}\right\|_{L^\infty(B_1)}ds\right) \leq Cv_M(t)^{\delta^{p-1}}.$$

Hence we obtain

$$u(x,t) \leq Cv_M(t)^\delta \quad \text{in } B_{1/2}\times(\tau, T_M). \qquad (4.6)$$

since $\phi \equiv 1$ on $B_{1/2}$. We repeat this manipulation again together with $\phi \in C_0^\infty(B_{1/2})$ satisfying $0 \leq \phi \leq 1$ and $\phi \equiv 1$ on $B_{1/4}$. Then we obtain

$$u(x,t) \leq C + \exp\left(\int_\tau^t \left\|\frac{f(u(s))}{u(s)}\right\|_{L^\infty(B_{1/2})}ds\right) \text{ in } B_{1/4}\times(\tau, T_M).$$

On the other hand, it follows from condition (GU) that

$$f(\eta\xi) \leq \eta^p f(\xi) \qquad \text{for } 0 < \eta < 1,\ \xi \geq \eta^{-1}\xi_0. \tag{4.7}$$

From (4.6) and (4.7), we see

$$\int_\tau^t \left\| \frac{f(u(s))}{u(s)} \right\|_{L^\infty(B_{1/2})} ds \leq \int_\tau^t \frac{f(Cv_M(s)^\delta)}{v_M(s)^\delta} ds$$
$$\leq C \int_\tau^t v_M(s)^{p(\delta-1)-\delta} f(v_M(s)) ds = C \int_{v_M(\tau)}^{v_M(t)} v^{p(\delta-1)-\delta} dv \leq C(M, p, \delta, \tau).$$

Thus we have proved a local boundedness of u; $u(x,t) \leq C$ in $B_{1/4} \times (\tau, T_M)$.

The proof of the local boundedness is almost the same for the case $p = 3$ although it is more involved.

We shall consider the case $1 < p < 3$. This case is more difficult to prove than the case $p \geq 3$. We shall return to the estimate (4.4). In this case, an elementary calculation yields

$$\int_\tau^t (t-s)^{-1/2} \|u(s)\|_{L^\infty(B_1)} ds \leq C(N,p)\delta (T_M - t)^{1/2-1/(p-1)}$$
$$\leq C(N,p)\delta \int_\tau^t (T_M - \sigma)^{-1/2-1/(p-1)} d\sigma$$

(see Lemma 2.2 of [9] for detail). We write $\tilde{\delta} = \delta^{p-1}$ for simplicity of notation. Using the Gronwall type inequality (Lemma 2.3 of [9]), we obtain

$$\begin{aligned}
&\|w(t)\|_{L^\infty(B_1)} \\
\leq& \Big[\|w(\tau)\|_{L^\infty(B_1)} + C(N,p)\delta \int_\tau^t (T_M - \sigma)^{-1/2-1/(p-1)} \\
&\qquad \times \exp\Big(-\tilde{\delta} \int_\tau^\sigma \frac{f(v(\xi))}{v(\xi)} d\xi \Big) d\sigma \Big] \exp\Big\{ \tilde{\delta} \int_\tau^t \frac{f(v(s))}{v(s)} ds \Big\} \\
\leq& C(M,N,p,\tau)\delta \Big[v(\tau) + \int_\tau^t (T_M - \sigma)^{-1/2-(1-\tilde{\delta})/(p-1)} d\sigma \Big] (T_M - t)^{-\tilde{\delta}/(p-1)} \\
\leq& C(M,N,p,\tau)(T_M - t)^{1/2-1/(p-1)}
\end{aligned}$$

provided that δ is chosen so that $1/2 - (1-\tilde{\delta})/(p-1) < 0$. It follows that

$$u(x,t) \leq C(M,N,p,\tau)(T_M - t)^{1/2-1/(p-1)} \ \text{ in } B_{1/2} \times (\tau, T_M).$$

We iterate this argument finitely many times. Take the smallest integer k such that $k > 1/(p-1)$. After k steps later, we have

$$u(x,t) \leq C_k(M,N,p,\tau)(T_M - t)^{-1/2(p-1)} \quad \text{in } B_{2^{-k}} \times (\tau, T_M),$$

and then $u(x,t) \le C_{k+1}(M,N,p,\tau)(T_M - t)^{-\tilde{\delta}/(p-1)}$ in $B_{2^{-k-1}} \times (\tau, T_M)$. We argue similarly to the case $p \ge 3$, so that we get a bound; $u(x,t) \le C$ in $B_{2^{-k-2}} \times (\tau, T_M)$. □

When $u_0 \not\equiv \|u_0\|_\infty$, one is able to prove the existence of δ as in the condition (4.3). The next lemma is reorganization of Lemmas 3.1 and 3.4 of [7].

Lemma 4.2.3. *Assume (A1) and (A2). Suppose that $u_0 \not\equiv \|u_0\|_\infty$. Then for any $\epsilon > 0$, there exist $\eta \in (0,1)$ and $\tau \in (0, T_M)$ such that*

$$u(x,t) \le \epsilon v(t) \quad \text{in } B_\eta \times (\tau, T_M). \tag{4.8}$$

Proof. Let $\hat{u}$ be a solution to the linear problem

$$\begin{cases} \hat{u}_t = \Delta \hat{u}, & \text{in } B_1 \times (0, T_M), \\ \hat{u} = u(x,t), & \text{on } \partial B_1 \times (0, T_M), \\ \hat{u}(x,0) = u_0(x), & \text{in } B_1. \end{cases}$$

Then it is not difficult to show that the assertion (4.8) for $\hat{u}$ (see Lemma 3.2 of [7] for detail). Let $t_1 \in (0, T_M)$. We claim that there exists $\delta \in (0,1)$ such that

$$u(x,t) \le \delta v_M(t) \quad \text{in } B_1 \times (t_1, T_M). \tag{4.9}$$

Let w be the solution of the heat equation $w_t = \Delta w$ in $\mathbf{R}^N \times (0,\infty)$ with initial data $w(x,0) = M^{-1}u_0(x)$. Applying the strong maximum principle yields $w < 1$ in $\mathbf{R}^N \times (0,\infty)$, so that there is $\delta \in (0,1)$ such that $w \le \delta$ in $B_1 \times (t_1, T_M]$. We set $\bar{u} := v_M w$ and observe that it is a supersolution to (1.1). Thus we see $u \le \bar{u} \le \delta v_M$ in $B_1 \times (t_1, T_M)$ and the claim (4.9) follows.

Let $G(x,y,t,s)$ be a fundamental solution of the heat equation for the Dirichlet problem in B_1. In view of positivity of G, there is a constant $c > 0$ such that $c \le \int_{B_1} G(x,y,t,s)dy \le 1$ for $x \in B_\eta$. Then if we choose t sufficiently close to T_M, we have

$$\begin{aligned} u - \hat{u} &= \int_0^t \int_{B_1} G(x,y,t,s) f(u(y,s)) dy ds \\ &\le \int_0^{t_1} f(v(s)) ds \int_{B_1} G(x,y,t,s) dy + \int_{t_1}^t f(\delta v(s)) ds \int_{B_1} G(x,y,t,s) dy \\ &\le (\delta^p + \epsilon) \int_{t_1}^t f(v(s)) ds \int_{B_1} G(x,y,t,s) dy \le (\delta^p + \epsilon) v(t) \quad \text{for } x \in B_\eta. \end{aligned}$$

Hence, choosing ϵ so small as to satisfy $\delta^p + 2\epsilon < \delta$, we get a better estimate: $u(x,t) \le (\delta^p + 2\epsilon)v(t)$ in $B_\eta \times (t_1, T_M)$. Iterating this manipulation finitely many times yields the assertion (4.8). □

The proof of Theorem 3.3. We shall prove Theorem 3.3(i). Let a be a point in $\mathbf{R}^N$. Take $R > 0$ sufficiently large so that $a \in B_R$. It is proved that there exists $\delta \in (0,1)$ such that

$$u(x,t) \le \delta v_M(t) \quad \text{in } B_R \times (0, T_M). \tag{4.10}$$

in the same way to derive (4.9). Actually, it holds with any $\epsilon > 0$ replacing δ by Lemma 4.2.3. Therefore, we see that the solution u is locally bounded around the point a at $t = T_M$ via Lemma 4.2.1.

Note that if $u(z, t_1) \le L' < M$ for some $t_1 \in [0, T_M)$, $L < L' < M$ and $z \in \mathbf{R}^N$, one is able to construct supersolution w independently of z as in Part 1. The constant $\delta \in (0,1)$ in (4.10) exsits independently of z. Then the statement (ii) is obtained similarly to that of Part 1 since one is able to show Lemma 4.1.2 under the assumptions $(A1)$, $(A2)$ and (GU) by using Lemma 4.2.1 and Lemma 4.2.3 instead of Lemma 4.1.1. □

5. Some examples

We shall demonstrate some examples of directional blow-up. Let u be a solution to the Cauchy problem (1.1) which blows up at minimal blow-up time. We denote by $S(u_0)(\subset S^{N-1})$ the set of all blow-up directions. It is a closed set in S^{N-1}. Indeed, for any $\varphi \in S^{N-1} \setminus S$, there exist a constant $C > 0$ and an open neighborhood $D \subset S^{N-1}$ of φ such that $\sup_{x/|x| \in D} u(x,t) \le C$. It follows that there are no blow-up directions in D. Thus $S^{N-1} \setminus S(u_0)$ is an open set in S^{N-1}.

Example 1. (Single directional blow-up) We give an example of initial datum such that $S(u_0)$ consists of a single direction. Let $\psi \in S^{N-1}$ be a direction and set $D = \{r\psi \ ; \ r \ge 0\}$. For a point $x \in \mathbf{R}^N$, take a point $p_x \in D$ such that $\operatorname{dist}\{x, D\} = |x - p_x|$. Let P be a paraboloid defind by $P = \{x \in \mathbf{R}^N \ ; \ |x - p_x| \le \sqrt{|p_x|}\}$ and let $Q = \{x \in \mathbf{R}^N \ ; \ |x - p_x| \le 2\sqrt{|p_x|}\}$. We choose an initial datum $u_0 \in BC(\mathbf{R}^N)$ so as to fulfill

$$\begin{cases} u_0(x) = M \ \text{ if } x \in P, \ |x| \ge 1; \\ 0 \le u_0(x) < M \ \text{ if } x \in Q \setminus P; \\ u_0(x) = 0 \ \text{ if } x \notin Q. \end{cases}$$

Let u be a solution of (1.1) with initial datum u_0. Take a sequence $x_n = n\psi$ $(n = 1, 2, ...)$. Clearly, the sequence $\{x_n\}$ satisfies condition $(A3_\psi)$, so that ψ is a direction of mean convergence of u_0. Thanks to Theorem 3.3, it is a blow-up direction of u. Let $\varphi \neq \psi$ be another direction. Then there are an open neighborfood G of φ in S^{N-1} and a constant $r_0 > 0$ such that $\{x \in \mathbf{R}^N \ ; \ |x| \geq r_0, \ x/|x| \in G\} \cap Q = \emptyset$. It follows that

$$\sup\{u_0(x) \ ; \ |x| \geq r_0, \ x/|x| \in G\} \leq L < M$$

for some $L \in (0, M)$. Consequently, φ cannot be a direction of mean convergence of u_0 to M. From Theorem 3.3, we observe that this φ is not a blow-up direction of u. Thus we obtain $S(u_0) = \{\psi\}$.

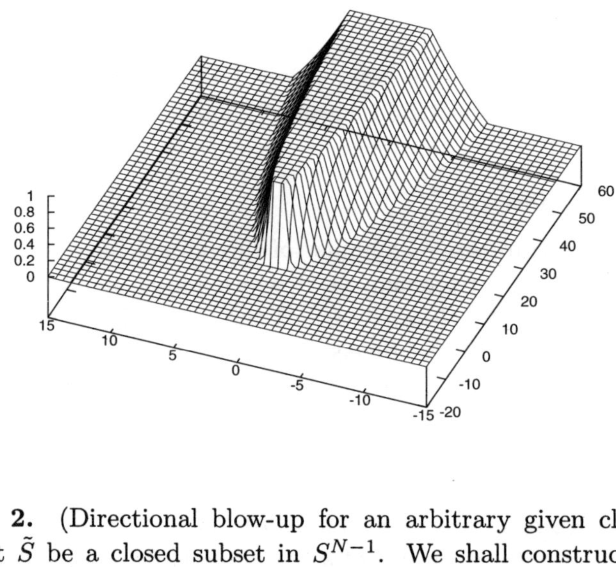

Example 2. (Directional blow-up for an arbitrary given closed set in S^{N-1}) Let $\tilde{S}$ be a closed subset in S^{N-1}. We shall construct an initial datum u_0 such that $S(u_0)$ coincides with $\tilde{S}$. We set

$$D = \cup_{r \geq 0} r\tilde{S} = \cup_{r \geq 0}\{x \in \mathbf{R}^N \ ; \ x = r\psi, \psi \in \tilde{S}\}.$$

For a point $x \in \mathbf{R}^N$, denote by $p_x \in D$ a point in D such that the distance of x and D is achieved; $\operatorname{dist}\{x, D\} = |x - p_x|$. Let $P = \cup\{x \in \mathbf{R}^N \ ; \ |x - p_x| \leq \sqrt{|p_x|}\}$ and let $Q = \cup\{x \in \mathbf{R}^N \ ; \ |x - p_x| \leq 2\sqrt{|p_x|}\}$, where the notation "$\cup$" runs for all such p_x. We define an initial datum $u_0 \in BC(\mathbf{R}^N)$ so as to fulfill

$$\begin{cases} u_0(x) = M \ \text{ if } x \in P, \ |x| \geq 1; \\ 0 \leq u_0(x) < M \ \text{ if } x \in Q \setminus P; \\ u_0(x) = 0 \ \text{ if } x \notin Q. \end{cases}$$

Then it is proved by the way similar to Example 1 that $S(u_0) = \tilde{S}$.

Example 3. (No direction of mean convergence of almost periodic initial data) Let us restrict ourselves to one-dimensional problem. It is easily seen that if the initial data is a non-constant periodic function, then it has no direction of mean convergence, so that the corresponding solution does not blow up at minimal blow-up time. In fact, this is also true for almost periodic functions. Here a function F defined for $-\infty < x < \infty$ is called almost periodic, if for any $\epsilon > 0$ there exists a trigonometric polynomial $T_\epsilon(x)$ such that

$$|F(x) - T_\epsilon(x)| < \epsilon, \qquad -\infty < x < +\infty.$$

Proof. Let u_0 be an almost periodic function in $\mathbf{R}$ which is not a constant and let M be its maximum. Contrary to the conclusion, suppose that u_0 had a direction of mean convergence to M. Then there exists a sequence $\{x_n\} \subset \mathbf{R}$ such that $u_0(x + x_n) \to M$ a.e. in $\mathbf{R}$ as $n \to \infty$. From a characteristic property of almost periodic functions, one can extract a subsequence, which is also denoted by $\{x_n\}$, such that the convergence $u_0(x + x_n) \to M$ is uniform (see Chapter 1 of [2]). Take a point $y \in \mathbf{R}$ such that $m := u_0(y) < M$. Define $H := M - m$ and $z_n := y - x_n$. Then $|u_0(z_n + x_n) - M| = H$ for all n. This means that there is no subsequence $\{x_{n'}\} \subset \{x_n\}$ for which the sequence $u_0(x + x_{n'})$ is uniformly convergent. This is a contradiction. □

Acknowledgement. The second author is grateful to Mr. Tsuyoshi Yoneda for his discussion on almost periodic functions in Example 3. The work of the first author was partly supported by the Grant-in-Aid for Scientific Reserch, No.17654037, No.18204011, the Japan Society of the Promotion of Science (JSPS) and by COE "Mathematics of Nonlinear Structures via Singularities" (Hokkaido University) sponsored by JSPS. The works of the second and third authors were supported by the 21st century COE Program "Base for New Development of Mathematics to Science and Technology" sponsored by JSPS at Graduate School of Mathematical Sciences, the University of Tokyo.

References

[1] Y.-G. Chen, *On blow-up solutions of semilinear parabolic equations; Analytical and numerical studies*, Thesis(Ph.D.)-University of Tokyo, 1988.

[2] C. Corduneanu, *Almost Periodic Functions*, Interscince Publishers, New York, 1968.

[3] E. DiBenedetto, *Continuity of weak solutions to a general porous medium equation*, Indiana Univ. Math. J. **32**(1983), 83-118.

[4] H. Fujita and Y.-G. Chen, *On the set of blow-up points and asymptotic behaviours of blow-up solutions to a semilinear parabolic equation*, Analyse Mathématique et Applications, Contrib. Honneur Jaques-Louis Lions, (1988), 181-201.

[5] A. Friedman and B. McLeod, *Blow-up of positive solutions of semilinear heat equations*, Indiana Univ. Math. J. **34**(1985), 425-447.

[6] Y. Giga and N. Umeda, *On blow-up at space infinity for semilinear heat equations*, J. Math. Anal. Appl. **316**(2006), 538-555.

[7] Y. Giga and N. Umeda, *Blow-up directions at space infinity for solutions of semilinear heat equations*, Bol. Soc. Paran. Mat. **23**(2005), 9-28.

[8] Y. Giga and N. Umeda, *Correction to "Blow-up directions at space infinity for solutions of semilinear heat equations"*, Bol. Soc. Paran. Mat., **24**(2006), 9–28.

[9] Y. Giga and R.V. Kohn, *Nondegeneracy of blow-up for semilinear heat equations*, Comm. Pure Appl. Math. **42**(1989) 845-884.

[10] A. A. Lacey, *The form of blow-up for nonlinear parabolic equations*, Proc. R. Soc. Edinb., Sect. A **98**(1984), 183-202.

[11] K. Mochizuki and R. Suzuki, *Blow-up sets and asymptotic behavior of interfaces for quasilinear degenerate parabolic equations in* $\boldsymbol{R}^N$, J. Math. Soc. Japan **44** (1992), 485-504.

[12] M. H. Protter and H. F. Weinberger, *Maximum principles in Differential Equations*, Englewood Cliffs, N.J. Prentice-Hall, 1967.

[13] Y. Seki, *On directional blow-up for quasilinear parabolic equations with fast diffusion*, J. Math. Anal. Appl. **338**(2008), 572-587.

[14] Y. Seki, R. Suzuki and N. Umeda, *Blow-up directions for quasilinear parabolic equations*, Proc. R. Soc. Edinb., Sect. A, in press.

DIFFUSION MEDIATED TRANSPORT WITH A LOOK AT MOTOR PROTEINS

STUART HASTINGS

Department of Mathematics, University of Pittsburgh, Pittsburgh, PA 15260, USA, email: sph+@pitt.edu

DAVID KINDERLEHRER

Center for Nonlinear Analysis and Department of Mathematical Sciences, Carnegie Mellon University, Pittsburgh, PA 15213, email: davidk@cmu.edu
Research supported by NSF DMS 0305794 and DMS 0405343

J. BRYCE MCLEOD

Department of Mathematics, University of Pittsburgh, Pittsburgh, PA 15260 email: mcleod@pitt.edu

Abstract: In this note we discuss transport properties of weakly coupled parabolic systems of evolution equations. These arise in the study of molecular motors, like conventional kinesin, which are responsible for eukaryotic intracellular transport. It falls under the rubric of what we call diffusion mediated transport. Diffusion mediated transport generally concerns directed transport or oriented fluctuations of a system with a high degree of randomness and requires special collaboration among its various elements to achieve. We discuss how this plays out in multiple state systems.

1. Introduction

Intracellular transport in eukarya is attributed to motor proteins that transduce chemical energy into conformational changes which lead to directed mechanical motion. Nanoscale motors like kinesins tow organelles and other cargo on microtubules or filaments, have a role separating the mitotic spindle during the cell cycle, and perform many other functions. The simplest description gives rise to a weakly coupled system of evolution equations. The transport process, to the mind's eye, is comparable to a biased coin toss. This intuition may be confirmed by a careful analysis of the cooperative effects among the conformational changes and the potentials. Two models illustrating this are a two state system and a larger system intended to account for the neck linker apparatus. We then discuss how collaboration may fail when connectivity among the elements comprising the network is disrupted, as in the case of spacers added to the neck-linker. Most of these remarks are about [11].

Suppose that $\rho_1, ..., \rho_n$ are partial probability densities defined on the unit interval $\Omega = (0,1)$ satisfying

$$\begin{aligned} \frac{d}{dx}(\sigma\frac{d\rho_i}{dx} + \psi_i'\rho_i) + \sum_{j=1,\dots,n} a_{ij}\rho_j &= 0 \text{ in } \Omega \\ \sigma\frac{d\rho_i}{dx} + \psi_i'\rho_i &= 0 \text{ on } \partial\Omega,\ i = 1, ...n, \\ \rho_i \geqq 0 \text{ in } \Omega,\ \int_\Omega (\rho_1 + \cdots + \rho_n)dx &= 1. \end{aligned} \tag{1.1}$$

Here $\sigma > 0$, $\psi_1, ..., \psi_n$ are smooth non-negative functions of period $1/N$, and $A = (a_{ij})$ is a smooth rate matrix of period $1/N$, that is, a_{ij} are $1/N-$periodic functions with

$$\begin{aligned} &a_{ii} \leqq 0,\ a_{ij} \geqq 0 \text{ for } i \neq j \text{ and} \\ &\sum_{i=1,..,n} a_{ij} = 0,\ j = 1, ..., n. \end{aligned} \tag{1.2}$$

Note that for $\tau > 0$ small, the matrix $P = \mathbf{1} + \tau A$ is a probability matrix. The system (1.1) is the stationary equations of the evolution system

$$\begin{aligned} \frac{\partial\rho_i}{\partial t} = \frac{\partial}{\partial x}(\sigma\frac{\partial\rho_i}{\partial x} + \psi_i'\rho_i) + \sum_{j=1,\dots,n} a_{ij}\rho_j &\text{ in } \Omega,\ t > 0, \\ \sigma\frac{\partial\rho_i}{\partial x} + \psi_i'\rho_i = 0 &\text{ on } \partial\Omega,\ t > 0,\ i = 1, ...n, \\ \rho_i \geqq 0 \text{ in } \Omega,\ \int_\Omega (\rho_1 + \cdots + \rho_n)dx = 1,&\ t > 0. \end{aligned} \tag{1.3}$$

Before discussing the result further, let us discuss what we intend by transport. In a chemical or conformational change process, a reaction coordinate (or coordinates) must be specified. This is the independent variable. In a mechanical system, it is usually evident what this coordinate must be. In our situation, even though both conformational change and mechanical effects are present, it is natural to specify the distance along the motor track, the microtubule, here the interval Ω, as the independent variable.

We interpret the migration of density during the evolution to one end of the track as evidence of transport.

Transport results from functional relationships in this system. A straightforward way to approach this issue is inspection of the dissipation principle which gives an implicit scheme for its solution. This implicit scheme is based on a Monge-Kantorovich-Wasserstein metric, which, we recall, may be defined by

$$d(f,f^*)^2 = \inf_{\mathbf{P}} \int_{\Omega\times\Omega} |x-y|^2 dp(x,y) \tag{1.4}$$

where $\mathbf{P}$ denotes the set of joint distributions for f and f^*, nonnegative densities with the same total mass. Let

$$F(\eta) = \int_\Omega \sum_{i=1,\dots,n} \{\psi_i\eta_i + \sigma\eta_i \log\eta_i\}dx \tag{1.5}$$

$$\eta_i \geqq 0 \text{ and } \int_\Omega \sum_{i=1,\dots,n} \eta_i dx = 1$$

denote the free energy. Now given a state ρ^*, determine its successor state ρ by resolving the variational principle

$$\begin{gathered}\frac{1}{2\tau}\sum_{i=1,\dots,n} d(\rho_i,(P\rho^*)_i)^2 + F(\rho) = \min, \\ \int_\Omega \rho_i dx = \int_\Omega (P\rho^*)_i dx,\end{gathered} \tag{1.6}$$

where P is the probability matrix above. Below we address why we call this a dissipation principle. For the moment, determine an implicit scheme by the rule: given $\rho^{(k-1)}$, set $\rho^* = \rho^{(k-1)}$ and $\rho^k = \rho$ in (1.6), then set

$$\rho^\tau = \rho^k,\ k\tau \leqq t < (k+1)\tau.$$

Finally, let $\tau \to 0$. Then $\rho^\tau \to \rho$, the solution of (1.3) in $\Omega\times(0,T)$ for any $T<\infty$.

The interpretation of (1.6) as a dissipation principle is given in [5] and is based in part on the Benamou and Brenier [3] result that in (1.4)

$$\frac{1}{\tau}d(f,f^*)^2 = \min \int_0^\tau \int_\Omega v^2 f dxdt, \tag{1.7}$$

where the minimum is taken over families $f(x,t), 0 \leqq t \leqq \tau$, of deformations satisfying

$$\frac{\partial f}{\partial t} + \frac{\partial}{\partial x}(vf) = 0 \text{ in } \Omega \times (0,\tau),$$
$$f(x,0) = f^*(x) \text{ and } f(x,\tau) = f(x).$$

The right hand side of (1.7) is the minimum dissipation of an ensemble of highly damped particles initially distributed by f^* and terminally distributed by f, expressed in an eulerian frame. So the minimum energy budget in moving the system from ρ^* to ρ is given by the variational problem (1.6). It has the merit of isolating the free energy, the dissipation, and the conformational change. This is modulo some modelling of the entropic contribution, for which we have adopted combinatorial indeterminacy, the simplest possible choice. We may also interpret this as representing a collection of molecular motors as a conformation changing ensemble of spring-mass-dashpots. Here we are glossing over the many issues present in modelling small scale systems, where mechanics and chemistry or conformational changes themselves operate at disparate time and length scales.

Our first statement is that the stationary solution of the system (1.1), which we denote by $\rho^\sharp$, is globally stable: given any solution $\rho(x,t)$ of (1.3),

$$\rho(x,t) \to \rho^\sharp(x) \text{ as } t \to \infty \tag{1.8}$$

So the migration of density, constituting transport, we referred to previously may be ascertained by inspection of $\rho^\sharp$. In the sequel, we simply set $\rho = \rho^\sharp$.

We shall characterize transport properties of (1.1) by decay properties of ρ. If the preponderance of mass of ρ is distributed at one end of the track, then transport is present. Our main result, stated precisely later, is that with suitable collaboration among the potentials $\psi_1, ..., \psi_n$ and the rate matrix A, there are constants K and M, independent of σ, such that

$$\sum_{i=1}^{n} \rho_i(x + \frac{1}{N}) \leqq K e^{-\frac{M}{\sigma}} \sum_{i=1}^{n} \rho_i(x), \; x \in \Omega, \; x < 1 - \frac{1}{N} \tag{1.9}$$

for sufficiently small $\sigma > 0$.

Our objective in this note is to explain what we have found to be the suitable collaboration mentioned above. A first possibility is to ask simply for the equilibrium configuration, associated to (1.5), that is, its minimum energy configuration, given by

$$\begin{aligned}\rho_i^\sharp(x) &= \frac{1}{Z}e^{-\frac{\psi_i(x)}{\sigma}},\ x\in\Omega,\ i=1,...,n\\ Z &= \int_\Omega (e^{-\frac{\psi_1(x)}{\sigma}} + \cdots + e^{-\frac{\psi_n(x)}{\sigma}})dx,\end{aligned} \tag{1.10}$$

which is $1/N-$periodic in Ω. This solution does not exhibit transport and for it to be a solution of (1.1) means that the equations decouple and

$$A\rho = 0 \text{ in } \Omega.$$

This is a detailed balance situation, separately for the Markov Process described by the n individual Fokker-Planck Equations and for the Markov chain with transition matrix P. So, to favor transport, equilibrium should not be attained and detailed balance broken. But failure of detailed balance is far from sufficient to produce transport.

We illustrate this in Figure 1 where solutions of a two species system for two arrangements of symmetric potential wells are shown. The matrix A was chosen constant, which favors a maximum amount of transport. Detailed balance fails for the solutions. Asymmetry of the potentials was suggested early in the study of motor proteins, cf. [2], [17], and it is known that the microtubules and actin filiments which host motors are polarized, is suspected to play a role here. Asymmetry itself is insufficient. In Figure 2, asymmetric potentials differ from each other by a slight shift, actually 1/8 period, and there is no noticable transport in the solution.

If we adopt the pragmatic notion that in a two species system, the two species function in the same way, we are led to interdigitated potentials ψ_j of the form in Figure 3. This is not a reason, of course. We discuss this further below.

We are led to the intriguing question of the relationship between the ψ_j and A. Even under the most most propitious circumstances, one may always add to the system independent uncoupled equations. So it is necessary, in view of (1.2), that

$$a_{ii} \not\equiv 0 \text{ in } \Omega$$

But where and how? What are the possibilities here? The basic mechanism of diffusional transport is that mass is transported to specific sites determined by minima and local minima of the potential. For directed transport, to the left toward $x = 0$, for example, in any subinterval of a period interval, there should be some ψ_i which is increasing. This explains the

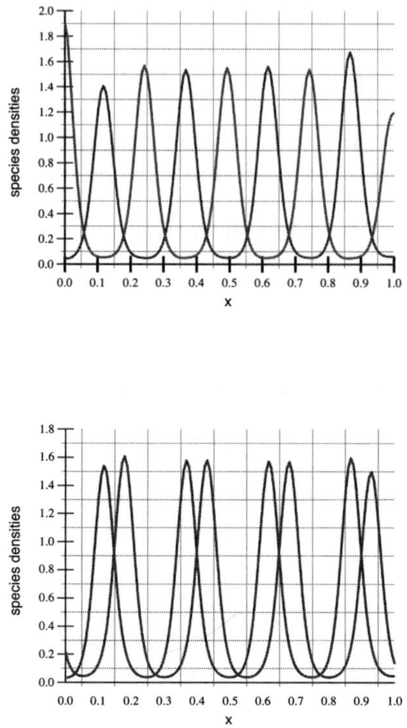

Figure 1. Solutions of (1.1) for two state systems with symmetric potentials placed symmetrically (left) and symmetric potentials placed asymmetrically (right) showing lack of transport of density. A was chosen constant to optimize the possibility of transport. Detailed balance is not satisfied by the solutions

result shown in Figure 2, where the potentials are asymmetric and transport is not present. Moreover, some interchange must take place: mass in states associated to each of the ψ_j which is decreasing should have the opportunity to change to the ith−state. This is reminiscent of an ergodic hypothesis. It does not say that all states are connected, but it will be a very strong condition since it will be required to hold near all the minima of all of the potentials. In the the neck linker example we have mentioned, the condition fails and so does the conclusion of our theorem. We give a more precise description below.

We have, in the above discussion, tacitly assumed the existence of a

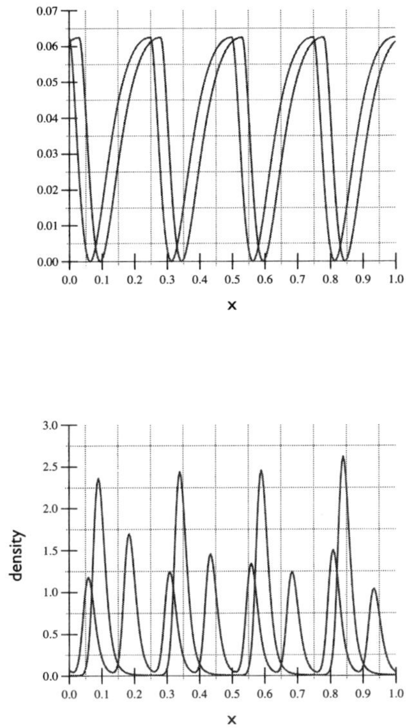

Figure 2. Slightly shifted asymmetric wells, left, and the solutions of (1.1), right, illustrating lack of transport. The matrix A was chosen to optimize transport possibilities.

positive solution of (1.1). In fact, under suitable conditions on the a_{ij}, we can assert not only the existence of a solution of (1.1), without assuming a priori its positivity, but also that it is unique and necessarily positive. Sufficient conditions for this are (1.2) together with the hypothesis that, with $P_\tau = \mathbf{1} + \tau A$ the probability matrix already mentioned, that

$$\begin{aligned} &a_{ij} \not\equiv 0 \text{ in } \Omega, \text{ or more generally,} \\ &Q_\tau = \int_\Omega P_\tau(x)dx \text{ is ergodic,} \end{aligned} \tag{1.11}$$

i.e., Q_τ^k has all positive entries for some integer k.

Further, we can prove that the positive stationary solution, $\rho^\dagger(x)$ say, of (1.1) is globally stable in the sense that given any initial data $f(x)$ with

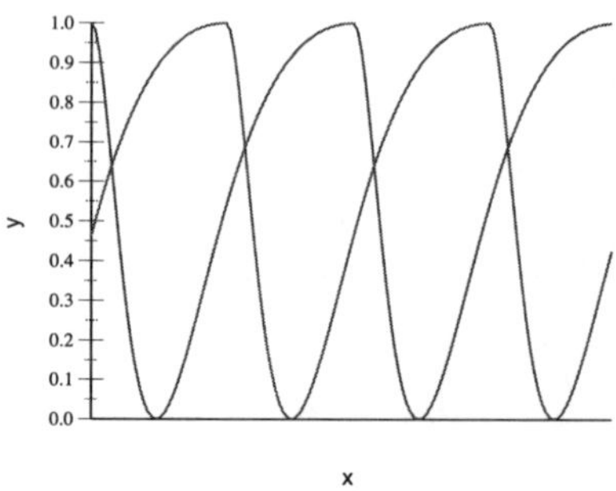

Figure 3. Asymmetric periodic potentials symmetrically interdigitated, the configuration which promotes transport

corresponding solution $\rho(x,t)$ of (1.3) there exist $c > 0$ and $\omega > 0$ such that

$$\rho(x,t) = c\rho^\dagger(x) + O(e^{-\omega t}) \text{ as } t \to \infty \tag{1.12}$$

There are various ways of proving this, they all depend on ideas from positive operators. In these frameworks, we define the operator e^S by expressing the solution ρ of (1.3) by

$$\rho(x,t) = e^{tS} f(x), \ \ x \in \Omega, t > 0.$$

From the maximum principle, eg. [25], this is a positive operator. This enables us to apply a variant of the Krein-Rutman Theorem, the appropriate generalization of the Perron-Frobenius Theorem for matrices, to understand the spectrum of S. We refer to [4], [11] for additional details. In [23], Perthame succeeds in constructing an appropriate entropy function for the system, at least when the number of species $n = 2$.

We take this opportunity to thank our collaborators Michel Chipot and Michal Kowalczyk for their help. We also thank Bard Ermentrout, Michael Grabe, and Jonathan Rubin, and the Mathematical Biology Seminar of the University of Pittsburgh for generosity of spirit and their interest.

2. Main transport result

Here we state our main result about transport in multiple state systems.

Theorem 2.1. *Suppose that ρ is a positive solution of* (1.1), *where the coefficients a_{ij}, $i,j = 1,...,n$ and the ψ_i, $i = 1,...,n$ are smooth and 1/N-*

periodic in $\overline{\Omega}$. Suppose that (1.2) holds and also that the following conditions are satisfied.

(i) *Each ψ_i' has only a finite number of zeros in $\overline{\Omega}$.*
(ii) *There is some interval in which $\psi_i' > 0$ for all $i = 1, ..., n$.*
(iii) *In any interval in which no ψ_i' vanishes, $\psi_j' > 0$ in this interval for at least one j.*
(iv) *If $I, |I| < 1/N$, is an interval in which $\psi_i' > 0$ for $i = 1, .., p$ and $\psi_i' < 0$ for $i = p+1, .., n$, and a is a zero of at least one of the ψ_k' which lies within ϵ of the right-hand end of I, then for ϵ sufficiently small, there is at least one index i, $i = 1, ..., p$, with $a_{ij} > 0$ in $(a - \eta, a)$ for some $\eta > 0$, all $j = p+1, .., n$.*

Then, there exist positive constants K, M independent of σ such that

$$\sum_{i=1}^{n} \rho_i(x + \frac{1}{N}) \leqq Ke^{-\frac{M}{\sigma}} \sum_{i=1}^{n} \rho_i(x), \ x \in \Omega, \ x < 1 - \frac{1}{N} \tag{2.1}$$

for sufficiently small σ.

We give a sketch of the proof intended to highlight the role of (iv). Let $\lambda = 1/N$ denote the period. Adding the equations in (1.1) gives

$$\frac{d}{dx} \sum_{i=1,\dots,n} (\sigma \frac{d\rho_i}{dx} + \psi_i' \rho_i) = 0 \text{ in } \Omega$$

and by the boundary condition

$$\sum_{i=1,\dots,n} (\sigma \frac{d\rho_i}{dx} + \psi_i' \rho_i) = 0 \text{ in } \Omega \tag{2.2}$$

This suggests application of the Gronwall Lemma. It will be successful only in subintervals where all the ψ_i' are positive, and there are some by (ii). So let us consider, for a fixed index ν, the νth-equation of the system,

$$\sigma \rho_\nu'' + \psi_\nu' \rho_\nu' + \psi_\nu'' \rho_\nu + a_{\nu\nu}\rho_\nu + \sum_{j \neq \nu} a_{\nu j}\rho_j = 0 \text{ in } I \tag{2.3}$$

Equation (2.3) represents a balance between ρ_ν and the other ρ_j. As seen below in Step 3, since items in the $\sum$ are nonnegative, they can be discarded and (2.3) can be employed to find an upper bound for ρ_ν when ψ_ν is increasing. We can then exploit it to impede the growth of the other $\{\rho_j\}$. Namely, $\{\rho_j\}$ cannot be too large without forcing ρ_ν negative. But this can only be assured if the coupling is really there, namely if $a_{\nu j} > 0$. This is the motivation for the ergodic type hypothesis in (iv).

Given $\xi_0, 0 < \xi_0 < 1 - \lambda$, fix the period interval $\Lambda = [\xi_0, \xi_0 + \lambda]$. We now limit our attention to Λ.

Select intervals $a + [-\delta, \delta]$ about zeros a of the ψ_i'. On the complement, there is a $k(\delta) > 0$ with

$$\psi_i' \geqq k(\delta) \text{ or } \psi_i' \leqq -k(\delta).\ i = 1, ..., n \tag{2.4}$$

Step 1: By (ii), there is at least one interval $I_0 = x_0 + [-L_0, L_0] \subset \Lambda$ with

$$\psi_i' \geqq k(\delta) \text{ in } I_0 \tag{2.5}$$

Then

$$\frac{d}{dx} \sum_{i=1,\ldots,n} \rho_i \leqq -\frac{K_0}{2\sigma} \sum_{i=1,\ldots,n} \rho_i \text{ in } I_0$$

and by Gronwall,

$$\sum_{i=1,\ldots,n} \rho_i(x_0 + L_0) \leqq e^{-\frac{K_0}{\sigma}} \sum_{i=1,\ldots,n} \rho_i(x_0 - L_0) \tag{2.6}$$

This is the exponential decay we are seeking. The remainder of the argument is to prove that (2.6) is not compromised when (2.5) does not hold.

Step 2: We check the zeros a of the ψ_i'. Although $\sum \rho_i$ may grow exponentially in these intervals, they are finite, say N, in number and we may choose them of small length. So restricting δ,

$$\sum_{i=1,\ldots,n} \rho_i(a + \delta) \leqq e^{\frac{M_0}{\sigma}} \sum_{i=1,\ldots,n} \rho_i(a - \delta)$$

with, eg.,

$$NM_0\delta < K_0 L_0 \tag{2.7}$$

Also assume that $\delta \leqq \epsilon$ of (iv). Now δ is fixed and in the sequel we suppress dependence of various constants on it.

Step 3: Both inequalities hold in (2.4) and we must exploit the coupling. Let $I = [\alpha, \beta] \subset \Lambda$ be an interval where

$$\begin{aligned} &\psi_i' \geqq k,\ i = 1, ..., p, \\ &\psi_j' \leqq -k,\ j = p + 1, ..., n, \\ &|\beta - a| < \delta \leqq \epsilon \text{ for some zero } a \text{ of the } \psi_k',\ k = 1, ..., n. \end{aligned} \tag{2.8}$$

Choose a favorable $\nu, \nu = 1, ..., p$, from (iii), and assume that (iv) holds, i.e., $a_{\nu j} > 0, j = p + 1, ..., n$ in $\beta - \eta \leqq x \leqq \beta$, and consider the equation

(2.3) for ρ_ν. We can integrate this for a more convenient form, which gives us

$$\begin{aligned}\frac{d\rho_\nu}{dx}(x) =&\frac{d\rho_\nu}{dx}(\alpha)e^{-\frac{1}{\sigma}(\psi_\nu(x)-\psi_\nu(\alpha))}\\ &-\frac{1}{\sigma}\int_\alpha^x \{(a_{\nu\nu}+\psi_\nu')\rho_\nu + \sum_{j\neq\nu} a_{\nu j}\rho_j\}e^{-\frac{1}{\sigma}(\psi_\nu(s)-\psi_\nu(\alpha))}ds \text{ in } I\end{aligned} \tag{2.9}$$

We shall use this in the two ways described at the beginning of the proof sketch. First, the sum in the right hand side of (2.9) is nonnegative, so we can omit it. Note also that ψ_ν is increasing so $\psi_\nu(x) - \psi_\nu(\alpha) > 0$ in I. After some manipulation and applications of Gronwall, this leads to the upper bound

$$\rho_\nu(x) \leqq K(\rho_\nu(\alpha) + \sigma|\rho_\nu'(\alpha)|), \; x \in I \tag{2.10}$$

Consequently, with

$$C(\alpha) = \sum_{\nu=1,\dots,p} (\rho_\nu(\alpha) + \sigma|\rho_\nu'(\alpha)|),$$

we obtain an upper bound for the favorable states which reads

$$\sum_{\nu=1,\dots,p} \rho_\nu(x) \leqq KC(\alpha), \; x \in I \tag{2.11}$$

where K depends only on δ and the problem parameters.

Returning to (2.2), we then find the estimate

$$\frac{d}{dx}(\rho_{p+1}+\cdots+\rho_n) \geqq -\frac{K_1}{\sigma}C(\alpha) + \frac{K_2}{\sigma}(\rho_{p+1}+\cdots+\rho_n), \text{ in } I. \tag{2.12}$$

This will tell us that if $\rho_{p+1}+\cdots+\rho_n$ is large at some x^*, it becomes exponentially larger for $x \geqq x^*$. We shall then show that this leads to a contradiction. In fact, we claim

$$\rho_{p+1}+\cdots+\rho_n \leqq \frac{K_1}{K_2}C(\alpha) \text{ in } I. \tag{2.13}$$

This will conclude the proof. Suppose that

$$(\rho_{p+1}+\cdots+\rho_n)(x^*) \geqq \frac{K_1}{K_2}AC(\alpha) \text{ with } A > 1, \text{ for some } x^* \in I. \tag{2.14}$$

Integrating (2.12) then gives us that

$$\begin{aligned}(\rho_{p+1}+\cdots+\rho_n)(x) \geqq&\frac{K_1}{K_2}C(A-1)e^{\frac{K_2}{\sigma}(x-x^*)}\\ &+\frac{K_1}{K_2}C, \; x^* < x < \beta\end{aligned} \tag{2.15}$$

$$C = C(\alpha).$$

Now, as foretold, we return to (2.9). Precisely, from (iv), $a_{\nu j} > 0$ for $j = p+1, ..., n$, so there is a $\mu > 0$ such that

$$a_{\nu p+1}\rho_{p+1} + \cdots + a_{\nu n}\rho_n \geqq \mu(\rho_{p+1} + \cdots + \rho_n) \text{ in } [\beta - \delta, \beta].$$

Using this and integrating (2.9), we find numbers M, M_1 and M_2 such that

$$0 \leqq \frac{\rho_\nu(x)}{C(\alpha)} \leqq M_1 - (A-1)M_2\delta\frac{K_1}{K_2}e^{\frac{M\delta}{\sigma}}, \text{ in } [\beta - \delta, \beta]. \tag{2.16}$$

Now for $A > 1$, (2.16) cannot hold for small σ since the right hand side becomes negative. Thus (2.13) holds for σ sufficiently small.

The theorem now follows by concatening the three Steps.

3. Correlated and uncorrelated heads in a three state system

One of the simplest systems we may consider is a two state system for the unknown $\rho(x,t) = (\rho_1(x,t), \rho_2(x,t))$, a solution of (1.3) with $n = 2$. Let us assume for this a configuration of potentials resembling (3) and a conformation change matrix

$$A = \begin{pmatrix} -\alpha_1 & \alpha_2 \\ \alpha_1 & -\alpha_2 \end{pmatrix} \tag{3.1}$$

where the support of the α_i is assumed to be a neighborhood of the minima of the potentials ψ_j. The conditions of Theorem 2.1 are satisfied, in particular, the ergodic-type condition (iv). The result of a sample simulation is given in Figure 4 and is a standard way to model conventional kinesin, cf [1], [4], [5]. The two heads are correlated.

We may attempt a slightly increased degree of sophistication. The two heads of conventional kinesin are known to be connected by a rigid protein structure, called the neck linker, [29]. Hackney *et al.* in a recent experiment describe the decrease of transport properties of a version of kinesin (a chimera) fashioned by extending the neck linker through insertion of a spacer, rendering it more flexible, [9]. The two kinesin heads then fail to be correlated. Abstracting this situation, we may consider a three state system for conventional kinesin consisting of head-one, head-two, and the neck linker states. The neck linker does not participate in transport, so, in our framework, it is accounted for only in the conformation change matrix A. This will be a novel feature of the system, since we shall have only an ordinary differential equation for ρ_3. The two heads are not connected

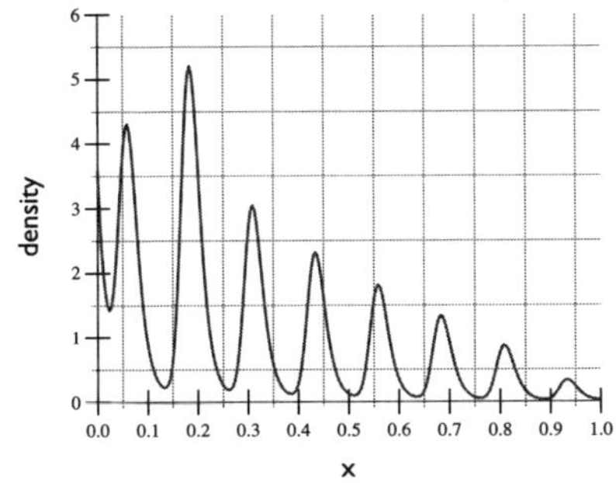

Figure 4. Computed solution for two state rachet for interdigitated asymmetric potentials, period 4. Plot shows summed density $\rho_1 + \rho_2$.

directly to each other but only to the neck linker. The system has the form

$$\begin{aligned}
&\frac{\partial \rho_i}{\partial t} = \frac{\partial}{\partial x}(\sigma\frac{\partial \rho_i}{\partial x} + \psi_i'\rho_i) + \sum_{j=1,\ldots,3} a_{ij}\rho_j \text{ in } \Omega,\ t > 0, i = 1,2 \\
&\frac{d\rho_3}{dt} = \sum_{j=1,..,3} a_{3j}\rho_j \text{ in } \Omega, t > 0 \\
&\sigma\frac{\partial \rho_i}{\partial x} + \psi_i'\rho_i = 0 \text{ on } \partial\Omega,\ t > 0,\ i = 1,2, \\
&\rho_i \geqq 0\ in\ \Omega,\ \int_\Omega (\rho_1 + \rho_2 + \rho_3)dx = 1,\ t > 0.
\end{aligned} \tag{3.2}$$

where the 3×3 matrix A has the generic representation

$$A = \begin{pmatrix} * & 0 & * \\ 0 & * & * \\ * & * & * \end{pmatrix}. \tag{3.3}$$

The stationary system for (3.2) is just

$$\begin{aligned}
\frac{d}{dx}(\sigma\frac{d\rho_i}{dx} + \psi_i'\rho_i) + &\sum_{j=1,\ldots,3} a_{ij}\rho_j = 0 \text{ in } \Omega, i = 1,2 \\
&\sum_{j=1,\ldots,3} a_{3j}\rho_j = 0 \text{ in } \Omega \\
&\sigma\frac{d\rho_i}{dx} + \psi_i'\rho_i = 0 \text{ on } \partial\Omega,\ i = 1,2,
\end{aligned} \tag{3.4}$$

Let us begin by revisiting the result about correlated two state systems in the present context. Thus we assume that all of the $*$ items in (3.3) do not vanish identically. In particular we assume that, in accord with (1.2),

$$A = \begin{pmatrix} a_{11} & 0 & a_{13} \\ 0 & a_{22} & a_{23} \\ a_{31} & a_{32} & a_{33} \end{pmatrix} \tag{3.5}$$

and

$$\mathit{supp}\ a_{ij} = \text{a neighborhood of the minima of } \psi_1 \text{ and } \psi_2 \tag{3.6}$$

There are only 4 independent quantities in (3.5) owing to (1.2) and we assume that the supports of these functions are all equal. We may suppose that $\rho_3 = 0$ outside *supp* a_{33} so that the last equation in (3.4) always makes sense. This can be guaranteed in the evolution system (3.2) by choice of suitable initial data. Eliminating ρ_3, we now find a new system for ρ_1 and ρ_2 alone of the form

$$\begin{aligned} &\frac{d}{dx}(\sigma\frac{d\rho_1}{dx} + \psi_1'\rho_1) - \alpha_1\rho_1 + \alpha_2\rho_2 = 0 \text{ in } \Omega, i = 1,2 \\ &\frac{d}{dx}(\sigma\frac{d\rho_2}{dx} + \psi_2'\rho_2) + \alpha_1\rho_1 - \alpha_2\rho_2 = 0 \text{ in } \Omega, i = 1,2 \\ &\qquad\qquad \sigma\frac{d\rho_i}{dx} + \psi_i'\rho_i = 0 \text{ on } \partial\Omega,\ i = 1,2, \end{aligned} \tag{3.7}$$

where, with the excruciating details, forms of α_1 are

$$\begin{aligned} \alpha_1 &= -\frac{a_{23}}{a_{33}}a_{31} = \frac{a_{23}}{a_{13} + a_{23}}a_{31} \\ &= -a_{11}\frac{a_{23}}{a_{13} + a_{23}} = -a_{11}(1 - \frac{a_{13}}{a_{13} + a_{23}}) \geqq 0 \end{aligned} \tag{3.8}$$

and similarly for α_2. The new system (3.7) for just ρ_1 and ρ_2 satisfies the conditions of Theorem 2.1. Thus the conclusion of the theorem applies, and we recapture our prior result. The result of a simulation is given in Figure 5 on the left.

We can decorrellate the heads. The idea, obviously, is to choose the $\{a_{ij}\}$ so that (3.4) decouples. So assume in (3.5) that

$$\begin{aligned} &\Lambda_1 = \mathit{supp}\ a_{11} = \mathit{supp}\ a_{13} = \text{a neighborhood of the minima of } \psi_1 \\ &\Lambda_2 = \mathit{supp}\ a_{22} = \mathit{supp}\ a_{23} = \text{a neighborhood of the minima of } \psi_2\text{, and} \\ &\Lambda_1 \cap \Lambda_2 = \emptyset \end{aligned} \tag{3.9}$$

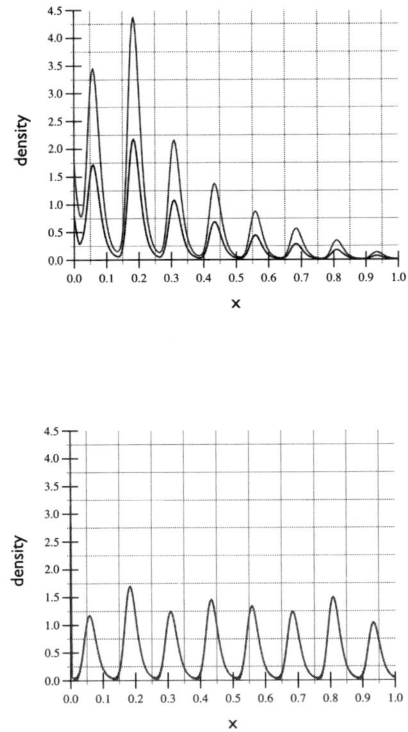

Figure 5. Two state rachet with neck linker with heads correlated (left) and heads uncorrelated (right). Red plot is the sum of the head densities $\rho_1 + \rho_2$ and gray plot is the neck linker density ρ_3. The left plot illustrates transport. In the uncorrelated example, there is no transport and it turns out that $\rho_3 = \rho_1 + \rho_2$.

Note now that from (3.8), $\alpha_1 = 0$ and the same for α_2.

We find for the system (3.4),that

$$\begin{aligned} &\frac{d}{dx}(\sigma\frac{d\rho_i}{dx} + \psi_i'\rho_i) = 0 \text{ in } \Omega, i = 1, 2 \\ &\rho_3 = \rho_1 \text{ or } \rho_2 \text{ or } 0 \text{ in } \Omega \\ &\sigma\frac{d\rho_i}{dx} + \psi_i'\rho_i = 0 \text{ on } \partial\Omega,\ \ i = 1, 2, \end{aligned} \tag{3.10}$$

which is the same except for mass fractions, cf. (1.10), as writing that

$$\rho_1 = \rho_1^\sharp \text{ and } \rho_2 = \rho_2^\sharp$$

where there is no transport. The result of a simulation is given in Figure 5 on the right.

References

[1] AJDARI, A. AND PROST, J. (1992) Mouvement induit par un potentiel périodique de basse symétrie: dielectrophorese pulse, C. R. Acad. Sci. Paris t. **315**, Série II, 1653.

[2] ASTUMIAN, R.D. (1997) Thermodynamics and kinetics of a Brownian motor, Science **276** (1997), 917–922.

[3] BENAMOU, J.-D. AND BRENIER, Y. (2000) A computational fluid mechanics solution to the Monge-Kantorovich mass transfer problem, Numer. Math. **84** , 375–393.

[4] CHIPOT, M., HASTINGS, S., AND KINDERLEHRER, D., Transport in a molecular motor system (2004) Math. Model. Numer. Anal. (M2AN) **38** no. 6, 1011–1034

[5] CHIPOT, M., KINDERLEHRER, D. AND KOWALCZYK, M. (2003) A variational principle for molecular motors, Meccanica, **38**, 505–518

[6] DOERING, C., ERMENTROUT, B. AND OSTER, G. (1995) Rotary DNA motors. Biophys. J. **69(6)**, 2256-67

[7] DOLBEAULT, J., KINDERLEHRER, D., AND KOWALCZYK, M. Remarks about the flashing rachet, to appear Proc. PASI 2003

[8] HACKNEY, D.D. (1996) The kinetic cycles of myosin, kinesin, and dynein, Ann. Rev. Physiol., **58**, 731 - 750

[9] HACKNEY, D.D., STOCK, M. F., MOORE, J., AND PATTERSON, R. (2003) Modulation of kinesin half-site ADP release and kinetic processvity by a spacer between the head grounps, Biochem., **42**, 12011 – 12018

[10] HASTINGS, S. AND KINDERLEHRER, D. (2005) Remarks about diffusion mediated transport: thinking about motion in small systems, Nonconvex Optim. Appl., Springer, 79, 497-511

[11] HASTINGS, S., KINDERLEHRER, D. AND MCLEOD, J.B. Diffusion mediated transport in multiple state systems, to appear

[12] HOWARD, J. (2001) Mechanics of Motor Proteins and the Cytoskeleton, Sinauer Associates, Inc.

[13] HUXLEY, A.F. (1957) Muscle structure and theories of contraction, Prog. Biophys. Biophys. Chem. **7**, 255–318.

[14] JORDAN, R., KINDERLEHRER, D. AND OTTO, F. (1998) The variational formulation of the Fokker-Planck equation, SIAM J. Math. Anal. Vol. **29** no. 1, 1–17.

[15] KINDERLEHRER, D. AND KOWALCZYK, M (2002) Diffusion-mediated transport and the flashing ratchet, Arch. Rat. Mech. Anal. **161**, 149–179.

[16] KINDERLEHRER, D. AND WALKINGTON, N. (1999) Approximation of parabolic equations based upon Wasserstein's variational principle, Math. Model. Numer. Anal. (M2AN) **33** no. 4, 837–852.

[17] MAGNASCO, M. (1993) Forced thermal rachets, Phys. Rev. Lett. **67**, 1477–1481

[18] OKADA, Y. AND HIROKAWA, N. (1999) A processive single-headed motor: kinesin superfamily protein KIF1A, Science Vol. **283**, 19

[19] OKADA, Y. AND HIROKAWA, N. (2000) Mechanism of the single headed processivity: diffusional anchoring between the K-loop of kinesin and the C terminus of tubulin, Proc. Nat. Acad. Sciences **7** no. 2, 640–645.

[20] OTTO, F. (1998) Dynamics of labyrinthine pattern formation: a mean field theory, Arch. Rat. Mech. Anal. **141**, 63-103

[21] OTTO, F. (2001) The geometry of dissipative evolution equations: the porous medium equation, Comm. PDE **26**, 101-174

[22] PARMEGGIANI,A., JÜLICHER, F., AJDARI, A. AND PROST, J. (1999) Energy transduction of isothermal ratchets: generic aspects and specific examples close and far from equilibrium, Phys. Rev. E, **60** no. 2, 2127–2140.

[23] PERTHAME, B. The general relative entropy principle

[24] PESKIN, C.S.. ERMENTROUT, G.B. AND OSTER, G.F. (1995) *The correlation ratchet: a novel mechanism for generating directed motion by ATP hydrolysis*, in Cell Mechanics and Cellular Engineering (V.C Mow et.al eds.), Springer, New York

[25] PROTTER, M. AND WEINBERGER, H.(1967) Maximum principles in differential equations, Prentice Hall, Englewood Cliffs, N. J.

[26] PURCELL, E. M.,(1971) Life at low Reynolds' Number Am. J. Phys. 45, 3-11

[27] REIMANN, P. (2002) Brownian motors: noisy transport far from equilibrium, Phys. Rep. **361** nos. 2–4, 57–265.

[28] SCHLIWA, M., ED (2003) Molecular Motors, Wiley-VCH Verlag, Wennheim

[29] VALE, R.D. AND MILLIGAN, R.A. (2000) The way things move: looking under the hood of motor proteins, Science **288**, 88–95.

[30] VILLANI, C (2003) Topics in optimal transportation, AMS Graduate Studies in Mathematics vol. 58, Providence

[31] ZEIDLER, E (1986) Nonlinear functional analysis and its applications, I Springer, New York, N.Y.

MATHEMATICAL STUDY OF SMOLDERING COMBUSTION UNDER MICRO-GRAVITY

KOTA IKEDA

Mathematical Institute, Tohoku University, Sendai 980-8578, Japan

MASAYASU MIMURA

Department of Mathematics / Institute for Mathematical Sciences school of Science and Technology Meiji University, Kawasaki 214-8571, Japan

Abstract: Various finger-like smoldering patterns are observed in experiments under micro-gravity. For theoretical understanding of such pattern phenomena, a model of reaction-diffusion system has been proposed. In this paper, we consider a large-time behavior of solutions and show nonexistence results of traveling wave solutions. We also consider how the solution behaves if a parameter is large.

1. Introduction

It is shown in [4] that thin solid, for an example, paper, cellulose dialysis bags and polyethylene sheets, burning against oxidizing wind develops *finger-like patterns* or *fingering patterns*. The oxidizing gas is supplied in a uniform laminar flow, opposite to the directions of the front propagation and the authors of [4] control the flow velocity of oxygen, denoted by V_{O2}. When V_{O2} is decreased below a critical value, the smooth front develops a structure which marks the onset of instability. As V_{O2} is decreased further, the peaks are separated by cusp-like minima and a fingering pattern is formed. In addition, thin solid is stretched out straight onto the bottom plate and they also control the adjustable vertical gap, denoted by a parameter h, between top and bottom plates. Experimentally, fingering pattern occurs for small h, which implies that fingering pattern appears in the absence of natural convection. Similar phenomena have been observed in a micro-gravity experiment in space (see [2]).

Here we propose a phenomenological model described by the following reaction-diffusion system for the (dimensionless) temperature u, the density of paper v, the concentration of the mixed gas w.

$$(\mathrm{RD})\begin{cases}\dfrac{\partial u}{\partial t}=\mathrm{Le}\Delta u+\lambda\dfrac{\partial u}{\partial x}+\gamma k(u)vw-au^m, & (x,y)\in I\times\Omega, t>0,\\ \dfrac{\partial v}{\partial t}=-k(u)vw, & (x,y)\in I\times\Omega, t>0,\\ \dfrac{\partial w}{\partial t}=\Delta w+\lambda'\dfrac{\partial w}{\partial x}-k(u)vw, & (x,y)\in I\times\Omega, t>0,\end{cases}$$

where the constants Le, called *Lewis number*, a and γ are positive parameters, λ and λ' are nonnegative parameters, $m\geq 1$, $k(u)$ is a nonlinear term called *Arrhenius kinetics* and defined by $k(u)=\exp(-1/u)$, $I\subset(-\infty,\infty)$ is a bounded interval $(0,l_x)$ or a whole line $(-\infty,\infty)$, $\Omega\subset\mathbb{R}^n$ is a bounded domain, and $\Delta=\partial^2/\partial x^2+\sum_{i=1}^n\partial^2/\partial y_i^2$ is Laplacian as usual. We suppose that if $I=(0,l_x)$, u,w satisfy

$$\frac{\partial u}{\partial x}(0,y,t)=\frac{\partial u}{\partial x}(l_x,y,t)=0,\quad \frac{\partial w}{\partial x}(0,y,t)=0,\quad w(l_x,y,t)=w_r>0$$

for any $y\in\Omega$ and $t>0$, and if $I=(-\infty,\infty)$,

$$\lim_{|x|\to\infty}u(x,y,t)=0,\quad \lim_{x\to\infty}w(x,y,t)=w_r,\quad \lim_{x\to-\infty}w(x,y,t)=w_l\geq 0$$

for any $y\in\Omega$ and $t>0$. In both cases we also suppose that u,w satisfy

$$\frac{\partial u}{\partial\nu}(x,y,t)=\frac{\partial w}{\partial\nu}(x,y,t)=0$$

for $x\in I$, $y\in\partial\Omega$ and $t>0$, where ν is the unit exterior normal vector on $\partial\Omega$. We suppose that initial values u_0,v_0 and w_0 satisfy

$$\begin{aligned}u(x,y,0)&=u_0(x,y)\geq 0,\\ v(x,y,0)&=v_0(x,y)\geq 0,\\ w(x,y,0)&=w_0(x,y)\geq 0.\end{aligned}$$

In numerical simulations, we take $\lambda=0$ and λ' as a controlled parameter. If λ' is large, a smooth flame front is observed (see Figure 1 (a)). When λ' is decreased, the instability of a smooth flame front occurs. As λ' is decreased further, a fingering pattern is formed (see Figure 1 (b), (c)). Numerical simulations suggest that the model (RD) exhibits a qualitative agreement with the experimental results. This motives us to discuss analytically (RD) from pattern formation viewpoints. As the first step, we will show the existence and uniqueness of global solution of (RD) and to study the asymptotic behavior of the global solution.

This paper is organized as follows; In Section 2, we show the global existence and uniqueness of a solution of (RD) (Theorem 2.1). Furthermore

(a) (b) (c)

large $\longleftarrow$ λ' $\longrightarrow$ small

Figure 1. various patterns in (RD)

we have the upper bound of a solution of (RD) (Lemma 2.1). In Section 3, we consider the asymptotic behavior of a global solution given in Section 2 (Theorem 3.1). In Section 4, we obtain the nonexistence results of a traveling wave solution (Lemmas 4.1, 4.2). From experiments and simulations, we expect that there is a stable traveling wave solution if λ' is large. Then we would like to prove the existence of a traveling wave solution and in general, however, it is difficult. Hence it is necessary to obtain such conditions as there are no traveling wave solutions. We prove that there are no traveling wave solutions if a or λ is large. In addition, we also obtain the upper bound of the wave speed of a traveling wave solution in Lemma 4.2. In Section 5, we have the asymptotic behavior of solutions of (RD) as $\lambda' \to \infty$. If λ' is large, oxidizing gas supplied immediately spreads throughout the domain and a smooth front occurs, as described previously. We consider the limit of $\lambda' \to \infty$ and show that the three-components system (RD) can be reduced to other two-components system (Theorem 5.1).

2. Existence and uniqueness of a global solution

In this section, we prove the existence and uniqueness of a global solution. We replace w by z such as $w = z + \omega$, where $\omega = \omega(x)$ is a smooth positive function and satisfies $\omega(l_x) = w_r$ and $\omega'(0) = 0$ if $I = (0, l_x)$, or $\omega \to w_r$ as $x \to \infty$ and $\omega \to w_l$ as $x \to -\infty$ if $I = (-\infty, \infty)$. Then we consider the following system derived from (RD) with respect to (u, v, z);

$$\begin{cases} \dfrac{\partial u}{\partial t} = \mathrm{Le}\Delta u + \lambda \dfrac{\partial u}{\partial x} + \gamma k(u) v(z+\omega) - a u^m, & (x,y) \in I \times \Omega, t > 0, \\ \dfrac{\partial v}{\partial t} = - k(u) v(z+\omega), & (x,y) \in I \times \Omega, t > 0, \\ \dfrac{\partial z}{\partial t} = \Delta z + \lambda' \dfrac{\partial z}{\partial x} - k(u) v(z+\omega) + \omega'' + \lambda' \omega', & (x,y) \in I \times \Omega, t > 0. \end{cases} \tag{2.1}$$

The initial values u_0, v_0 and z_0 are

$$\begin{aligned} u(x,y,0) = u_0(x,y) \geq 0, \quad v(x,y,0) = v_0(x,y) \geq 0, \\ z(x,y,0) = w_0(x,y) - \omega(x) \equiv z_0(x,y) \end{aligned} \tag{2.2}$$

for $x \in I$ and $y \in \Omega$. We suppose that u satisfies

$$\begin{aligned} \frac{\partial u}{\partial x}(0,y,t) = \frac{\partial u}{\partial x}(l_x,y,t) = 0, \quad \text{if } I = (0, l_x), \\ \lim_{|x| \to \infty} u(x,y,t) = 0, \qquad \text{if } I = (-\infty, \infty) \end{aligned} \tag{2.3}$$

for $y \in \Omega$ and $t > 0$ and z does

$$\begin{aligned} \frac{\partial z}{\partial x}(0,y,t) = z(l_x,y,t) = 0, \quad \text{if } I = (0, l_x), \\ \lim_{|x| \to \infty} u(x,y,t) = 0, \qquad \text{if } I = (-\infty, \infty) \end{aligned} \tag{2.4}$$

for $y \in \Omega$ and $t > 0$. In addition we suppose that u, z satisfy

$$\frac{\partial u}{\partial \nu}(x,y,t) = 0, \quad \frac{\partial z}{\partial \nu}(x,y,t) = 0 \tag{2.5}$$

for $x \in I$, $y \in \partial\Omega$ and $t > 0$.

We easily prove the existence and uniqueness of a global solution of the above system. In the proof, we shall use the standard theory of an analytic semigroup and prove the existence of the following integral equation;

$$\Phi(t) = T(t)\Phi_0 + \int_0^t T(t-s) F(\Phi(s)) ds, \tag{2.6}$$

where $\Phi = (u, v, z)^t$, $\Phi_0 = (u_0, v_0, z_0)^t$, $T(t)$ is a semigroup generated by a differential operator A defined by

$$A = \begin{pmatrix} \mathrm{Le}\Delta + \lambda \dfrac{\partial}{\partial x} & 0 & 0 \\ 0 & 0 & 0 \\ 0 & 0 & \Delta + \lambda' \dfrac{\partial}{\partial x} \end{pmatrix}$$

and

$$F(\Phi) = \begin{pmatrix} \gamma k(u)v(\omega + z) - au^m \\ -k(u)v(\omega + z) \\ -k(u)v(\omega + z) + \omega'' + \lambda'\omega' \end{pmatrix}.$$

We consider the integral equation (2.6) in the functional space X defined by

$$X = L^p(I \times \Omega) \times L^\infty(I \times \Omega) \times L^p(I \times \Omega)$$

for $p > n+1$. And the domain of A, denoted by $D(A)$, is defined by

$$D(A) = W_N^{2,p}(I \times \Omega) \times L^\infty(I \times \Omega) \times W_{N,0}^{2,p}(I \times \Omega),$$

where $W_N^{2,p}(I \times \Omega)$ is defined by

$$W_N^{2,p}(I \times \Omega) = \{u \in W^{2,p}(I \times \Omega) \mid \frac{\partial u}{\partial \nu} = 0 \text{ for } x \in I,\ y \in \partial\Omega \text{ and } u \text{ satisfies (2.3)}\}$$

and $W_{N,0}^{2,p}(I \times \Omega)$ is defined by

$$W_{N,0}^{2,p}(I \times \Omega) = \{z \in W^{2,p}(I \times \Omega) \mid \frac{\partial z}{\partial \nu} = 0 \text{ for } x \in I,\ y \in \partial\Omega \text{ and } z \text{ satisfies (2.4)}\}.$$

The functional space $W^{2,p}(I \times \Omega)$ is a usual Sobolev space.

We assume that $u_0 \in D(L_u^\alpha)$, $v_0 \in C^\kappa(I \times \Omega)$ and $z_0 \in D(L_z^\alpha)$ for $1/2 < \alpha < 1$ and $0 < \kappa < 1$. The functional spaces $D(L_u^\alpha)$ and $D(L_z^\alpha)$ are called *fractional spaces* where $L_u = -\mathrm{Le}\Delta - \lambda'\partial/\partial x$ and $L_z = -\Delta - \lambda\partial/\partial x$ (see Section 2.6 of [3]). And $C^\kappa(I \times \Omega)$ is the Hölder space with a Hölder exponent κ. Then we have the following theorem for existence of a global solution.

Theorem 2.1. *Assume that $p > n+1$, $1/2 < \alpha < 1$, $0 < \kappa < 1$, and $\partial\Omega \in C^2$. In addition, suppose that the function ω has the second order continuous derivatives in $x \in I$ and they belong to $L^p(I \times \Omega) \cap C^\kappa(I \times \Omega)$. Then, for any $(u_0, v_0, z_0) \in D(L_u^\alpha) \times C^\kappa(I \times \Omega) \times D(L_z^\alpha)$, there exists a unique global classical solution (u, v, z) of (2.1), (2.2), (2.3), (2.4) and (2.5).*

To prove that the solution exists globally, we need to obtain a priori estimate.

Lemma 2.1. *Let (u, v, z) be a solution given in Theorem 2.1 and set $w = z + \omega$. Then there exists a constant $R > 0$, depending on initial values u_0, v_0 and w_0, such that for any $(x, y) \in I \times \Omega$, $t > 0$,*

$$0 \le u \le R, \quad 0 \le v \le R, \quad 0 \le w \le R.$$

3. Asymptotic behavior of u, v and w

In this section we consider the asymptotic behavior of classical solutions of (RD).

Theorem 3.1. *Set $I = (0, l_x)$ and let (u, v, z) be a solution given in Theorem 2.1 and $w = z + \omega$. Then u, v and w have the following asymptotic behavior (i), (ii), (iii) as $t \to \infty$:*

(i) For any $(x, y) \in I \times \Omega$, $\lim_{t\to\infty} u(x, y, t) = 0$.
(ii) There exists $v_\infty(x, y) \in L^\infty(I \times \Omega)$ such that $\lim_{t\to\infty} v(x, y, t) = v_\infty(x, y)$ for $(x, y) \in I \times \Omega$ and the function v_∞ has a positive value at any points $(x, y) \in I \times \Omega$ where $v_0(x, y) > 0$.
(iii) For any $(x, y) \in I \times \Omega$, $\lim_{t\to\infty} w(x, y, t) = w_r$.

We need lemmas to prove Theorem 3.1. As the first step of the proof of Theorem 3.1, we consider the reaction term $k(u)vw$ and prove that it approaches 0 as $t \to \infty$.

Lemma 3.1. *Let (u, v, z) be a solution given in Theorem 2.1 and set $w = z + \omega$. Then it holds that $k(u)vw \to 0$ as $t \to \infty$ at any $(x, y) \in I \times \Omega$.*

Proof. It is easy to see that there exists $v_\infty(x, y)$ such that $v(x, y) \to v_\infty(x, y)$ as $t \to \infty$ because v decreases monotonically. Hence we can show that $v_t \to 0$ as $t \to \infty$. at any $(x, y) \in I \times \Omega$. From the second equation of (RD), we complete the proof. □

The previous lemma implies that u, v or w approach 0. In fact, the function u must tend to 0 as $t \to \infty$.

Lemma 3.2. *Let (u, v, z) be a solution given in Theorem 2.1. Set $I = (0, l_x)$ and $w = z + \omega$. Then it holds that $u \to 0$ as $t \to \infty$ at any $(x, y) \in I \times \Omega$.*

Proof. From Lemma 3.1, there exists $T > 0$ such that $\gamma k(u)vw \le au^m/2$ for any $t > T$ and $(x, y) \in I \times \Omega$. Then u satisfies

$$u_t \le \mathrm{Le}\Delta u + \lambda \frac{\partial u}{\partial x} - \frac{a}{2} u^m$$

for any $t > T$ and $(x, y) \in I \times \Omega$. Now we use a constant R given in Lemma 2.1 and define $q = q(t)$ by a solution of

$$q' = -\frac{a}{2} u^m, \quad t > T, \qquad q = R, \quad t = T. \tag{3.1}$$

Using $u \leq R$ and applying the comparison principle to u and q, we have $u \leq q$ for $t > T$. It is easy to see that $q \to 0$ as $t \to \infty$. Hence u approaches 0, which completes the proof. □

Now we are in position to prove the asymptotic behaviors (ii), (iii) of Theorem 3.1.

Proof. Let $q = q(t)$ be a function given in the proof of Lemma 3.2. As stated previously, we have $0 \leq u \leq q$ for any $t > T$ and $(x, y) \in I \times \Omega$. Let M be a constant defined by $M \equiv \sup_{u>0} k(u)/u^m$. Then it follows from $w \leq R$, as stated in Lemma 2.1, and the second equation of (RD) that

$$\begin{aligned} v(x,y,t) &\geq v(x,y,T)\exp(-MR\int_T^t u^m ds) \\ &= v(x,y,T)\exp(-\frac{2MR}{a}(q(T)-q(t))) \\ &\geq v(x,y,T)\exp(-\frac{2MR}{a}q(T)). \end{aligned} \tag{3.2}$$

Here we have an estimate of $v(x, y, T)$ such as

$$v(x,y,T) = v_0(x,y)\exp(-\int_0^T k(u)wds) \geq v_0(x,y)\exp(-RT), \tag{3.3}$$

because of $w \leq R$ and $k(u) < 1$. Therefore it follows from (3.2) and (3.3) that $v_\infty(x, y) > 0$ if $v_0(x, y) > 0$.

Next we prove (iii) of our theorem. In the third equation of (RD), we set $\phi = w - w_r$. Then ϕ satisfies the equation

$$\frac{\partial \phi}{\partial t} = \Delta\phi + \lambda' \frac{\partial \phi}{\partial x} - k(u)vw, \tag{3.4}$$

and the boundary conditions (2.4) and $\partial\phi/\partial\nu = 0$ for $x \in I, y \in \partial\Omega$ and $t > 0$. Using the analytic semigroup $T_z(t)$ generated by $L_z = \Delta + \lambda'\partial/\partial x$, we rewrite (3.4) to the following integral equation

$$\phi(t) = T_z(t-\tilde{T})\phi(\tilde{T}) - \int_{\tilde{T}}^t T_z(t-s)k(u(s))v(s)w(s)ds$$

for $t > \tilde{T}$, where $\tilde{T} > T$ is sufficiently large and we omit the space variable of functions for the convenience of readers. Here T was taken in the proof of Lemma 3.2. In fact we can show that there exists a constant $c_0 > 0$ such that

$$\|T_z(t)\psi\|_{L^\infty(I\times\Omega)} \leq c_0 \min\{1, \frac{1}{\sqrt{t}}\}\|\psi\|_{L^\infty(I\times\Omega)}$$

for any $\psi \in L^\infty(I \times \Omega)$. Let $\Phi = \Phi(t)$ be a fundamental solution of

$$\frac{\partial \Phi}{\partial t} = \frac{\partial^2 \Phi}{\partial x^2} + \lambda' \frac{\partial \Phi}{\partial x}, \quad x \in (-\infty, \infty), t > 0,$$

which has a singularity at $(x,t) = (0,0)$. Then we can show that $\Phi * \|\psi\|_{L^\infty(I\times\Omega)}$ is a super-solution of $T_z(t)\psi$, where $\Phi * \|\psi\|_{L^\infty(I\times\Omega)}$ is a convolution of Φ and $\|\psi\|_{L^\infty(I\times\Omega)}$ as usual. Similarly, $-\Phi * \|\psi\|_{L^\infty(I\times\Omega)}$ is a sub-solution of $T_z(t)\psi$. Applying usual $L^\infty - L^1$ and $L^\infty - L^\infty$ estimates to $\Phi * \|\psi\|_{L^\infty(I\times\Omega)}$, we obtain

$$\|T_z(t)\psi\|_{L^\infty(I\times\Omega)} \le \|\Phi * 1\|_{L^\infty(I\times\Omega)}\|\psi\|_{L^\infty(I\times\Omega)} \le c_0 \min\{1, \frac{1}{\sqrt{t}}\}\|\psi\|_{L^\infty(I\times\Omega)}.$$

Hence it follows that

$$\begin{aligned}
\|\phi(t)\|_{L^\infty(I\times\Omega)} &\le \frac{c_0}{\sqrt{t-\tilde{T}}}\|\phi(\tilde{T})\|_{L^\infty(I\times\Omega)} + c_0 R^2 \int_{\tilde{T}}^t \|k(u(s))\|_{L^\infty(I\times\Omega)} ds \\
&\le \frac{c_0}{\sqrt{t-\tilde{T}}}\|\phi(\tilde{T})\|_{L^\infty(I\times\Omega)} + c \int_{\tilde{T}}^t a(q(s))^m ds \\
&= \frac{c_0}{\sqrt{t-\tilde{T}}}\|\phi(\tilde{T})\|_{L^\infty(I\times\Omega)} + 2c(q(\tilde{T}) - q(t))
\end{aligned}$$

for a constant $c > 0$, from which we have

$$\limsup_{t\to\infty} \|\phi(t)\|_{L^\infty(I\times\Omega)} \le 2cq(\tilde{T}).$$

Since $\tilde{T}$ is any large constant and $q \to 0$ as $t \to \infty$, $q(\tilde{T})$ tends to 0 as $\tilde{T}$ goes to ∞. Therefore we have $\|\phi\|_{L^\infty(I\times\Omega)} = \|w - w_r\|_{L^\infty(I\times\Omega)} \to 0$ as $t \to \infty$. □

4. Nonexistence of a traveling wave solution

From the results of experiments and simulations, we expect that there exists a traveling wave solution independent of y-direction and moving opposite to the flow of the oxidizing gas, which is a positive solution of (RD) and can be written such as $u(x,t) = U(x-ct)$, $v(x,t) = V(x-ct)$ and $w(x,t) = W(x-ct)$, where $c > 0$ is a wave speed. Here we set $z = x - ct$. Then our equations satisfied by (U, V, W) have the forms

$$\begin{cases} -cU' = \mathrm{Le}U'' + \lambda U' + \gamma k(U)VW - aU^m, \\ -cV' = -k(U)VW, \\ -cW' = W'' + \lambda' W' - k(U)VW, \end{cases} \tag{4.1}$$

where the prime $'$ is d/dz, and boundary conditions are

$$U(\pm\infty) = 0, \qquad V(+\infty) = v_r > 0,$$
$$W(+\infty) = w_r > 0,\ W(-\infty) = w_l(< w_r).$$

In this paper, we consider the conditions of the parameters as there does not exist a traveling wave solution, although we would like to prove the existence of a traveling wave solution (U, V, W, c). Recalling the constant M, defined in the previous section.

Lemma 4.1. *If $a \geq \gamma M v_r w_r$, there exists no traveling wave solution.*

Proof. We assume that there exists a traveling wave solution (U, V, W, c) for $a \geq \gamma M v_r w_r$. It is easy to see that $V \leq v_r$ and $W \leq w_r$. From the boundary condition, the function U must have the maximum at some point. Then we have $U' = 0$ there and from the first equation of (4.1)

$$\mathrm{Le}U'' = aU^m - \gamma k(U)VW \geq U^m(a - \gamma M v_r w_r) \geq 0,$$

which is a contradiction. □

Next we show that there does not exist traveling wave solutions if λ is large.

Lemma 4.2. *Set $m = 1$. Then there exists a constant c^* independent of Le, λ such that (4.1) possesses no traveling wave solutions with the wave speed c satisfying $c > \sqrt{Le}c^* - \lambda$.*

From this lemma, we can state that the wave speed of traveling wave solutions must satisfy $c \leq \sqrt{\mathrm{Le}}c^* - \lambda$ if exists. This is the upper bound of the wave speed.

Proof. From Lemma 4.1, we can assume $a < \gamma M v_r w_r$ without loss of generality. Here we set $f(U) = \gamma k(U) v_r w_r - aU$. Then $f(U)$ has zeros at $U = 0, U_1, U_2$ for $0 < U_1 < U_2$. Moreover it holds that $f(U) < 0$ for $0 < U < U_1$ and $f(U) > 0$ for $U_1 < U < U_2$, which implies that $f(U)$ is a nonlinear term of a bistable type. Therefore there exists a traveling wave solution $Q = Q(\tilde{z})$ with the wave speed c^* independent of Le, λ such that

$$-\sqrt{\mathrm{Le}}c^* Q' = \mathrm{Le}Q'' + f(Q), \quad Q(-\infty) = U_2, \quad Q(+\infty) = 0,$$

where $\tilde{z} = x - (\sqrt{\mathrm{Le}}c^* - \lambda)t$ and the prime $' = d/d\tilde{z}$ (see Theorem 4.2 of [1]). Setting $q(x,t) = Q(x - (\sqrt{\mathrm{Le}}c^* - \lambda)t)$, we know that q satisfies

$$\frac{\partial q}{\partial t} = \mathrm{Le}\frac{\partial^2 q}{\partial x^2} + \lambda\frac{\partial q}{\partial x} + f(q).$$

Now we assume that there exists a traveling wave solution (U, V, W) with the wave speed $c > \sqrt{\mathrm{Le}c^*} - \lambda$. We have obtained $V \leq v_r$ and $W \leq w_r$ in Lemma 4.1. Then $u(x,t) = U(x - ct)$ satisfies

$$\frac{\partial u}{\partial t} \leq \mathrm{Le}\frac{\partial^2 u}{\partial x^2} + \lambda\frac{\partial u}{\partial x} + f(u)$$

and $u < U_2$ for $-\infty < x < \infty$. Hence for some $h \in \mathbb{R}$, we have $u(x,t) < q(x - h, t)$ for any $-\infty < x < \infty$ and $t > 0$, that is, $U(x - ct) < Q(x - h - (\sqrt{\mathrm{Le}c^*} - \lambda)t)$. However, since $c > \sqrt{\mathrm{Le}c^*} - \lambda$, U must reach Q at some point in a finite time, which is a contradict. □

5. Behavior of solutions for large λ'

If λ' is sufficiently large, we expect that flame uniformly spreads like the figure (a) of Figure 1. Now we formally set $\lambda' \to \infty$ in (RD) and have $\partial w/\partial x = 0$. Then, from the boundary condition of w, we obtain $w \equiv w_r$ and (RD) is reduced to the new system (SS) as follows;

$$\text{(SS)}\begin{cases} \dfrac{\partial u}{\partial t} = \mathrm{Le}\Delta u + \lambda\dfrac{\partial u}{\partial x} + \gamma k(u)vw_r - au^m, & (x,y) \in I \times \Omega, t > 0, \\ \dfrac{\partial v}{\partial t} = -k(u)vw_r, & (x,y) \in I \times \Omega, t > 0. \end{cases}$$

We expect that the solution of (RD) is close to that of (SS) if λ' is sufficiently large. In the following, we assume that the initial value of (RD), denoted by $(u_0^{\lambda'}, v_0^{\lambda'})$, depends on the parameter λ' and is close to that of (SS), denoted by (u_0, v_0), as follows:

$$\|u_0^{\lambda'} - u_0\|_\alpha \to 0, \quad \|v_0^{\lambda'} - v_0\|_{L^\infty(I\times\Omega)} \to 0, \tag{5.1}$$

where $\|\cdot\|_\alpha \equiv \|\cdot\|_{L^p(I\times\Omega)} + \|L_u^\alpha \cdot\|_{L^p(I\times\Omega)}$ for $p > n+1$. On the other hand, we do not need to assume that the initial value of w of (RD), denoted by $w_0^{\lambda'}$, is close to w_r.

Theorem 5.1. *Let $(u^{\lambda'}, v^{\lambda'}, z^{\lambda'})$ be a solution of (RD) given in Theorem 2.1 with an initial value $(u_0^{\lambda'}, v_0^{\lambda'}, z_0^{\lambda'}) \in D(L_u^\alpha) \times C^\kappa(I \times \Omega) \times D(L_z^\alpha)$ for $1/2 < \alpha < 1$ and $0 < \kappa < 1$ which depends on the parameter λ', and set $w^{\lambda'} = z^{\lambda'} + \omega$. Let (u, v) be a solution of (SS) with an initial value $(u_0, v_0) \in D(L_u^\alpha) \times C^\kappa(I \times \Omega)$. Assume that $(u_0^{\lambda'}, v_0^{\lambda'})$ and (u_0, v_0) satisfy*

(5.1). Then it holds that

$$\sup_{0\le t\le T}\|u^{\lambda'}(\cdot,\cdot,t)-u(\cdot,\cdot,t)\|_\alpha\to 0,$$
$$\sup_{0\le t\le T}\|v^{\lambda'}(\cdot,\cdot,t)-v(\cdot,\cdot,t)\|_{L^\infty(I\times\Omega)}\to 0,$$
$$\sup_{\delta\le t\le T}\|w^{\lambda'}(\cdot,\cdot,t)-w_r\|_{L^\infty(I\cap(-K,\infty)\times\Omega)}\to 0$$

as $\lambda'\to\infty$ for any $\delta, T, K>0$.

In the following, we prove the only case of $I=(0,l_x)$. Setting $\phi = w-w_r$, ϕ is a solution of

$$\begin{cases} \dfrac{\partial\phi}{\partial t}=\Delta\phi-k(u)vw+\lambda'\dfrac{\partial\phi}{\partial x}, & (x,y)\in I\times\Omega, t>0,\\ \phi(x,y,0)=w_0(x,y)-w_r, & (x,y)\in I\times\Omega,\\ \dfrac{\partial\phi}{\partial x}(0,y,t)=0,\quad \phi(l_x,y,t)=0, & y\in\Omega, t>0,\\ \dfrac{\partial\phi}{\partial\nu}(x,y,t)=0, & (x,y)\in I\times\partial\Omega, t>0, \end{cases}\tag{5.2}$$

where we abbreviate $u^{\lambda'}, v^{\lambda'}$ and $w^{\lambda'}$ as u, v and w for simplicity. At first we study the property of the semigroup $T_z(T)$, generated by $L_z=\Delta+\lambda'\partial/\partial x$.

Lemma 5.1. *For any $\psi\in L^\infty(I\times\Omega)$ and $t\ge 0$,*

$$\|T_z(t)\psi\|_{L^\infty(I\times\Omega)}\le\|\psi\|_{L^\infty(I\times\Omega)}.$$

Moreover, for any $\psi\in L^\infty(I\times\Omega)$ and $t\ge 4l_x/\lambda'$,

$$\|T_z(t)\psi\|_{L^\infty(I\times\Omega)}\le\exp(-\frac{(\lambda')^2}{8}t)\|\psi\|_{L^\infty(I\times\Omega)}.$$

Proof. The first statement has already been proved in Section 3.

Let $\tilde\psi=\tilde\psi(x,t)$ denoted by

$$\tilde\psi(x,t)=\exp(-\frac{(\lambda')^2}{4}t+\frac{\lambda'}{2}(l_x-x))\|\psi\|_{L^\infty(I\times\Omega)}.$$

Then $\tilde\psi$ satisfies

$$\begin{cases} \dfrac{\partial\tilde\psi}{\partial t}=\dfrac{\partial^2\tilde\psi}{\partial x^2}+\lambda'\dfrac{\partial\tilde\psi}{\partial x}, & x\in I, t>0,\\ \tilde\psi(x,0)\ge\|\psi\|_{L^\infty(I\times\Omega)}, & x\in I,\\ \dfrac{\partial\tilde\psi}{\partial x}(0,t)\le 0,\quad \tilde\psi(l_x,t)>0, t>0, \end{cases}$$

from which we have $\|T_z(t)\psi\|_{L^\infty(I\times\Omega)} \leq \|\tilde{\psi}\|_{L^\infty(I)}$. Hence we have

$$\begin{aligned}\|T_z(t)\psi\|_{L^\infty(I\times\Omega)} &\leq \exp(-\frac{(\lambda')^2}{4}t + \frac{\lambda'}{2}l_x)\|\psi\|_{L^\infty(I\times\Omega)} \\ &\leq \exp(-\frac{(\lambda')^2}{8}t)\|\psi\|_{L^\infty(I\times\Omega)}\end{aligned}$$

for any $\psi \in L^\infty(I \times \Omega)$ and $t \geq 4l_x/\lambda'$. □

We rewrite (5.2) to an integral equation

$$\phi(t) = T_z(t)(w_0 - w_r) - \int_0^t T_z(t-s)k(u)vwds.$$

From Lemmas 2.1 and 5.1, we have

$$\begin{aligned}\|\phi\|_{L^\infty(I\times\Omega)} \leq& \exp(-\frac{(\lambda')^2}{8}t)\|w_r - w_0\|_{L^\infty(I\times\Omega)} \\ &+ \int_0^{t-4l_x/\lambda'} \exp(-\frac{(\lambda')^2}{8}(t-s))\|k(u)vw\|_{L^\infty(I\times\Omega)}ds \\ &+ \int_{t-4l_x/\lambda'}^t \|k(u)vw\|_{L^\infty(I\times\Omega)}ds \to 0\end{aligned}$$

as $\lambda' \to \infty$. On the other hand, we readily show that (u, v) also satisfies the property of Theorem 5.1 because of the assumption for the initial value and the standard argument with Gronwall's inequality, which completes the proof.

Acknowledgement. Special thanks go to Professor Eiji Yanagida for many stimulating discussions and continuous encouragement.

References

[1] D.G.Aronson and H.F.Weinberger, *Nonlinear diffusion in population genetics, combustion, and nerve pulse propagation*, Partial differential equations and related topics, Lecture Notes in Math. **446** Springer (1975) 5–49.

[2] S.L.Olson, H.R.Baum and T.Kashiwagi, *Finger-like smoldering over thin cellulosic sheets in microgravity*, The Combustion Institute (1998) 2525–2533.

[3] A.Pazy, *Semigroups of linear operators and applications to partial differential equations*, Springer-Verlag New York **44** (1983).

[4] O.Zik, Olami and E.Moses, *Fingering Instability in Combustion*, Phys. Rev. Lett. **81** 3836 (1998).

MOTION OF NON-CONVEX POLYGON BY CRYSTALLINE CURVATURE FLOW AND ITS GENERALIZATION

TETSUYA ISHIWATA

Gifu university, Yanagido 1-1, Gifu, 501-1193, JAPAN

Abstract: The motion of polygons by crystalline curvature is discussed. We also mention the "convexity phenomena" and show examples of non-existence of non-convex self-similar solutions.

1. Introduction

In this short paper we discuss a motion of polygons $\Omega(t)$ in the plane by a crystalline curvature flow:

$$V_j = H_j \tag{CCF}$$

and its generalization:

$$V_j = g(\theta_j, H_j) \tag{GCCF}$$

where V_j, $\theta_j \in S^1 = \mathbb{R}/2\pi\mathbb{Z}$ and H_j denote the normal velocity, the normal angle and the crystalline curvature of the j-th edge of the solution polygon, respectively. Here the function g is a continuous function from $S^1 \times \mathbb{R}$ to $\mathbb{R}$ and solution polygons are considered in the special class which is characterized by an interfacial energy density. The polygon belonging to this class is called "admissible polygon." The definitions and assumptions are mentioned in the next section. This framework of the interface motion is introduced by S. Angenent and M.E. Gurtin [3] and J.E. Taylor [14], independently (history: see [1]).

If the initial polygon is convex, solution polygon keeps its convexity and any edges never disappear as long as an enclosed area of solution polygon is positive. The asymptotic behavior of solution polygons, especially extinction rate and limiting shape, is well studied (see [13, 2, 11, 12] and the references in them). However, there are few results for non-convex case. K. Ishii and H. M. Soner [8] discussed a motion of admissible polygon by $V_j = H_j$. M.-H. Giga and Y. Giga [5] treat more general case and show that a degenerate pinching singularity never occur under an assumption on the function g. And they also mentioned "convexity phenomena" : Nonconvex curve becomes convex in a finite time before the extinction time. Figure 1 shows a numerical example of this phenomena. We can see that solution polygon becomes convex in a finite time and then shrinks to a single point. The convexity phenomena are well known for a classical curvature flow $v = \kappa$ [6] and its anisotropic version [4]. But this assertion is not valid for (GCCF) since there exist non-convex self-similar solutions for some cases [10]. At present, conditions for the convexity phenomena and behaviors of solution polygons are not clear.

In this paper, we assume some conditions on the function g and the interfacial energy density and show that (i) the solution polygon belongs to the admissible class as long as the solution exists, (ii) a crystalline curvature of each edges becomes non-negative before the extinction time and (iii) the solution polygon also becomes star-shaped before the extinction time. In the final section, we discuss some examples of a non-existence of non-convex self-similar solutions for (CCF).

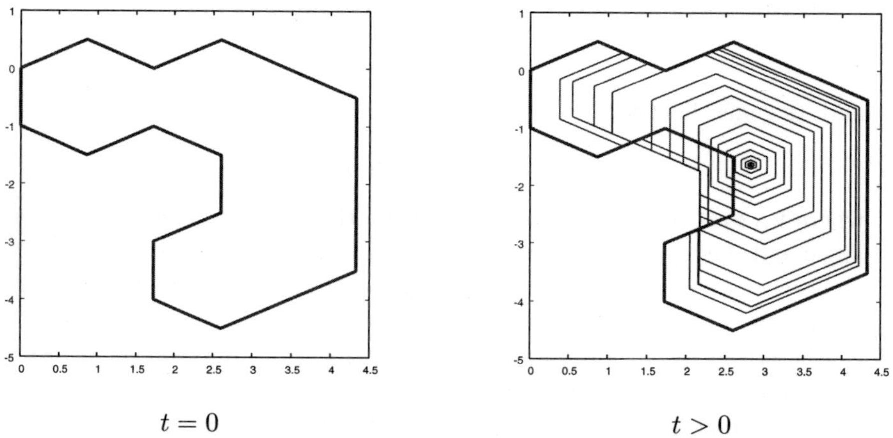

Figure 1. Numerical example of the convexity phenomena for $V_j = H_j$. The left is $\Omega(0)$ and the right is a time evolution of $\Omega(t)$.

2. Motion of admissible polygons by crystalline curvature

We first mention the Wulff shape which plays important role in definitions of an admissibility of a polygon and a crystalline curvature. Let $\sigma = \sigma(\theta)$

be a positive continuous function defined on S^1, which describes an interfacial energy density in the direction θ of a crystal. It is well-known that the equilibrium shape of crystal is given as the shape which attains the minimum of the total interfacial energy on some curve Γ , $\int_\Gamma f(\boldsymbol{n})ds$ (Here $f(\boldsymbol{n}) = \sigma(\theta), \boldsymbol{n} = -(\cos\theta, \sin\theta)$). This minimizing problem is called the Wulff problem and the solution figure is called the Wulff shape, which is given by

$$\mathcal{W}_\sigma = \bigcap_{\theta\in S^1} \{(x,y) \in \mathbb{R}^2 \,|\, x\cos\theta + y\sin\theta \le \sigma(\theta)\}.$$

Note that $\mathcal{W}_\sigma$ is always convex. For example, if $\sigma(\theta) \equiv 1$, then the Wulff shape is a unit circle. If $\mathcal{W}_\sigma$ is a polygon, then the energy σ is called "crystalline energy." In this paper, we only treat this case.

Let $\mathcal{W}_\sigma$ and φ_i be N_σ-sided polygon and normal angle of i-th edge of $\mathcal{W}_\sigma$ for $i = 0, 1, \cdot, N_\sigma - 1$, respectively. We also define a set of normal angles by

$$\Theta_{\mathcal{W}_\sigma} = \{\varphi_0, \varphi_1, \cdots, \varphi_{N_\sigma - 1}\}$$

Here, there is a following order : $\varphi_0 < \varphi_1 < \cdots < \varphi_{N_\sigma-1} < \varphi_{N_\sigma} := \varphi_0 + 2\pi$.

We next introduce a notion of admissible polygon. Let Ω and $\mathcal{P}$ be an N-sided polygon and its boundary. We denote j-th vertex of Ω, the j-th edge and a normal angle of the j-th edge for $j = 0, 1, \dots, N-1$ by $p_j = (x_j, y_j)$, $\mathcal{S}_j$ and θ_j in anticlockwise order, respectively. We also define $p_N (= (x_N, y_N)) := p_0$ and $\theta_N := \theta_0$. Then the j-th edge and the boundary of Ω are described by $\mathcal{S}_j := \{(1-t)(x_j, y_j) + t(x_{j+1}, y_{j+1}) \,|\, 0 \le t \le 1\}$ and $\mathcal{P} = \bigcup_{j=0}^{N-1} \mathcal{S}_j$, respectively. Let Θ_Ω $\boldsymbol{N}_j$ and $\boldsymbol{T}_j$ be a set of normal angles: $\Theta_\Omega = \{\theta_0, \theta_1, \cdots, \theta_{N-1}\}$, an inward and a tangential normal vectors of $\mathcal{S}_j$, which are given by $\boldsymbol{N}_j = -(\cos\theta_j, \sin\theta_j)$ and $\boldsymbol{T}_j = (-\sin\theta_j, \cos\theta_j)$, respectively.

If Ω satisfies the following two conditions, we call Ω an admissible polygon: (1) $\Theta_{\mathcal{W}_\sigma} = \Theta_\Omega$ and (2) $s\theta_j + (1-s)\theta_{j+1} \notin \Theta_{\mathcal{W}_\sigma}$ for any $0 < s < 1$ and $j = 0, 1, \cdots, N-1$. The admissibility is the analogical notion of smoothness of a closed curve: Ω has all normal angles in $\Theta_{\mathcal{W}_\sigma}$ and if $\theta_j = \varphi_k \in \Theta_{\mathcal{W}_\sigma}$, then $\theta_{j+1}, \theta_{j-1} \in \{\varphi_{k-1}, \varphi_{k+1}\}$, that is, normal angle θ_j change "smoothly" in $\Theta_{\mathcal{W}_\sigma}$.

Here and hereafter, we only consider admissible polygons. The quantity defined for admissible polygon Ω by

$$H_j = \chi_j \frac{l_\sigma(\theta_j)}{d_j}$$

is called crystalline curvature. Here d_j and $l_\sigma(\theta_i)$ be the length of $\mathcal{S}_j$ and the length of the edge of the Wulff shape having normal angle θ_j, respectively. The quantity χ_j is called transition number which is given by

$$\chi_j = \frac{1}{2}(\mathrm{sgn}(\det(\boldsymbol{N}_{j-1}, \boldsymbol{N}_j)) + \mathrm{sgn}(\det(\boldsymbol{N}_j, \boldsymbol{N}_{j+1}))).$$

This takes $+1$ (resp. -1) if $\mathcal{P}$ is convex (resp. concave) at $\mathcal{S}_j$ in the N_j direction; otherwise it takes zero. Note that if Ω is convex, then $\chi_j = +1$ for all j. Let $N^0(\Omega) := \#\{j|\chi_j = 0\}, N^+(\Omega) := \#\{j|\chi_j = 1\}$ and $N^-(\Omega) := \#\{j|\chi_j = -1\}$. Then we have $N = N^0(\Omega) + N^+(\Omega) + N^-(\Omega)$ and $N^+(\Omega) - N^-(\Omega) = N_\sigma$. Note that a crystalline curvature of each edge of the Wulff shape is always one. This means that the Wulff shape plays a role like as the unit circle in the usual sense.

In this paper we consider (GCCF) as the generalization of (CCF) and assume the following conditions on the function g:

Assumption (G1): $\lambda \mapsto g(\theta_j, \lambda)$ is Lipschitz continuous on $\mathbb{R} \setminus \{0\}$ for all $\theta_j \in \Theta_{\mathcal{W}_\sigma}$.

Assumption (G2): $g(\theta_j, \lambda)$ is monotone non-decreasing on λ and satisfies $g(\theta_j, 0) = 0$, $g(\theta_j, \lambda) > 0$ $(\lambda > 0)$, $g(\theta_j, -\lambda) = -g(\theta_j, \lambda)$ and $\lim_{\lambda\to\infty} g(\theta_j, \lambda) = +\infty$ for all $\theta_j \in \Theta_{\mathcal{W}_\sigma}$.

Let the initial polygon $\Omega(0)$ be N-sided admissible polygon. Then, time-local solution polygon exists uniquely in the class of N-sided admissible polygons and we can obtain the time-local solution as long as $\Omega(t)$ is N-sided admissible polygon because (GCCF) can be written as the following:

$$\dot{d}_j(t) = (\cot(\theta_{j+1} - \theta_j) + \cot(\theta_j - \theta_{j-1}))V_j - \frac{1}{\sin(\theta_j - \theta_{j-1})}V_{j-1}$$
$$-\frac{1}{\sin(\theta_{j+1} - \theta_j)}V_{j+1} \qquad \text{for } j = 0, 1, \ldots, N-1. \tag{2.1}$$

However, by assumption (G2), we can easily show that at least one edge disappear in a finite time. Therefore, the solution polygon does not exist globally in time in the class of N-sided admissible polygons. But the edge-disappearing does not always lead a breakdown of the time evolution of solution polygon. If the limiting shape which is obtained after some edges disappear is still admissible, we can continue a time evolution of solution polygon in the class of admissible polygons beyond the edge-disappearing. In this case, the number of edges of solution polygon becomes smaller. That is, the system size of the problem (2.1) may change. But, there is a possibility that the limiting shape is not admissible and thus we can no longer continue the time evolution of polygons in the class of admissible polygon.

On the other hand, self-contacting of the boundary $\mathcal{P}(t) = \partial\Omega(t)$ may occur before the edge-disappearing. Indeed, some examples and numerical simulations of split-type self-contacting are shown in [7]. When the solution polygon splits into some polygons, if each polygons are still admissible, we can continue to track the motions of each polygons. However, some results in [7] show that we can not use this procedure in general.

In the next section, we discuss the above two types of singularities and clarify the conditions which guarantee to track the motion of solution polygon in the class of admissible polygon till the area of solution polygon becomes zero.

3. Beyond the singularities

We add the following three assumptions:

(W1): $\sigma(\theta+\pi) = \sigma(\theta)$ for any $\theta \in S^1$.
(G3): $g(\theta_j+\pi,\lambda) = g(\theta_j,\lambda)$ for any $\theta_j \in \Theta_{\mathcal{W}_\sigma}$ and $\lambda \in \mathbb{R}$.
(G4): For any $\theta_j \in \Theta_{\mathcal{W}_\sigma}$, the function $g(\theta,\lambda)$ satisfies

$$\int_{\lambda_0}^{\infty} g(\theta_j,\lambda)\lambda^{-2}d\lambda = \infty.$$

The assumption (W1) means that the Wulff shape $\mathcal{W}_\sigma$ is point symmetric. For example, regular hexagon and octagon satisfy (W1). The condition (G3) means a symmetry of mobility. For an anisotropic crystalline curvature flow, i.e., $g(\theta,\lambda) = M(\theta)\lambda$ ($M(\theta)$ is a positive function called "the mobility."), if $M(\theta+\pi) = M(\theta)$, then (G3) holds. The assumption (G4) seen in [5] and means that the growth rate of $g(\cdot,\lambda)$ as $\lambda \to \pm\infty$ is linear or superlinear. If (G4) fails, the degenerate pinching may occur (see also [2, 7]).

Under all assumptions, we obtain the following result:

Theorem 3.1. *([9]) Let $\Omega(0)$ be a N-sided admissible polygon and assume that (G1,2,3,4) and (W1) hold. Then, there exists a finite time $T_1 > 0$ such that $\Omega(t)$ is N-sided admissible polygon for $0 \le t < T_1$ and there exists k such that $\lim_{t\to T_1} d_k(t) = 0$.*

Moreover, one of the following two phenomena occurs exclusively at $t = T_1$:

(1) $\Omega(t)$ shrinks to a single point, that is, $\lim_{t\to T_1} d_j(t) = 0$ for all j.

$t = 0$ $t = T$

Figure 2. Example motion when (W1) does not hold: The two center above edges have zero crystalline curvature and thus never move in each normal direction. The bottom edge moves upward and eventually touches the above vertex. The solution polygon splits into two particles and each polygon is not admissible.

$t = 0$ $t = T$

Figure 3. Example motion when (G3) does not hold: If $g(\pi/2, \cdot) \ll g(3\pi/2, \cdot)$, the bottom edge moves upward faster than the above edge and eventually the both edges touch each other. The solution polygon splits into two particles and each polygon is not admissible.

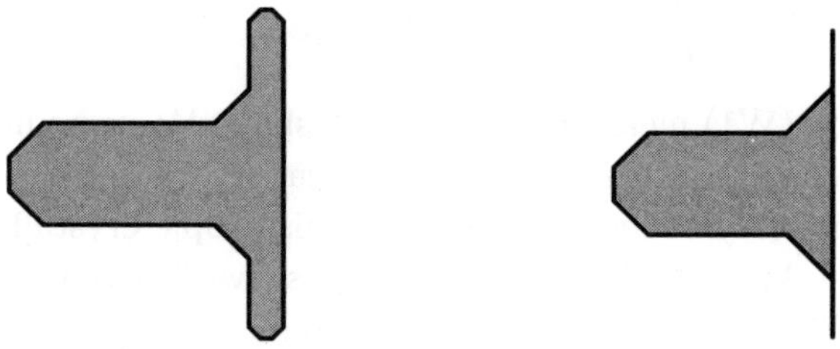

$t = 0$ $t = T$

Figure 4. Example motion when (G4) does not hold: Because of the lack of (G4), the degenerate pinching occurs. The limiting shape is not admissible.

(2) There exists j_0 such that $\inf_{0 \le t < T_1} d_{j_0}(t) > 0$. Then, the transition number of the edge which disappears at $t = T_1$ is zero and the limiting shape $\Omega^* = \lim_{t \to T_1} \Omega(t)$ is N'-sided admissible polygon ($N' < N$).

This result says that the self-contacting never occur and we can continue a time evolution of admissible polygon beyond the edge-disappearing.

Remark: These three assumptions (G3,4) and (W1) do not come of a technical reason for the proof. If one of them fails, then we can find counter examples. Figures 2, 3 and 4 show typical examples.

4. Almost convexity phenomena

In this section we always assume all condition in Theorem 3.1 and consider a behavior of solution polygons near the extinction time.

Since a number of edges of the initial polygon $\Omega(0)$ is finite and new edges are not generated during the time evolution. By Theorem 3.1, edge-disappearing must happen in a finite time and thus $N(\Omega(t))$ is monotone non-increasing in time. Hence, we have finite sequence of edge-disappearing times:

$$0 < T_1 < T_2 < \cdots < T_m < \infty.$$

Let $T_0 = 0$. Note that by Theorem 3.1, the solution polygon shrinks to a single point at the extinction time T_m.

The following result characterizes shapes of solution polygons near the extinction time.

Theorem 4.1. *([9]) Assume the same conditions in the previous theorem. Then, we have $N^-(\Omega(t)) = 0$ and $N^0(\Omega(t)) = 0$ or 2 for $t \in [T_{m-1}, T_m)$. Moreover, in the case where $N^0(\Omega(T_{m-1})) = 2$, these two edges which have zero transition number are adjacent each other.*

In the above theorem, the case where $N^0(\Omega(T_{m-1})) = 0$ means that the solution polygon become convex at $t = T_{m-1}$. The non-convex self-similar solutions are typical solutions for the case where $N^0(\Omega(T_{m-1})) = 2$ with $m = 1$.

From this theorem, we easily obtain the following corollaries.

Corollary 4.1. *(Almost convexity phenomena)*

(1) There exists $T_* \leq T_{m-1}$ such that $H_j(t) \geq 0$ for all j and any $t \geq T_*$.

(2) There exists $T^* \leq T_{m-1}$ such that the solution polygon $\Omega(t)$ is star-shaped for any $t \geq T^*$.

Here, the order of T_* and T^* depends on the initial shape.

Corollary 4.2. *(Sufficient condition for convexity phenomenon) If the initial polygon $\Omega(0)$ is point symmetric, then the solution polygon becomes convex at $t = T_{m-1}$.*

5. Discussion

In this final section we only consider the case $V_j = H_j$ and $\mathcal{W}_\sigma$ is a even sided regular polygon. This setting satisfies all assumptions in the previous theorems.

It is known that the convexity phenomena does not always appear for (GCCF) because there are some examples of non-convex self-similar solutions (see [10]). However, it is shown in [8] that the motion of polygonal curve by $V_j = H_j$ approximates the motion of smooth curve by curve-shortening flow and also well-known that the convexity phenomena always appear in the later motion (see [6]). Therefore, we expect that the convexity phenomena always appear for $V_j = H_j$ when the number of edges of $\mathcal{W}_\sigma$ is large.

It is obvious that non-existence results of non-convex self-similar solutions lead a necessary condition for the convexity phenomena. We here show two examples of them.

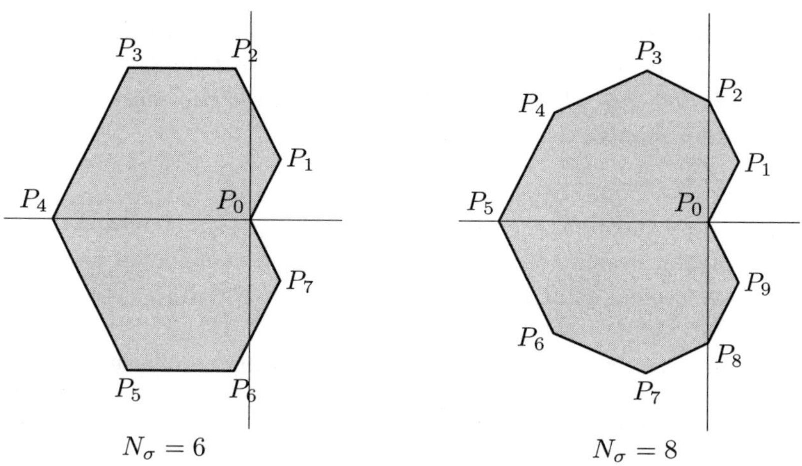

Figure 5. Non-convex self-similar solutions (left:$N_\sigma = 6$, right:$N_\sigma = 8$).

Example 5.1. Let $N_\sigma = 6$ and $l_\sigma(\theta_j) = 1$ for $\theta_j \in \Theta_{\mathcal{W}_\sigma} := \{\pi/6, \pi/2, 5\pi/6, 7\pi/6, 3\pi/2, 11\pi/6\}$. By theorem 4.1, the non-convex self-similar solution has 8 edges and only two adjacent edges have zero crystalline curvature (see Figure 5). We set $P_0 = (0,0)$ and $d_j = \lambda_j d_0$ for $j = 1, 2, \cdots, 7$. Here λ_j's are positive constants since the solution is self-similar. Let consider the upper curve from P_0 to P_4, i.e., $C_1 := \cup_{j=0}^{3} \mathcal{S}_j$. By geometry, $P_j = (x_j, y_j)$ are given by $x_1 = Ad_0, y_2 = Bd_0, x_2 = Ad_0(1-\lambda_1), y_2 = Bd_0(1+\lambda_1), x_3 = Ad_0(1-\lambda_1-2\lambda_2), y_3 = Bd_0(1+\lambda_1), x_4 = Ad_0(1-\lambda_1-2\lambda_2-\lambda_3), y_4 = Bd_0(1+\lambda_1-\lambda_3)$ where $A = \cos(\pi/3)$ and $B = \sin(\pi/3)$. By $V_0 = 0$ and $V_1 = 1/\lambda_1 d_0$, we have $\dot{d}_0 = -1/B\lambda_1 d_0$ since $\boldsymbol{N}_1 \cdot \boldsymbol{T}_0 = -B$. Next let see a motion of P_2. The time derivatives of x_2 and y_2 are given by $\dot{x}_2 = A\dot{d}_0(1-\lambda_1) = -A(1-\lambda_1)/B\lambda_1 d_0$ and $\dot{y}_2 = B\dot{d}_0(1+\lambda_1) = -(1+\lambda_1)/\lambda_1 d_0$. On the other hand, $V_j = 1/\lambda_j d_0$ $(j = 1, 2)$ yield that $\dot{x}_2 = (\dfrac{A}{\lambda_2} - \dfrac{1}{\lambda_1})\dfrac{1}{Bd_0}$ and $\dot{y}_2 = -\dfrac{1}{\lambda_2 d_0}$. Thus we have $\lambda_2 = \lambda_1/(1+\lambda_1)$. To consider a motion of P_3, we similarly have $\lambda_3 = (1+\lambda_1)/(2+\lambda_1)$. Then we have $y_4 = Bd_0(\lambda_1 + 1/(2+\lambda_1)) > 0$ for any $d_0 > 0$ and $\lambda_1 > 0$. Thus the curve C_1 does not touch or cross the x-axis except at the origin. We can also show in the same way that the lower curve $C_2 := \cap_{j=4}^{7} \mathcal{S}_j$ does not touch or cross the x-axis except at the origin. This contradicts to the closedness of $\mathcal{P}(t)$. Therefore, the non-convex self-similar solution does not exist.

Example 5.2. Let $N_\sigma = 8$ and $l_\sigma(\theta_j) = 1$ for $\theta_j \in \Theta_{\mathcal{W}_\sigma} := \{\pi/8, 3\pi/8, 5\pi/8, 7\pi/8, 9\pi/8, 11\pi/8, 13\pi/8, 15\pi/8\}$. The non-convex self-similar solution has 10 edges and only two adjacent edges have zero crystalline curvature (see Figure 5). Let λ_j be positive constants. We set $P_0 = (0,0)$ and $d_j = \lambda_j d_0$ for $j = 1, 2, \cdots, 9$. Let consider the upper curve from P_0 to P_5. We can make the relations between λ_j and λ_1 for $j = 2, 3, 4$ in the same manner in the previous example and get $\lambda_2 = \lambda_1/(\lambda_1 + \sqrt{2})$, $\lambda_3 = \lambda_1(\lambda_1 + \sqrt{2})/(\sqrt{2}\lambda_1^2 + 4\lambda_1 + \sqrt{2})$ and $\lambda_4 = (\sqrt{2}\lambda_1^3 + 6\lambda_1 + 5\sqrt{2}\lambda_1 + 2)/(\sqrt{2}\lambda_1^3 + 9\lambda_1^2 + 11\sqrt{2}\lambda_1 + 6)$. By these relations and $y_5 = d_0[(1+\lambda_1-\lambda_4)\cos(\pi/8) + (\lambda_2 - \lambda_3)\sin(\pi/8)]$, we can show $y_5 > 0$ for any $d_0 > 0$ and $\lambda_1 > 0$. Thus, the non-convex self-similar solution does not exist.

Acknowledgment

The part of this work was done while the author was visiting the Department of Pure and Applied Mathematics, University of L'Aquila. The author is partially supported by Grant-in-Aid for Encouragement of Young Scientists (No. 18740048).

References

[1] F. Almgren and J.E. Taylor, *Flat flow is motion by crystalline curvature for curves with crystalline energies*, J. Diff. Geom. **42** (1995) 1–22.

[2] B. Andrews, Singularities in crystalline curvature flows, Asian J. Math. **6** (2002) 101–122.

[3] S. Angenent and M.E. Gurtin, Multiphase thermomechanics with interfacial structure, 2. Evolution of an isothermal interface, Arch. Rational Mech. Anal. **108** (1989) 323–391.

[4] K.S. Chou and X.P. Zhu, A convexity theorem for a class of anisotropic flows of plane curves, Indiana Univ. Math. J. **48** (1999) 139–154.

[5] M.-H. Giga and Y. Giga, Crystalline and level set flow – convergence of a crystalline algorithm for a general anisotropic curvature flow in the plane, Free boundary problems: theory and applications, I (Chiba, 1999), GAKUTO Internat. Ser. Math. Sci. Appli., **13**, Gakkōtosho, Tokyo, (2000) 64–79.

[6] M. Grayson, The heat equation shrinks embedded plane curves to round points, J. Differential Geometry 26 (1987), 285-314.

[7] C. Hirota, T. Ishiwata and S. Yazaki, Numerical study and examples on singularities of solutions to anisotropic crystalline curvature flows of nonconvex polygonal curves, to appear in ASPM.

[8] K. Ishii and H. M. Soner, Regularity and convergence of crystalline motion, SIAM J. Math. Anal. **30** (1999) 19–37 (electronic).

[9] T. Ishiwata, Motion of non-convex polygons by crystalline curvature and almost convexity phenomena, preprint.

[10] T. Ishiwata, T.K. Ushijima, H. Yagisita and S. Yazaki, Two examples of nonconvex self-similar solution curves for a crystalline curvature flow, Proc. Japan Acad. Vol. 80, Ser. A, No. 8 (2004) 151–154.

[11] T. Ishiwata and S. Yazaki, On the blow-up rate for fast blow-up solutions arising in an anisotropic crystalline motion, J. Comp. App. Math. **159** (2003) 55–64.

[12] T. Ishiwata and S. Yazaki, A fast blow-up solution and degenerate pinching arising in an anisotropic crystalline motion, submitted.

[13] A. Stancu, Asymptotic behavior of solutions to a crystalline flow, Hokkaido Math. J. **27** (1998) 303–320.

[14] J. E. Taylor, Constructions and conjectures in crystalline nondifferential geometry, Proceedings of the Conference on Differential Geometry, Rio de Janeiro, Pitman Monographs Surveys Pure Appl. Math. **52** (1991) 321–336, Pitman London.

FUNDAMENTAL PROPERTIES OF SOLUTIONS TO A SCALAR-FIELD TYPE EQUATION ON THE UNIT SPHERE

YOSHITSUGU KABEYA*

Department of Mathematical Sciences,
Osaka Prefecture University
1-1 Gakuencho, Sakai, 599-8531, Japan

HIROKAZU NINOMIYA†

Department of Applied Mathematics and Informatics,
Ryukoku University
Seta, Otsu, 520-2194, Japan

Abstract: A scalar-field type equation on a spherical cap in the unit sphere under the homogeneous Dirichlet condition is considered and the existence of an azimuthal solution is discussed. The scalar-field type equation under the azimuthal condition is reduced to a specific ordinary differential equation and the analysis of this ODE is crucial. To understand properties of solutions, behaviors of a solution to the linearized equation are also investigated. The associated Legendre functions play key roles.

1. Introduction and Results

In this note, we consider the nonlinear elliptic equation

$$\Lambda u - \lambda u + u_+^p = 0, \quad \text{in } \Omega \subset \mathbb{S}^n \tag{1.1}$$

under the homogeneous Dirichlet boundary condition, where Λ is the Laplace-Beltrami operator on the sphere, $n \geq 3$, $p > 1$, $\lambda > 0$, $\mathbb{S}^n \subset \mathbb{R}^{n+1}$ is the standard unit sphere, $\Omega \subset \mathbb{S}^n$ is a geodesic ball (called a "spherical cap") centered at the south pole $(0, \dots, 0, -1)$. Actually, let $(y_1, y_2, \cdots, y_{n+1}) \in$

*Supported in part Grant-in-Aid for Scientific Research (C)(No. 19540224), Japan Society for the Promotion of Science.

†Supported in part Grant-in-Aid for Scientific Research (C)(No. 18540147), Japan Society for the Promotion of Science.

$\Omega \subset \mathbb{S}^n$. We express y_i in the polar coordinates:

$$\begin{cases} y_k = \Big(\prod_{j=1}^{k} \sin\theta_j\Big)\cos\theta_{k+1}, \quad k = 1, 2, \dots, n-2, \\ y_{n-1} = \Big(\prod_{j=1}^{n-1} \sin\theta_j\Big)\cos\phi, \\ y_n = \Big(\prod_{j=1}^{n-1} \sin\theta_j\Big)\sin\phi, \\ y_{n+1} = \cos\theta_1. \end{cases}$$

Note that Ω can be expressed in the polar coordinates as

$$\Omega = \Omega(\theta_1^*) = \Big\{(\theta_1, \theta_2, \dots, \theta_{n-1}, \phi)\Big| \quad 0 < \theta_1^* < \theta_1 \le \pi, \\ 0 \le \theta_i \le \pi, (i = 2, 3, \dots, n-1),\ 0 \le \phi \le 2\pi\Big\}$$

with some $\theta_1^* > 0$. Then Λ is written as

$$\Lambda u = \sum_{k=1}^{n-1} (\sin\theta_1 \dots \sin\theta_{k-1})^{-2} (\sin\theta_k)^{k-n} \frac{\partial}{\partial\theta_k}\Big\{(\sin\theta_k)^{n-k}\frac{\partial u}{\partial\theta_k}\Big\} \\ + \Big(\prod_{k=1}^{n-1}\sin\theta_k\Big)^{-2}\frac{\partial^2 u}{\partial\phi^2}.$$

For example, see Section 3 in Chapter 2 of Shimakura [15] by applying the decomposition of $(n+1)$ dimensional Laplacian. Here we can consider (1.1) in the class of radial functions, that is, functions depending only on the azimuthal angle θ_1 (="latitude"). For a function v depending only on θ_1, Λ can be rewritten as

$$\Lambda v = \frac{1}{\sin^{n-1}\theta_1}\frac{\partial}{\partial\theta_1}\left\{\left(\sin^{n-1}\theta_1\right)\frac{\partial v}{\partial\theta_1}\right\}, \tag{1.2}$$

which is denoted by Λ_{θ_1}. Then (1.1) becomes

$$\Lambda_{\theta_1} v - \lambda v + v_+^p = 0. \tag{1.3}$$

Although the equation (1.1) is defined on a domain of the unit sphere, we can project the problem stereographically on the Euclidean space. In

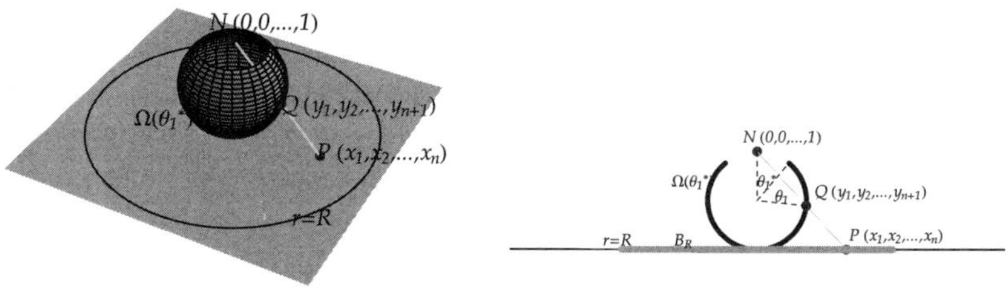

(a) (b)

Figure 1. Stereographic projection from the north pole to the Euclidean space. (a) a bird's-eye view of the domain, (b) the cross-section at the hyperplane.

view of the stereographic projection from the north pole $(0,\ldots,0,1)$, any point $(y_1, y_2, \ldots, y_{n+1}) \in \Omega$ corresponds to

$$(x_1, x_2, \ldots, x_n) = \left(\frac{y_1}{1-y_{n+1}}, \frac{y_2}{1-y_{n+1}}, \ldots, \frac{y_n}{1-y_{n+1}}\right)$$

on $\mathbb{R}^n$. Expressing $(y_1, y_2, \ldots, y_{n+1})$ in terms of x_i, we have

$$(y_1, y_2, \ldots, y_{n+1}) = \left(\frac{2x_1}{|x|^2+1}, \frac{2x_2}{|x|^2+1}, \ldots, \frac{2x_n}{|x|^2+1}, \frac{|x|^2-1}{|x|^2+1}\right).$$

See Figure 1. Since $\Omega(\theta_1^*)$ is a geodesic ball with its radius $\pi - \theta_1^*$, its stereographically projected image is also a ball B_R in $\mathbb{R}^n$ centered at the origin with its radius $R = \cot(\theta_1^*/2)$.

The operator Λ can be also rewritten as a weighted differential operator of the elliptic type as expressed below. After tedious calculations, we obtain

$$\begin{aligned}\Lambda u &= \Big(\frac{|x|^2+1}{2}\Big)^2 \Delta u - (n-2)\frac{|x|^2+1}{2}(y \cdot \nabla u) \\ &= \Big(\frac{1+|x|^2}{2}\Big)^n \operatorname{div}\left(\Big(\frac{2}{|x|^2+1}\Big)^{n-2} \nabla u\right).\end{aligned}$$

Thus, in the case of radial function $w(r)$, Λ can be written as

$$\Lambda w = \frac{(r^2+1)^n}{4r^{n-1}} \frac{\partial}{\partial r}\left(\frac{r^{n-1}}{(r^2+1)^{n-2}} \frac{\partial w}{\partial r}\right), \tag{1.4}$$

which we denote by $\Lambda_r w$. Finally, (1.1) yields

$$\Lambda_r w - \lambda w + w_+^p = 0. \tag{1.5}$$

Although the linear elliptic problems have been considered on the sphere since 19th century (the Legendre era), nonlinear elliptic problems on the unit sphere or the spherical caps are now getting attention. For example, Bandle and Peletier [4], Bandle and Benguria [3] are basic results. In [3], the problem (1.1) with $\lambda < 0$, $n = 3$ and the homogeneous Dirichlet condition is treated. They showed the existence and the non-existence results by using the variational methods as are done in Brezis and Nirenberg [8]. They emphasized that different phenomena would occur if Ω contains the "southern hemisphere" (note that they project the spherical cap from the south pole and in their sense, this word should be "northern hemisphere"). Also related results concerning the existence of a minimizer for the best Sobolev constants were discussed by Bandle and Peletier [4].

Later, Stingelin [16] considered (1.1) with $\lambda > 0$ and the homogeneous Dirichlet condition, and showed numerically the bifurcation diagram, which seemed as if imperfect bifurcations occur (see Figure 2). His diagrams look very like what we obtained in Kabeya, Morishita and Ninomiya [10]. There must be the similar mechanism in behind in this problem. We would like to assure the bifurcation diagram mathematically rigorously. Especially, what will happen to the bifurcation diagram if we let $\theta_1^* \to 0$?

Recently, Brezis and Peletier [4] studied this problem further than what Stingelin [16] did and they showed several properties of the bifurcation diagram for large $\lambda > 0$. Moreover, very recently, Bandle and Wei [5, 6, 7] studied intensively on this subject from the singular perturbation point of view, and arrived to almost the same height as in Ambrosetti, Malchiodi and Ni [1, 2] and Malchiodi, Ni and Wei [13]. Various concentration phenomena have been observed in [5, 6, 7].

We would like to see what the structure of solutions to (1.5) is. Unfortunately, (1.5) cannot be transformed to a Matsukuma type equation or the canonical form as discussed in Yanagida and Yotsutani [18, 19] and theorems in Kabeya, Yanagida and Yotsutani [11] are not applicable.

We should note that (1.5) on $\mathbb{S}^n$ has a constant solution $\lambda^{1/(p-1)}$. Although the constant solution is never a solution to the Dirichlet problem, as in Section 4 of [4], the analysis of the linearized problem is important.

As a first step to analyze the problem, we here consider the linear eigenvalue problem: for given $R > 0$, find eigenvalues such that

$$\begin{cases} \Lambda_r w + (p-1)\lambda w = 0, \\ w_r(0) = 0, \ w(R) = 0, \end{cases}$$

has a nontrivial solution. However, as for the linear problem, we should better consider the original problem on the spherical cap and find azimuthal eigenvalues. That is, for given $\theta_1^* > 0$, we consider the problem

$$\begin{cases} \Lambda_{\theta_1} v + (p-1)\lambda v = 0 \\ v_{\theta_1}(\pi) = 0, \ v(\theta_1^*) = 0 \end{cases} \tag{1.6}$$

and find suitable eigenvalue $\lambda_{k,\theta_1^*}^D > 0$ and a solution $v = \varphi_{k,\theta_*}^D$ with $k-1$ zeros in (θ_1^*, π) $(k = 1, 2, 3, \dots)$. This is equivalent to find k-th azimuthal eigenvalue of $-\Lambda$, i.e., to find a nontrivial function Φ in $H^1_{0,\mathrm{az}}(\Omega(\theta_1^*))$ such that

$$\int_{\Omega(\theta_1^*)} |\Lambda^{1/2}\Phi|^2 \, dS = (p-1)\lambda_k^D(\theta_1^*) \int_{\Omega(\theta_1^*)} \Phi^2 \, dS,$$

where

$$H^1_{0,\mathrm{az}}(\Omega(\theta_1^*)) = \left\{ \Phi = \Phi(\theta_1) \;\middle|\; \int_{\Omega(\theta_1^*)} |\Lambda^{1/2}\Phi|^2 \, dS < \infty, \ \Phi\Big|_{\theta_1 = \theta_1^*} = 0 \right\},$$

dS denotes the "spherical element" and $\Lambda^{1/2}$ is defined as

$$\int_{\Omega(\theta_1^*)} |\Lambda^{1/2}u|^2 \, dS = -\int_{\Omega(\theta_1^*)} (\Lambda u) u \, dS$$

for $u \in C_0^\infty(\Omega(\theta_1^*))$ and is extended to $H^1_{0,\mathrm{az}}(\Omega(\theta_1^*))$ by completion. Note that for a function f depending only on θ_1, we have

$$\int_{\Omega(\theta_1^*)} f(\theta_1) \, dS = |\mathbb{S}^{n-1}| \int_{\theta_1^*}^{\pi} f(\theta_1)(\sin\theta_1)^{n-1} \, d\theta_1,$$

where $|\mathbb{S}^{n-1}|$ denotes the area of $\mathbb{S}^{n-1}$.

As a comparison, we also consider properties of a solution to the following eigenvalue problems

$$\begin{cases} \Lambda_{\theta_1} v + (p-1)\lambda v = 0, \\ v_{\theta_1}(\pi) = v_{\theta_1}(0) = 0. \end{cases} \tag{1.7}$$

We say that λ_k $(k \in \mathbb{N})$ is the k-th eigenvalue if a nontrivial solution $v = \varphi_k$ to (1.7) changes its sign $(k-1)$ times in $[0, \pi)$. The first eigenvalue λ_1 is zero and the corresponding eigenfunction is a constant.

The explicit forms of the eigenvalues and the eigenfunctions will be discussed in Section 2. The eigenvalues and the eigenfunctions will play important roles to the analysis of our goal — to prove the imperfect bifurcations (the phenomena that diagrams of solutions look like a bifurcation diagram although there exists no bifurcating point. See Figure 2).

What are the behaviors of λ^D_{k,θ_1^*}? We can expect that λ^D_{k,θ_1^*} converges to the eigenvalue λ_k in the whole sphere case.

Theorem 1.1. *For any natural number k and for any sufficiently small $\theta_1^* > 0$, there holds $\lambda^D_{k,\theta_1^*} > \lambda_k$. Moreover, as $\theta_1^* \to 0$, $\lambda^D_{k,\theta_1^*} \to \lambda_k$ and $\varphi^D_{k,\theta_1^*} \to \varphi_k$ local uniformly in $(0,\pi]$.*

Remark 1.1. The convergence of eigenfunctions under the Dirichlet problem to the whole sphere ones might be strange. This would show that the solution to (1.1) very close to the constant solution have a boundary layer as the spherical cap approaches to the north pole. The bifurcation diagram of (1.1) would converge locally uniformly to that of the whole sphere case.

Using the information of eigenfunctions and eigenvalues, we have the existence theorem to the nonlinear problem, which would partially give a rigorous proof of numerically computed bifurcation diagrams (see Figure 2).

Theorem 1.2. *Let $n \geq 3$, $(n+2)/(n-2) \geq p > 1$ and $\lambda \neq \lambda_k (k = 1, 2, 3, \dots)$ but near some λ_j. Then, for sufficiently small $\theta_1^* > 0$, there exist $\alpha_* > 0$ and a solution $v(\theta; \theta_1^*, \alpha_*)$ to*

$$\begin{cases} \Lambda_{\theta_1} v - \lambda v + v^p = 0, & \theta_1^* < \theta_1 < \pi, \\ v(\theta_1^*) = 0, \quad v_{\theta_1}(\pi) = 0, & v(\pi) = \alpha_*. \end{cases} \tag{1.8}$$

The organization of this paper is as follows. Analysis on the linear problem will be done by using the Legendre associate functions, which will help us much, in Section 2. Sketches of proofs of Theorems 1.1 and 1.2 will be given in Section 3.

2. Investigation of the linearized problem

In this section, we investigate (1.6) and (1.7) and find exact solutions to these problems by using the associated Legendre functions.

Letting $t = \cos(\pi - \theta_1) = -\cos\theta_1 = -y_{n+1}$, we have

$$\frac{\partial}{\partial t}\left\{(1-t^2)^{n/2}\frac{\partial \psi}{\partial t}\right\} + (1-t^2)^{n/2-1}(p-1)\lambda\psi = 0, \tag{2.1}$$

and (2.1) is called a "hyper-sphere" equation. The fundamental solutions of (2.1) are expressed by the associated Legendre functions P^μ_ν, Q^μ_ν as

$$\psi = c_1(1-t^2)^{-\mu/2}P^\mu_\nu(t) + c_2(1-t^2)^{-\mu/2}Q^\mu_\nu(t), \tag{2.2}$$

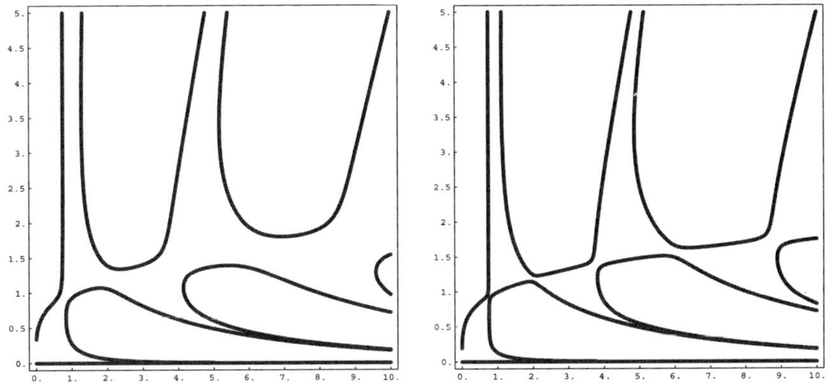

(a) $\theta_1^* = 0.02$ (b) $\theta_1^* = 0.002$

Figure 2. Bifurcation diagrams of (1.8) when $n = 3$ and $p = 5$. The horizontal axes stand for λ and the vertical axes do for $v(\pi)$.

where

$$\mu = \frac{n-2}{2}, \quad \nu = \frac{\sqrt{(n-1)^2 + 4(p-1)\lambda} - 1}{2}. \tag{2.3}$$

The associated Legendre functions P_ν^μ and Q_ν^μ are the independent solutions to the associated Legendre equation

$$\frac{d}{dt}\left\{(1-t^2)\frac{dP}{dt}\right\} + \left\{\nu(\nu+1) - \frac{\mu^2}{1-t^2}\right\} P = 0. \tag{2.4}$$

In case of $n = 2m - 1$, P_ν^μ has a singularity at $t = 1$ and hence $c_1 = 0$ must hold. Moreover, if

$$\lim_{t \to -1} (1-t^2)^{-\mu/2} Q_\nu^\mu(t)$$

is finite, then v corresponds to an eigenfunction and λ does to an eigenvalue. Since there holds

$$Q_\nu^\mu(-t) + \Big(\cos(\mu+\nu)\pi\Big) Q_\nu^\mu(t) = \frac{\pi}{2}\Big(\sin(\mu+\nu)\pi\Big) P_\nu^\mu(t),$$

$\mu + \nu$ should be an integer in this case. That is,

$$\frac{n-2}{2} + \frac{\sqrt{(n-1)^2 + 4(p-1)\lambda} - 1}{2} = \ell$$

for $\ell = n-2, n-1, \ldots$ when $n = 2m - 1$. Thus, we have

$$(p-1)\lambda = (\ell+1)(\ell+2-n). \tag{2.5}$$

Note also that

$$\lim_{t\to-1}\frac{d}{dt}\left\{(1-t^2)^{-\mu/2}Q_\nu^\mu(t)\right\} \tag{2.6}$$

is finite if $\mu+\nu$ is an integer.

Since $\mu\notin\mathbf{Z}$ when n is odd, we have

$$Q_\nu^\mu(t)=\frac{\pi}{2\sin\mu\pi}\left[(\cos\mu\pi)P_\nu^\mu(t)-\frac{\Gamma(\nu+\mu+1)}{\Gamma(\nu-\mu+1)}P_\nu^{-\mu}(t)\right]. \tag{2.7}$$

Also $Q_\nu^{m-3/2}(t)$ can be expressed in a finite sum of the combinations of the elementary functions by using

$$Q_\nu^{m-3/2}(t)=-\frac{\pi}{2}\frac{\Gamma(\nu+m-1/2)}{\Gamma(\nu-m+5/2)}P_\nu^{-(m-3/2)}(t),$$

which comes from (2.7),

$$P_\nu^{1/2}(\cos\theta)=i\sqrt{\frac{2}{\pi\sin\theta}}\cos\left(\nu+\frac{1}{2}\right)\theta,$$

$$P_\nu^{-1/2}(\cos\theta)=-\frac{2i}{2\nu+1}\sqrt{\frac{2}{\pi\sin\theta}}\sin\left(\nu+\frac{1}{2}\right)\theta,$$

with $t=\cos\theta$ $(\theta=\pi-\theta_1)$ and the inductive relation

$$P_\nu^{\mu+2}(t)-2(\mu+1)t(1-t^2)^{-1/2}P_\nu^{\mu+1}(t)+(\nu-\mu)(\nu+\mu+1)P_\nu^\mu(t)=0.$$

We choose a pure imaginary constant so that v should be a real-valued function.

On the other hand, in case of $n=2m$, then Q_ν^μ has a singularity at $t=1$ and there must hold $c_2=0$. Similarly, if

$$\lim_{t\to-1}(1-t^2)^{-\mu/2}P_\nu^\mu(t)$$

is finite, then v becomes an eigenfunction. Since P_ν^{m-1} is also expressed as

$$P_\nu^\mu(t)=P_\nu^{m-1}(t)=(1-t^2)^{(m-1)/2}\frac{d^{m-1}}{dt^{m-1}}P_\nu(t)$$

where $P_\nu(t)$ is the Legendre function of the first kind, only in the case where

$$\left|\frac{d^{m-1}}{dt^{m-1}}P_\nu(t)\Big|_{t=-1}\right|<\infty,$$

$P_\nu^{m-1}(t)$ becomes an eigenfunction. Since $\lim_{t\to-1}|P_\nu(t)|=\infty$ for $\nu\notin\mathbf{Z}$, more precisely,

$$P_\nu(-1+2t)\sim\frac{1}{\pi}(\sin\nu\pi)\log t$$

near $t = 0$ for $\nu \notin \mathbf{Z}$, ν should be an integer, which is denoted by ℓ. That is, the eigenvalues λ are expressed as

$$(p-1)\lambda = \left(\ell + \frac{n}{2}\right)\left(\ell + 1 - \frac{n}{2}\right) \tag{2.8}$$

for $\ell = n/2 - 1, n/2, \ldots$ when $n = 2m$. Note that

$$\lim_{t \to -1} \frac{d}{dt}\left\{(1-t^2)^{-\mu/2} P_\nu^\mu(t)\right\} \tag{2.9}$$

if ν is an integer and that $P_\nu^{m-1}(t)$ becomes a polynomial for $\nu \in \mathbf{Z}$.

As a summary in view of (2.5) and (2.8), we see that the eigenvalue λ_k to (1.7) is expressed as

$$\lambda_k = (k-1)(k+n-2)$$

for $k = 1, 2, \ldots$, well-known eigenvalues for $-\Lambda$. The case $k = 1$ corresponds to the constant eigenfunction. The corresponding eigenfunction $\varphi_k(\theta_1)$ is

$$\varphi_k(\theta_1) = \frac{1}{\sin^{(n-2)/2}\theta_1} Q^{(n-2)/2}_{k-1+(n-2)/2}(-\cos\theta_1)$$

when n is odd, and is

$$\varphi_k(\theta_1) = \frac{1}{\sin^{(n-2)/2}\theta_1} P^{(n-2)/2}_{k-1+(n-2)/2}(-\cos\theta_1)$$

when n is even. The formulae of the (associated) Legendre functions used here are seen in Chapters IV and V of [14]. The eigenfunction corresponding to λ^D_{k,θ_1^*} is denoted by φ^D_{k,θ_1^*}.

Remark 2.1. Consider the case of $n = 3$. By (2.3) we see that

$$\nu = \sqrt{(p-1)\lambda + 1} - \frac{1}{2}$$

and that a solution ψ to (2.1) is written as

$$\begin{aligned}\psi &= c_3(1-t^2)^{-1/4} Q_\nu^{1/2}(t) \\ &= \frac{c_4}{\sqrt{\sin\theta_1}} P_\nu^{-1/2}(-\cos\theta_1) = \frac{c_5 \sin\{\sqrt{(p-1)\lambda+1}(\pi-\theta_1)\}}{\sin(\pi-\theta_1)}\end{aligned}$$

with some constants c_j $(j = 3, 4, 5)$. If follows from $\psi(\theta_1^*) = 0$ that we have the eigenvalues

$$\lambda^D_{k,\theta_1^*} = \frac{1}{p-1}\left\{\left(\frac{k\pi}{\pi-\theta_1^*}\right)^2 - 1\right\}.$$

Thus, the solution to (1.6) is explicitly expressed as

$$\varphi^D_{k,\theta_1^*}(\theta_1) = \frac{c_6}{\sqrt{\sin\theta_1}} P_\nu^{-1/2}(-\cos\theta_1) = \frac{c_7}{\sin\theta_1} \sin\frac{k\pi(\pi-\theta_1)}{\pi-\theta_1^*}$$

where c_6 and c_7 are normalizing constants. The convergence of λ^D_{k,θ_1^*} is readily seen. See also [4].

Remark 2.2. All the eigenfunctions, not restricted to the azimuthal ones, are expressed by the Gegenbauer polynomials. See, e.g. Sections 2 and 4 of Chapter 2 in [15] or Section 4 of Chapter 8 in Taylor [17].

Hence we obtain the following.

Proposition 2.1. *The solution ψ to* (2.1) *satisfies $|\psi(t)| \to \infty$ as $t \to -1$ if $\lambda \neq \lambda_k$ $(k = 1, 2, 3, \dots)$.*

Remark 2.3. This is different from the usual Laplacian case on the Euclidean space, under which the solution of the linearized equation converges to 0 changing its sign infinitely many times as $r \to \infty$. Indeed, a unique solution to

$$\begin{cases} \dfrac{1}{r^{n-1}}(r^{n-1}u_r)_r + \mu^2 u = 0, \\ u(0) = 1 \end{cases}$$

with $\mu > 0$ is explicitly expressed as

$$u(r) = C_n \frac{J_{(n-2)/2}(\mu r)}{r^{(n-2)/2}}$$

with $C_n > 0$ being the normalizing constant. This may reflect the difference of properties of solutions caused by the compactness or non-compactness of the domain.

3. Sketch of Proofs

We give sketches of proofs of Theorems 1.1 and 1.2 by using the properties of the associated Legendre functions as discussed in Section 2.

Sketch of a proof of Theorem 1.1. Let $\psi(t;\lambda)$ be a solution of (2.1) and $\psi(1;\lambda) = 1$. By (2.2) and (2.3), $P(t;\lambda) = (1-t^2)^{\mu/2}\psi = c_1 P_\nu^\mu(t) + c_2 Q_\nu^\mu(t)$ satisfies (2.4). Recall the locally uniform convergence on $(-1, 1]$ of a solution of (2.4) with respect to ν and the continuity of P_ν^μ and Q_ν^μ with respect to ν. More precisely, we consider the case of $n = 2m - 1$

$(m = 2, 3, \cdots)$. In this case, $P(t; \lambda) = c_2 Q_\nu^\mu(t)$ as stated in Section 2. If $\lambda = \lambda_k$, then $P(t; \lambda)$ vanishes at $t = \pm 1$. If $\lambda \in (\lambda_k, \lambda_{k+1})$ is close to λ_k, then due to the Sturm (or Prüfer) comparison principle, $Q_\nu^\mu(t)$ has k zeros in $(-1, 1)$. We take the smallest zero t_1^* of $Q_\nu^\mu(t)$ in $(-1, 1)$. This means that such λ is an eigenvalue of the homogeneous Dirichlet problem (1.6) with $\theta_1^* = \cos^{-1}(-t_1^*)$. The Sturm comparison principle and the uniqueness of solutions to (2.4) with $P(1; \lambda) = 0$ also implies that the correspondence between λ and t_1^* is one-to-one.

Note that t_1^* tends to -1 as $\lambda \to \lambda_k$. Since $Q_\nu^\mu \to Q_\ell^\mu$ as $\lambda \to \lambda_k$ locally uniformly in $(-1, 1]$ and in view of (2.6), the eigenvalue converges to that of the whole sphere case.

Similarly, in case of $n = 2m$ if $\lambda (> \lambda_k)$ is sufficiently close to λ_k (this implies that ν is a little bigger than a natural number), then $P_\nu^\mu(t)$ has the zero t_2^*, which are close to $t = -1$ and

$$\lim_{t \to -1} (1 - t^2)^{-\mu/2} |P_\nu^\mu(t)| = \infty.$$

Hence, the same conclusion holds for this case due to (2.9).

Letting $\theta_1^* \to 0$ means that finding λ closer to λ_k so that the smallest zero is closer to $t = -1$.

In either case, the convergence of the norm of $\varphi_{k,\theta_1^*}^D$ to that of φ_k is deduced easily since $\varphi_{k,\theta_1^*}^D$ has no singularities. □

Remark 3.1. As in the proof of Theorem 1.1, the smallest zero of the associated Legendre function converges to the smallest zero $t = -1$ of $P(t; \lambda_k)$ as $\lambda \to \lambda_k$. Thus, from the view point of the associated Legendre function, the eigenvalues of the Dirichlet problem for (2.4) converges to the eigenvalues of that at $t = -1$. However, as a solution on the sphere, due to the weight $(1 - t^2)^{-\mu/2}$, the zero of $P(t; \lambda_k)$ at $t = -1$ is cancelled out for $\lambda = \lambda_k$. The eigenvalues of the Dirichlet problem on the spherical cap converges to those of the whole sphere (Nuemann) problem as the cap becomes wider and wider to cover the whole sphere except the north pole.

Sketch of proof of Theorem 1.2. We construct a solution in $H_{0,\mathrm{az}}^1(\Omega(\theta_1^*))$ by using the contraction mapping principle. We seek a solution of the form

$$v_{k,\theta_1^*} = w_{k,\theta_1^*}(\theta_1) + s\varphi_{k,\theta_1^*}^D(\theta_1) + h_{k,\theta_1^*}(\theta_1, s)$$

where $w_{k,\theta_1^*}(\theta_1)$ is an auxiliary function which converges to $\lambda^{1/(p-1)}$ as $\theta_1^* \to 0$. We can follow the argument in [10] to show that there exist $s \in \mathbb{R}$

and $h_{k,\theta_1^*} \in H^1_{0,\mathrm{az}}(\Omega(\theta_1^*))$ with $h_{k,\theta_1^*} \perp \varphi^D_{k,\theta_1^*}$ such that v_{k,θ_1^*} satisfies (1.8) for sufficiently small θ_1^*. □

Acknowledgment. The authors thank Professor Catherine Bandle for introducing Dr. Stingelin's work and useful private communications. Also they thank Professor Juncheng Wei for informing us of his results with Professor Bandle during the conference.

References

[1] A. Ambrosetti, A. Malchiodi, and W.-M. Ni, *Singularly perturbed elliptic equations with symmetry: existence of solutions concentrating on spheres, Part I*, Comm. Math. Phys. **235** (2003), 427–466.

[2] A. Ambrosetti, A. Malchiodi, and W.-M. Ni, *Singularly perturbed elliptic equations with symmetry: existence of solutions concentrating on spheres, Part II*, Indiana Univ. Math. J. **53** (2004), 297–329.

[3] C. Bandle and R. Benguria, *The Brézis-Nirenberg problem on* $\mathbb{S}^3$, J. Differential Equations, **178** (2002), 264–279.

[4] C. Bandle and L. A. Peletier, *Best Sobolev constants and Emden equations for the critical exponent in* $\mathbf{S}^3$, Math. Ann. **313** (1999), 83–93.

[5] C. Bandle and J. Wei, *Multiple clustered layer solutions for semilinear elliptic problems on* $\mathbf{S}^n$, preprint.

[6] C. Bandle and J. Wei, *Nonradial clustered spike solutions for semilinear elliptic problems on* $\mathbf{S}^n$, preprint.

[7] C. Bandle and J. Wei, *Solutions with an interior bubble and clustered layers for elliptic equations with critical exponents on spherical caps of* $\mathbf{S}^n$, preprint.

[8] H. Brezis, and L. Nirenberg, *Positive solutions of nonlinear elliptic equations involving critical Sobolev exponents*, Comm. Pure Appl. Math., **36** (1983), 437–477.

[9] H. Brezis and L. A. Peletier, *Elliptic equations with critical exponents on spherical caps of* $\mathbf{S}^3$, J. Anal. Math. **98** (2006), 279–316.

[10] Y. Kabeya, H. Morishita and H. Ninomiya, *Imperfect bifurcations arising from elliptic boundary value problems*, Nonlinear Anal., **48** (2002), 663–684.

[11] Y. Kabeya, E. Yanagida and S. Yotsutani, *Global structure of solutions for equations of Brezis-Nirenberg type on the unit ball*, Proc. Royal Soc. Edinburgh **131A** (2001), 647–665.

[12] Y. Kabeya, E. Yanagida and S. Yotsutani, *Canonical forms and structure theorems for radial solutions to semi-linear elliptic problems*, Comm. Pure Appl. Anal. **1** (2002), 85–102.

[13] A. Malchiodi, W.-M. Ni and J. Wei, *Multiple clustered layer solutions for semilinear Neumann problems on a ball*, Ann. Inst. H. Poincaré Anal. Non-linéaire, **22** (2005), 143–163.

[14] S. Moriguchi, K. Udagawa and S. Hitotsumatsu, Mathematical Formulae III, Special Functions (1960), Iwanami Shoten, Tokyo (in Japanese).

[15] N. Shimakura, "Partial Differential Operators of Elliptic Type", AMS, Providence, 1992.

[16] S. Stingelin, *New numerical solutions for the Brezis-Nirenberg problem on* $\mathbf{S}^n$, Universität Basel preprint 2003-15, 2003.

[17] M. E. Taylor, "Partial Differential Equations II", Springer-Verlag, New York, 1996.

[18] E. Yanagida and S. Yotsutani, *Pohozaev identity and its applications*, RIMS Koukyuuroku **834** (1993), 80–90.

[19] E. Yanagida and S. Yotsutani, *A unified approach to the structure of radial solutions for semi-linear elliptic problems*, Japan J. Indust. Appl. Math. **18** (2001), 503–519.

EXISTENCE THEOREMS FOR ELLIPTIC QUASI-VARIATIONAL INEQUALITIES IN BANACH SPACES

RISEI KANO

Department of Mathematics, Graduate School of Science & Technology, Chiba University, 1-33 Yayoi-cho, Inage-ku, Chiba, 263-8522 Japan

NOBUYUKI KENMOCHI

Department of Mathematics, Faculty of Education, Chiba University, 1-33 Yayoi-cho, Inage-ku, Chiba, 263-8522 Japan

YUSUKE MURASE

Department of Mathematics, Graduate School of Science & Technology, Chiba University, 1-33 Yayoi-cho, Inage-ku, Chiba, 263-8522 Japan

Abstract. A class of quasi-variational inequalities (QVI) of the elliptic type is studied in Banach spaces. The concept of QVI was ealier introduced by A. Bensoussan and J. L. Lions [2] and its general theory was evolved by many mathematicians, for instance, see [7,9,13] and a monograph [1]. In this paper we give a generalization of the existence theorem due to J. L. Joly and U. Mosco [6,7] from not only the view-point of the nonlinear operator theory, but also the application to nonlinear variational inequalities including partial differential operators. In fact, employing the compactness argument based on the Mosco convergence (cf.[11]) for convex sets and the graph convergence for nonlinear operaors, we shall prove an abstract existence result for our class of QVI's. Moreover we shall give some new applications to QVI's arising in the material science.

1. Introduction

Let X be a real reflexive Banach space and X^* be its dual. We assume that X and X^* are strictly convex and denote by $< \cdot, \cdot >$ the duality pairing between X^* and X. Given a nonlinear operator A from X into X^*, an element $g^* \in X^*$ and a closed convex subset K of X, the variational inequality is formulated as a problem to find u in X such that

$$u \in K, \quad < Au - g^*, u - w > \leq 0, \quad \forall w \in K. \tag{1.1}$$

Variational inequality has been studied by many mathematicians, for instance see J. L. Lions and G. Stampacchia [10], F. Browder [5], H. Brézis [4] and their references.

The concept of quasi-variational inequality was introduced by A. Bensoussan and J. L. Lions [2] in order to solve some problems in the control theory. Given an operator $A : X \to X^*$, an element $g^* \in X^*$ and a family $\{K(v); v \in X\}$ of closed convex subsets of X, the quasi-variational inequality is a problem to find u in X such that

$$u \in K(u), \quad < Au - g^*, u - w > \leq 0, \quad \forall w \in K(u). \tag{1.2}$$

As is seen from (1.2), the constraint $K(u)$ for the quasi-variational inequality depends upon the unknown u, which causes one of main difficulties in the mathematical treatment of quasi-variational inequalities. The theory of quasi-variational inequality has been evolved for various classes of the mapping $v \to K(v)$ and the linear or nonlinear operator $A : X \to X^*$; see for instance [6,7,13], in which two approaches to quasi-variational inequalities were proposed. One of them is the so-called monotonicity method in Banach lattices X (cf. [13]), and for the mapping $v \to K(v)$ the monotonicity condition

$$\begin{aligned} &\min\{w_1, w_2\} \in K(v_1), \quad \max\{w_1, w_2\} \in K(v_2), \\ &\text{if } v_1, \ v_2 \in X \text{ with } v_1 \leq v_2, \end{aligned} \tag{1.3}$$

is required, and an existence result for (1.2) is proved with the help of a fixed point theorem in Banach lattices. Another is the so-called compactness method in which some compactness properties are required for the mapping $v \to K(v)$ such as $K(v_n)$ converges to $K(v)$ in the Mosco sense, if $v_n \to v$ weakly in X as $n \to \infty$. In this framework, an existence result for (1.2) was shown by J. L. Joly and U. Mosco [7]. However this result seems not enough from some points of applications. The objective of this paper is to generalize the result in [7] to the case that $A : X \to X^*$ is the multivalued pseudo-monotone operator, $Au = \tilde{A}(u, u)$, generated by a semimonotone operator $\tilde{A} : X \times X \to X^*$. In such a case our quasi-variational inequality is of the form: Find $u \in X$ and $u^* \in X^*$ such that

$$u \in K(u), \ u^* \in Au, \quad < u^* - g^*, u - w > \leq 0, \quad \forall w \in K(u). \tag{1.4}$$

This generalization (1.4) is new and enables us to apply it to the following quasi-variational inequality arising in the elastic-plastic torsion problem for visco-elastic material: Find $u \in H_0^1(\Omega)$ and $\tilde{u} \in L^2(\Omega)$ satisfying

$$\begin{cases} |\nabla u| \le k_c(u) \text{ a.e. on } \Omega, \ \tilde{u} \in \beta(u) \text{ a.e. on } \Omega, \\ \displaystyle\sum_{i,j=1}^{N} \int_\Omega a_{ij}(x,u) \frac{\partial u}{\partial x_i} \frac{\partial (u-w)}{\partial x_j} dx + \int_\Omega \tilde{u}(u-w)dx \\ \displaystyle\le \int_\Omega f(u-w)dx, \ \forall w \in H_0^1(\Omega) \text{ with } |\nabla w| \le k_c(u) \text{ a.e. on } \Omega, \end{cases} \tag{1.5}$$

where Ω is a bounded smooth domain in $\mathbf{R}^N$, f is given in $L^2(\Omega)$, $k_c(\cdot)$ is a positive, smooth and bounded function on $\mathbf{R}$ and $\beta(\cdot)$ is a maximal monotone graph in $\mathbf{R} \times \mathbf{R}$ with linear growth at $\pm\infty$. In this case our abstract result is applied to

$$K(v) := \{w \in H_0^1(\Omega); |\nabla w| \le k_c(u) \text{ a.e. on } \Omega\} \tag{1.6}$$

and

$$\tilde{A}(v,u) := -\sum_{i,j=1}^{N} \frac{\partial}{\partial x_j}\left(a_{ij}(x,v)\frac{\partial u}{\partial x_i}\right) + \beta(u); \tag{1.7}$$

it should be noted that the family $\{K(v); v \in H_0^1(\Omega)\}$ given by (1.6) does not satisfy the monotonicity condition (1.3), and that the term $\beta(u)$ of (1.7) is in general multivalued. Therefore, (1.5) is a new application in the respect that the differential form in (1.7) is quasi-linear and the additional term is multivalued in general.

2. Main results

Let X be a real Banach space and X^* be its dual space, and assume that X and X^* are strictly convex. We denote by $< \cdot, \cdot >$ the duality pairing between X^* and X, and by $|\cdot|_X$ and $|\cdot|_{X^*}$ the norms of X and X^*, respectively. For various general concepts on nonlinear multivalued operators from X into X^*, for instance, monotonicity and maximal monotonicity of operators, we refer to the monograph [1]. In this paper, we mean that operators are multivalued, in general. Given a general nonlinear operator A from X into X^*, we use the notations $D(A)$, $R(A)$ and $G(A)$ to denote its domain, range and graph of A. In this paper, we formulate quasi-variational inequalities for a class of nonlinear operators, which is called semimonotone, from $X \times X$ into X^*.

Definition 2.1. An operator $\tilde{A}(\cdot,\cdot) : X \times X \to X^*$ is called *semimonotone*, if $D(\tilde{A}) = X \times X$ and the following conditions (SM1) and (SM2) are satisfied:

(SM1) For any fixed $v \in X$ the mapping $u \to \tilde{A}(v,u)$ is maximal monotone form $D(\tilde{A}(v,\cdot)) = X$ into X^*.

(SM2) Let u be any element of X and $\{v_n\}$ be any sequence in X such that $v_n \to v$ weakly in X. Then, for every $u^* \in \tilde{A}(v,u)$ there exists a sequence $\{u_n^*\}$ in X such that $u_n^* \in \tilde{A}(v_n,u)$ and $u_n^* \to u^*$ in X^* as $n \to +\infty$.

Let $\tilde{A} : D(\tilde{A}) := X \times X \to X^*$ be a semimonotone operator. Then we define $A : D(A) = X \to X^*$ by putting $Au := \tilde{A}(u,u)$ for all $u \in X$, which is called the operator generated by $\tilde{A}$.

Now, for an operator A generated by semimonotone operator, any $g^* \in X^*$ and a mapping $v \to K(v)$ we consider a quasi-variational inequality, denoted by $P(g^*)$, to find $u \in X$ and $u^* \in X^*$ such that

$$P(g^*) \qquad u \in K(u), \ u^* \in Au, \quad < u^* - g^*, w - u > \leq 0, \quad \forall w \in K(u). \tag{2.1}$$

Our main results of this paper are stated as follows.

Theorem 2.1. *Let $\tilde{A} : D(\tilde{A}) = X \times X \to X^*$ be a bounded semimonotone operator and A be the operator generated by $\tilde{A}$. Let K_0 be a bounded, closed and convex set in X. Suppose that to each $v \in K_0$ a non-empty, bounded, closed and convex subset $K(v)$ of K_0 is assigned, and the mapping $v \to K(v)$ satisfies the following continuity properties* (K1) *and* (K2) *:*

(K1) If $v_n \in K_0$, $v_n \to v$ weakly in X (as $n \to \infty$), then for each $w \in K(v)$ there is a sequence w_n in X such that $w_n \in K(v_n)$ and $w_n \to w$ (strongly) in X.

(K2) If $v_n \to v$ weakly in X, $w_n \in K(v_n)$ and $w_n \to w$ weakly in X, then $w \in K(v)$.

Then, for any $g^* \in X^*$, the quasi-variational inequality $P(g^*)$ has at least one solution u.

The following theorem is a slightly general version of Theorem 2.1.

Theorem 2.2. *Let $\tilde{A} : D(\tilde{A}) = X \times X \to X^*$ be a bounded semimonotone operator and A be the operator generated by $\tilde{A}$. Suppose that to each $v \in X$ a non-empty, bounded, closed and convex subset $K(v)$ of X is assigned and* *there is a bounded, closed and convex subset* G_0 *of* X *such that*

$$K(v) \cap G_0 \neq \emptyset, \quad \forall v \in X, \tag{2.2}$$

and

$$\inf_{w^* \in Aw} \frac{< w^*, w - v >}{|w|_X} \to \infty \quad \text{as } |w|_X \to \infty \text{ uniformly in } v \in G_0. \tag{2.3}$$

Furthermore, the mapping $v \to K(v)$ *satisfies the following condition* (K'1) *and the same condition* (K2) *as in Theorem 2.1:*

(K'1) If $v_n \to v$ *weakly in* X*, then for each* $w \in K(v)$ *there is a sequence* w_n *in* X *such that* $w_n \in K(v_n)$ *and* $w_n \to w$ *in* X*.*

Then, for any $g^* \in X^*$*, the quasi-variational inequality* $P(g^*)$ *has at least one solution* u*.*

In our proof of Theorems 2.1 and 2.2 we use some results on nonlinear operators of monotone type, which are mentioned below.

Proposition 2.1. *Let* $\tilde{A} : D(\tilde{A}) = X \times X \to X^*$ *be a semimonotone operator and let* $A : X \to X^*$*. Then, the following two properties (a) and (b) are satisfied:*

(a) For any $v,\ u \in X$*,* $A(v, u)$ *is a non-empty, closed, bounded and convex subset of* X^**.*

(b) Let $\{u_n\}$ *and* $\{v_n\}$ *be sequences in* X *such that* $u_n \to u$ *weakly in* X *and* $v_n \to v$ *weakly in* X *(as* $n \to \infty$*). If* $u_n^* \in \tilde{A}(v_n, u_n)$*,* $u_n^* \to g$ *weakly in* X^* *and* $\limsup_{n\to\infty} < u_n^*, u_n > \leq < g, u >$*, then* $g \in \tilde{A}(v, u)$ *and* $\lim_{n\to\infty} < u_n^*, u_n > = < g, u >$*.*

Proof. Property (a) immediately follows from the maximal monotonicity of $\tilde{A}(v, \cdot)$ for each $v \in X$. Now we show (b). Assume that $\{u_n\}$, $\{v_n\}$ and $\{u_n^*\}$ are such as in the statement of (b), namely,

$$u_n \to u \text{ weakly in } X, \quad v_n \to v \text{ weakly in } X, \quad u_n^* \to g \text{ weakly in } X^* \tag{2.4}$$

and

$$u_n^* \in \tilde{A}(v_n, u_n), \quad \limsup_{n\to\infty} < u_n^*, u_n > \leq < g, u > . \tag{2.5}$$

Now we note from (SM1) that

$$< u_n^* - w_n^*, u_n - w > \geq 0, \quad \forall w \in X, \quad \forall w_n^* \in \tilde{A}(v_n, w). \tag{2.6}$$

For any $w \in X$ and any $w^* \in \tilde{A}(v, w)$, use (SM2) to choose a sequence $\{\tilde{w}_n^*\}$ with $\tilde{w}_n^* \in \tilde{A}(v_n, w)$ and $\tilde{w}_n^* \to w^*$ in X^*. Then, by substituting this sequence w_n^* and letting $n \to \infty$ in (2.6) we have with the help of (2.4) and (2.5)

$$< g - w^*, u - w > \geq 0, \quad \forall w \in X, \ \forall w^* \in \tilde{A}(v, w).$$

This implies that $g \in \tilde{A}(v, u)$, since $\tilde{A}(v, \cdot)$ is maximal monotone. Corresponding to this $g \in \tilde{A}(v, u)$, by (SM2) choose a sequence $g_n \in \tilde{A}(v_n, u)$ such that $g_n \to g$ in X^*. Then, by taking $w = u$ and $w_n^* = g_n$ and letting $n \to \infty$ in (2.6) we obtain

$$\liminf_{n\to\infty} < u_n^*, u_n - u > \geq \liminf_{n\to\infty} < g_n, u_n - u >= 0,$$

namely, $\liminf_{n\to\infty} < u_n^*, u_n > \geq < g, u >$. Therefore, on account of (2.5), it holds that

$$\lim_{n\to\infty} < u_n^*, u_n >=< g, u > .$$

Thus (b) has been seen. ◇

We consider a class of nonlinear operators $A : D(A) = X \to X^*$ satisfying the following properties (PM1) and (PM2):

(1) (PM1) For any $u \in X$, Au is a non-empty, closed, bounded and convex subset of X^*.
(2) (PM2) Let $\{u_n\}$ be a sequence in X such that $u_n \to u$ weakly in X. If $u_n^* \in Au_n$, $u_n^* \to g$ weakly in X^* and $\limsup_{n\to\infty} < u_n^*, u_n > \leq < g, u >$, then $g \in Au$ and $\lim_{n\to\infty} < u_n^*, u_n >=< g, u >$.

This class of nonlinear operaors A is called *pseudo-monotone*. The above proposition says that the operator generated by semimonotone $\tilde{A}$ is pseudo-monotone. As to pseudo-monotone operators we refer to [4,8] for fundamental results on their ranges.

Proposition 2.2. *Let $A_1 : D(A_1) \subset X \to X^*$ be a maximal monotone operator and $A_2 : D(A_2) = X \to X^*$ be a maximal monotone operator. Suppose that*

$$\inf_{v_1^* \in A_1 v, \ v_2^* \in A_2 v} \frac{< v_1^* + v_2^*, v - v_0 >}{|v|_X} \to \infty \quad \text{as} \quad |v|_X \to \infty, \ v_1 \in D(A_1).$$

Then $R(A_1 + A_2) = X^*$.

For a proof of Proposition 2.2, see [4,5,8].

3. Proof of main theorems

We begin with the proof of Theorem 2.1.

Proof of Theorem 2.1: The theorem is proved in the following two steps: (A) The case when $\tilde{A}(v, \cdot)$ is strictly monotone from X into X^* for every $v \in X$; (B) The general case as in the statement of Theorem 2.1.

(In the case of (A))
Let v be any element in K_0 and fix it. We consider the variational inequality with state constraint $K(v)$, namely, to find $u \in X$ and $u^* \in X^*$ such that

$$u \in K(v), \quad u^*(v) \in \tilde{A}(v, u), \quad < u^*(v) - g^*, u - v > \leq 0, \quad \forall w \in K(v). \tag{3.1}$$

This problem is written in the following form equivalent to (3.1):

$$g^* \in \tilde{A}(v, u) + \partial I_{K(v)}(u), \tag{3.2}$$

where $\partial I_{K(v)}(\cdot) : D(\partial I_{K(v)}) \to X^*$ is the subdifferential of the indicator function of $K(v)$, i.e.

$$I_{K(v)}(z) := \begin{cases} 0 & \text{if } z \in K(v), \\ \infty & \text{if } z \in X - K(v); \end{cases}$$

note that $\partial I_{K(v)}$ is maximal monotone. It follows from Proposition 2.2 that $R(\tilde{A}(v, \cdot) + \partial I_{K(v)}) = X^*$; in fact, the coerciveness condition of Proposition 2.2 for $A_1 := \partial I_{K(v)}$ and $A_2 := A(v, \cdot)$ is automatically satisfied, since $D(A_1) = K(v)$ is bounded in X. Hence there exists an element u which satisfies (3.2) (therefore (3.1)) for each $v \in K_0$. Moreover, the solution u is unique by the strict monotonicity of $\tilde{A}(v, \cdot)$ and $u \in K_0$. Using this fact, we define a mapping S from K_0 into itself which assigns to each $v \in K_0$ the solution $u \in K_0$ of (3.1), i.e. $u = Sv$.

Next, we show that S is weakly continuous in K_0. Let $\{v_n\}$ be any sequence in K_0 such that $v_n \to v$ weakly in X, and put $u_n = Sv_n$ $(\in K_0)$ for $n = 1, 2, \cdots$. Now, let $\{u_{n_k}\}$ be any weakly convergent subsequence of $\{u_n\}$ and denote by u the weak limit; note by condition (K2) that $u \in K(v)$. We are going to check that u is a unique solution of (3.1). To do so, first observe that there is $u_n^* \in \tilde{A}(v_n, u_n)$ such that

$$< u_n^* - g^*, u_n - w > \leq 0, \quad \forall w \in K(v_n) \tag{3.3}$$

Using condition (K1), we find a sequence $\{\tilde{u}_k\}$ such that $\tilde{u}_k \in K(v_{n_k})$ and $\tilde{u}_k \to u$ in X (as $k \to \infty$). By the boundedness of $\tilde{A}(\cdot, \cdot)$, we may assume that $u_{n_k}^* \to u^*$ weakly in X^* for some $u^* \in X^*$. Now, taking $n = n_k$ and $w = \tilde{u}_k$ in (3.3), we see that

$$\begin{aligned}\limsup_{k\to\infty} < u_{n_k}^*, u_{n_k} > &= \limsup_{k\to\infty}\{< u_{n_k}^*, u_{n_k} - \tilde{u}_k > + < u_{n_k}^*, \tilde{u}_k >\} \\ &\le \limsup_{k\to\infty}\{< g^*, u_{n_k} - \tilde{u}_k > + < u_{n_k}^*, \tilde{u}_k >\} \\ &= < u^*, u >\end{aligned}$$

Therefore it follows from Proposition 2.1 that

$$u^* \in \tilde{A}(v,u), \quad \lim_{k\to\infty} < u_{n_k}^*, u_{n_k} > = < u^*, u > . \tag{3.4}$$

We go back to (3.3) with $n = n_k$. For any $w \in K(v)$, we use (K1) to choose a sequence $w_k \in K(v_{n_k})$ such that $w_k \to w$ in X. Taking $n = n_k$ and $w = w_k$ in (3.3) and passing to the limit as $k \to \infty$ in (3.3), by (3.4) we obtain the variational inequality (3.1). Thus $u = Sv$, and S is weakly continuous in K_0.

Since K_0 is a weakly compact and convex set in X, we infer from the well-known fixed-point theorem for compact mappings that S has at least one fixed point in K_0. This fixed point u is clearly a solution of our quasi-variational ineqality P(g^*). ◇

(In the case of (B))

We approximate $\tilde{A}(v,u)$ by $\tilde{A}_\varepsilon(v,u) := \tilde{A}(v,u) + \varepsilon J(u)$ for any $u, v \in X$ and with parameter $\varepsilon \in (0,1]$; note that the duality mapping J from X into X^* is strictly monotone and hence $\tilde{A}_\varepsilon$ is a semimonotne operator such that $\tilde{A}_\varepsilon(v,\cdot)$ is strictly monotone for every $v \in X$. By the result of the case (A), for each $g^* \in X^*$ there exists a solution $u_\varepsilon \in K_0$ of the quasi-variational inequality

$$u_\varepsilon \in K(u_\varepsilon),\ u_\varepsilon^* \in Au_\varepsilon, \quad < u_\varepsilon^* + \varepsilon J u_\varepsilon - g^*, u_\varepsilon - w > \le 0, \ \ \forall w \in K(u_\varepsilon), \tag{3.5}$$

where A is the operator generated by $\tilde{A}$. Now, choose a sequence $\{\varepsilon_n\}$, with $\varepsilon_n \downarrow 0$, such that $u_n := u_{\varepsilon_n} \to u$ weakly in X for some $u \in K_0$. Then, by conditions (K1) and (K2), we see that $u \in K(u)$ and there is a sequence $\{\tilde{u}_n\}$ such that $\tilde{u}_n \in K(u_n)$ and $\tilde{u}_n \to u$ in X. Moreover, by the boundedness of $\{u_n^* := u_{\varepsilon_n}^*\}$ in X^*, we may assume that $u_n^* \to u^*$ weakly in X^* for some $u^* \in X^*$. Substitute u_n and $\tilde{u}_n$ for u_ε and w in (3.5) with

$\varepsilon = \varepsilon_n$, respectively, and pass to the limit as $n \to \infty$ to get

$$\begin{aligned}&\limsup_{n\to\infty} < u_n^*, u_n - u > \\ &= \limsup_{n\to\infty}\{< u_n^* + \varepsilon_n J u_n, u_n - \tilde{u}_n > + < u_n^* + \varepsilon_n J u_n, \tilde{u}_n - u >\} \\ &= \limsup_{n\to\infty}\{< g^*, u_n - \tilde{u}_n > + < u_n^*, \tilde{u}_n - u >\} \\ &\le 0.\end{aligned}$$

Since A is pseudo-monotone from X into X^* (cf. Proposition 2.1), it follows from the above inequality that

$$u^* \in Au, \qquad \lim_{n\to\infty} < u_n^*, u_n >=< u^*, u > . \tag{3.6}$$

Now, for each $w \in K(u)$, by (K1) we choose $\{\tilde{w}_n\}$ such that $\tilde{w}_n \in K(u_n)$ and $\tilde{w}_n \to w$ in X, and then substitute them for w in (3.5) with $\varepsilon = \varepsilon_n$ to have

$$< u_n^* + \varepsilon_n J u_n - g^*, u_n - \tilde{w}_n >\le 0. \tag{3.7}$$

By (3.6), letting $n \to +\infty$ in (3.7) yields that $< u^* - g^*, u - w >\le 0$. Thus u is a solution of our quasi-variational ineqality $\mathrm{P}(g^*)$. $\diamondsuit$

Next we proceed to the proof of Theorem 2.2.

Proof of Theorem 2.2: Put $d_1 := \sup_{w\in G_0} |w|_X$ and

$$d_2 := \sup\left\{|w|_X; w \in X, \inf_{w^*\in Aw} \frac{< w^*, w - v >}{|w|_X} \le |g^*|_{X^*}(1 + d_1), {}^\forall v \in G_0\right\}.$$

By condition (2.3), d_2 is finite. Also we put $M_0 := d_1 + d_2 + 1$, and for any number $M \ge M_0$ consider the closed ball $B_M := \{w \in X; |w|_X \le M\}$ as well as bounded closed and convex sets $K_M(v) := K(v) \cap B_M$ for all $v \in X$. Since $G_0 \subset B_M$, (2.2) implies that $K_M(v)$ is non-empty for every $v \in X$.

We now show that conditions (K1) and (K2) in Theorem 2.1 with $K_0 = B_M$ and $K(\cdot) = K_M(\cdot)$ are satisfied. The verification of (K2) is easy. We check condition (K1) for $K_0 = B_M$ and $K(\cdot) = K_M(\cdot)$ as follows. Let w be any element in $K_M(v)$. Then, by condition (K'1) for $K(\cdot)$, for a sequence $\{v_n\} \subset B_M$ weakly converging to v there is a sequence $\{w_n\}$ such that $w_n \in K(v_n)$ and $w_n \to w$ in X. In the case of $|w|_X < M$, we see that $|w_n|_X < M$ and hence $w_n \in K_M(v_n)$ for all large n. In the case of $|w|_X = M$, choose an element $v_0 \in K(v) \cap G_0$ and put

$$w^m := (1 - \frac{1}{m})w + \frac{1}{m}v_0, \quad m = 1, 2, \cdots . \tag{3.8}$$

Clealy $w^m \in K_M(v)$ and $|w^m|_X < M$. Therefore, according to the above argument, for each m there is a sequence $\{w_n^m\}_{n=1}^\infty$ such that $w_n^m \in K_M(v_n)$ and $w_n^m \to w^m$ in X as $n \to \infty$. For each m choose a number $n(m)$ so that $|w^m - w_n^m|_X \le \frac{1}{m}$ for all $n \ge n(m)$. We may choose $\{n(m)\}_{m=1}^\infty$ so that $n(m-1) < n(m)$ for all $m = 0, 1, \cdots$, where $n(0) = 1$. We put

$$w_n = w_n^m \quad \text{if} \quad n(m) \le n < n(m+1), \quad m = 0, 1, \cdots . \tag{3.9}$$

It is easy to see that $w_n \in K_M(v_n)$ and $w_n \to w$ in X.

By the above observation we can apply Theorem 2.1 to find an element u_M such that

$$\begin{aligned} &u_M \in K_M(u_M), \quad u_M^* \in Au_M, \quad < u_M^* - g^*, u_M - w > \le 0, \\ &\qquad \forall w \in K_M(u_M). \end{aligned} \tag{3.10}$$

Also, by condition (2.3), $\{u_M; M \ge M_0\}$ is bounded in X, so that there are a sequence $\{M_n\}$ with $M_n \uparrow \infty$ and elements $u \in X$, $u^* \in X^*$ such that $u_n := u_{M_n} \to u$ weakly in X and $u_n^* := u_{M_n}^* \to u^*$ weakly in X^* as $n \to \infty$. We note $u \in K(u)$ by (K2). It follows from (K'1) that for each $w \in K(u)$ there is a sequence $\{\tilde{w}_n\}$ such that $\tilde{w}_n \in K(u_n)$ and $\tilde{w}_n \to w$ in X. In particular, denote by $\{\tilde{u}_n\}$ the sequence $\{\tilde{w}_n\}$ corresponding to $w = u$. Here, we substitute M_n and $\tilde{u}_n$ for M and w in (3.10) to obtain $< u_n^* - g^*, u_n - \tilde{u}_n > \le 0$. Hence it follows that

$$\begin{aligned} &\limsup_{n\to\infty} < u_n^*, u_n - u > \\ &= \limsup_{n\to\infty} \{< u_n^* - g^*, u_n - \tilde{u}_n > + < u_n^*, \tilde{u}_n - u > + < g^*, u_n - \tilde{u}_n >\} \\ &\le 0. \end{aligned}$$

By the pseudo-monotonicity of A this implies that

$$u^* \in Au, \quad \lim_{n\to+\infty} < u_n^*, u_n > = < u^*, u > .$$

By making use of these properties with (K1) and passing to the limit as $n \to \infty$ in (3.10) with $M = M_n$ and $w = \tilde{w}_n$ as above, we see that $u \in K(u)$ and $< u^* - g^*, u - w > \le 0$ for all $w \in K(u)$. Thus u is a solution of our problem $\mathrm{P}(g^*)$. $\diamondsuit$

4. Application to obstacle problems

In this section, let Ω be a bounded domain in $\mathbf{R}^N$, $1 \le N < \infty$, with smooth boundary $\Gamma := \partial\Omega$, and put $X := W_0^{1,p}(\Omega)$ or $W^{1,p}(\Omega)$, $1 < p < \infty$. Let $a_i(x, \eta, \xi)$, $i = 1, 2, \cdots, N$, be functions on $\Omega \times \mathbf{R} \times \mathbf{R}^N$ such that (a1) for all $(\eta, \xi) \in \mathbf{R} \times \mathbf{R}^N$ the function $x \to a_i(x, \eta, \xi)$ is measurable on Ω for each $i = 1, 2, \cdot, N$;

(a2) for a.e. $x \in \Omega$ the function $(\eta, \xi) \to a_i(x, \eta, \xi)$ is continuous on $\mathbf{R} \times \mathbf{R}^N$ for each $i = 1, 2, \cdots, N$;

(a3) there are positive constants c_0, c_0' and c_1, c_1' such that

$$c_0|\xi|^{p-1} - c_0' \le a_i(x, \eta, \xi) \le c_1|\xi|^{p-1} + c_1', \ i = 1, 2, \cdots, N, \quad (4.1)$$

$$\text{a.e. } x \in \Omega, \ \forall \eta \in \mathbf{R}, \ \forall \ \xi = (\xi_1, \xi_2, \cdots, \xi_N) \in \mathbf{R}^N;$$

(a4) the following monotonicity property is satisfied:

$$\sum_{i=1}^{N} \left(a_i(x, \eta, \xi) - a_i(x, \eta, \bar{\xi})\right)(\xi_i - \bar{\xi}_i) \ge 0, \quad (4.2)$$

$$\text{a.e. } x \in \Omega, \ \forall \eta \in \mathbf{R}, \ \forall \xi = (\xi_1, \xi_2, \cdots, \xi_N),$$
$$\forall \bar{\xi} = (\bar{\xi}_1, \bar{\xi}_2, \cdots, \bar{\xi}_N) \in \mathbf{R}^N.$$

We define a mapping $\tilde{A}_0(\cdot, \cdot) : X \times X \to X^*$ by putting

$$< \tilde{A}_0(v, u), w >= \sum_{i=1}^{N} \int_\Omega a_i(x, v, \nabla u) \frac{\partial w}{\partial x_i} dx, \quad \forall u, v, w \in X, \quad (4.3)$$

and $A_0 u$ by $\tilde{A}_0(u, u)$ for every $u \in X$. Also, let β be a maximal monotone operator from $D(\beta) = \mathbf{R}$ into $\mathbf{R}$ such that the primitive $\hat{\beta}(r) := \int_0^r \beta(s) ds$ of β satisfies that

$$c_2|r|^p - c_2' \le \hat{\beta}(r) \le c_3|r|^p + c_3', \quad \forall r \in \mathbf{R}, \quad (4.4)$$

where c_2, c_2', c_3 and c_3' are positive constants. Now, we consider an operator $\tilde{A} : X \times X \to X^*$ given by $\tilde{A}(v, u) := \tilde{A}_0(v, u) + \beta(u)$ for all v, $u \in X$. It is easy to see from (4.3) and (4.4) that $\tilde{A}$ is a bounded and semimonotone operator from $X \times X$ into X^*.

Application 4.1. (Gradient obstacle problem)
Let $X = W_0^{1,p}(\Omega)$, and k_c be a Lipschitz continuous real function on $\mathbf{R}$ such that

$$0 < k_c(r) \le k^*, \quad \forall r \in \mathbf{R}, \quad (4.5)$$

where k^* is a positive constant, and put

$$K(v) := \{w \in X; |\nabla w| \le k_c(v) \ \text{ a.e. on } \Omega\}, \quad \forall v \in X. \quad (4.6)$$

Also, we set

$$K_0 := \{w \in X; |\nabla w| \le k^* \text{ a.e. on } \Omega\},; \tag{4.7}$$

note from the Sobolev imbedding theorem that K_0 is compact in $C(\bar{\Omega})$.

Lemma 4.1. *The family $\{K(v); v \in X\}$ and the set K_0, which are respectively given by (4.6) and (4.7), satisfy conditions (K1) and (K2).*

Proof. We prove (K2). Suppose that $v_n \in K_0$, $w_n \in K(v_n)$, $v_n \to v$ weakly in X and $w_n \to w$ weakly in X. Then, $v_n \to v$ in $C(\bar{\Omega})$ and hence $k_c(v_n) \to k_c(v)$ in $C(\bar{\Omega})$. Therefore, given $\varepsilon > 0$, there exists a positive integer n_ε such that

$$|k_c(v_n) - k_c(v)| \le \varepsilon \quad \text{on } \Omega, \quad \forall n \ge n_\varepsilon. \tag{4.8}$$

This shows that

$$|\nabla w_n| \le k_c(v_n) \le k_c(v) + \varepsilon \quad \text{a.e. on } \Omega, \quad \forall n \ge n_\varepsilon. \tag{4.9}$$

Clearly the set $K_\varepsilon(v) := \{w \in X; |\nabla w| \le k_c(v) + \varepsilon \text{ a.e. on } \Omega\}$ is bounded, closed and convex in X, so that $K_\varepsilon(v)$ is weakly compact in X. Now, we derive by letting $n \to +\infty$ in (4.9) that $w \in K_\varepsilon(v)$. Since $\varepsilon > 0$ is arbitrary, we have $w \in K(v)$. Thus (K2) holds.

Next we show (K1). Suppose that $v \in K_0$, $w \in K(v)$ and $\{v_n\} \subset K_0$ such that $v_n \to v$ weakly in X. By the compactness of K_0 in $C(\bar{\Omega})$ we have that $v_n \to v$ in $C(\bar{\Omega})$. Since $cw \in K(v)$ for all constant $c \in (0,1)$ and $cw \to w$ as $c \uparrow 1$ in X, it is enough to show the existence of a sequence $\{\tilde{w}_n\}$ such that $\tilde{w}_n \in K(v_n)$ and $\tilde{w}_n \to \tilde{w}$ in X, when $\tilde{w} = cw$ for any $c \in (0,1)$. In such a case, by conditions (4.5) and (4.8), we can take a small $\varepsilon > 0$ so that $|\nabla \tilde{w}| \le k_c(v) - \varepsilon$ a.e. on $\bar{\Omega}$. Furthermore, for this $\varepsilon > 0$ we can find a positive integer n_ε such that $k_c(v) \le k_c(v_n) + \varepsilon$ for all $n \ge n_\varepsilon$. This implies that $|\nabla \tilde{w}| \le k_c(v_n)$ a.e. on Ω, namely $\tilde{w} \in K(v_n)$ for all $n \ge n_\varepsilon$. Now we define $\{\tilde{w}_n\}$ by putting

$$\tilde{w}_n = \begin{cases} \tilde{w} & \text{for } n \ge n_\varepsilon, \\ \text{some function in } K(v_n) & \text{for } 1 \le n < n_\varepsilon. \end{cases}$$

Clearly this is a required sequence in condition (K1). ◇

According to Lemma 4.1, we can apply Theorem 2.1 to solve the following quasi-variational inequality:

$$\begin{cases} u \in X, \quad |\nabla u| \le k_c(u) \text{ a.e. on } \Omega, \quad u^* \in L^q(\Omega) \text{ with } u^* \in \beta(u) \text{ a.e. on } \Omega; \\ \displaystyle\sum_{i=1}^N \int_\Omega a_i(x,u,\nabla u)(u_{x_i} - w_{x_i})dx + \int_\Omega u^*(u-w)dx \le \int_\Omega f(u-w)dx, \\ \qquad\qquad \forall w \in X \text{ with } |\nabla w| \le k_c(u) \text{ a.e. on } \Omega, \end{cases} \tag{4.10}$$

where f is given in $L^q(\Omega)$, $\frac{1}{p} + \frac{1}{q} = 1$.

Application 4.2. (Interior obstacle problem)
Let Ω be a one-dimensional bounded open interval, say $(0,1)$, $X := W^{1,p}(0,1)$, $1 < p < \infty$, and let $a_1(x,\eta,\xi)$ be a function which satisfy (a1)-(a4) with $N = 1$. Let β be the same as above.

Also, let $k_c(\cdot)$ be a Lipschitz continuous real function on $\mathbf{R}$ such that

$$k_c(r) \le k^*, \quad \forall r \in \mathbf{R}, \tag{4.11}$$

where k^* is a constant, and put

$$K(v) := \{w \in X; w \ge k_c(v) \text{ on } (0,1)\}, \quad \forall v \in X. \tag{4.12}$$

Clearly, the constant function k^* belongs to $K(v)$ for all $v \in X$. Futhermore, it follows from (4.1) and (4.4) that

$$\int_0^1 \{a_1(x,v,v_x)v_x + v^*(v-k^*)\}\, dx \ge c_4|v|_X^p - c_4', \tag{4.13}$$

$$\forall v \in X, \quad \forall v^* \in L^q(0,1) \text{ with } v^* \in \beta(v) \text{ a.e. on } (0,1),$$

where c_4 and c_4' are positive constants.

Lemma 4.2. *The family $\{K(v); v \in X\}$ and the set $G_0 := \{k^*\}$ satisfy conditions (2.2), (2.3), (K'1) and (K2).*

Proof. We observe that X is compactly embedded in $C([0,1])$. On account of this fact, the verification of (K'1) and (K2) can be done in a way similar to that in the proof of Lemma 4.1. Also, (2.2) and (2.3) are immediately seen from (4.11), (4.12) and (4.13). ◇

Now, given $f \in L^q(0,1)$, by applying Theorem 2.2 we find functions u and u^* such that

$$\begin{cases} u \in X,\ u \geq k_c(u) \text{ on } (0,1),\ u^* \in L^q(0,1) \text{ with } u^* \in \beta(u) \text{ a.e. on } (0,1), \\ \displaystyle\int_0^1 \{a_1(x,u,u_x)(u_x - w_x) + u^*(u-w)\}\,dx \leq \int_0^1 f(u-w)dx, \\ \qquad\qquad \forall w \in X \text{ with } w \geq k_c(u) \text{ on } (0,1). \end{cases} \tag{4.14}$$

Application 4.3.(Boundary obstacle problem)
We consider a quasi-variational inequality with constraints on the boundary. Let $X := W^{1,p}(0,1)$, $1 < p < \infty$, and let $a_1(x,\eta,\xi)$ be a function satisfying condition (a1)-(a4) with $N = 1$. Let β be the same as above. Also, let $k_c^i(\cdot)$, $i = 0,1$, be two Lipschitz continuous real functions on $\mathbf{R}$ such that

$$k_c^i(r) \leq k^*, \quad \forall \in \mathbf{R},\ i = 0,1,$$

where k^* is a constant. We define

$$K(v) := \left\{w \in X; w(i) \geq k_c^i(v(i)),\ i = 0,1\right\}, \quad \forall v \in X, \tag{4.15}$$

and $G_0 = \{k^*\}$, being the singleton set of constant function k^*. Then it is easy to see that G_0 and the family $\{K(v)\}$ satisfy conditions (2.2), (2.3), (K'1) and (K2) in the statement of Theorem 2.2. Therefore, by Theorem 2.2, for each $f \in L^q(0,1)$, $\frac{1}{p} + \frac{1}{q} = 1$, we find functions u and u^* such that

$$\begin{cases} u \in X,\quad u(i) \geq k_c(u(i)),\ i = 0,1,\quad u^* \in L^q(0,1) \\ \qquad \text{with } u^* \in \beta(u) \text{ a.e. on } (0,1); \\ \displaystyle\int_0^1 \{a_1(x,u,u_x)(u_x - w_x) + u^*(u-w)\}\,dx \leq \int_0^1 f(u-w)dx, \\ \qquad\qquad \forall w \in X \text{ with } w(i) \geq k_c(u(i)),\ i = 0,1. \end{cases}$$

In Applications 4.2 and 4.3 we supposed that the space dimension of Ω is one, since we do not know whether condition (K1) or (K'1) holds or not in higher space dimensional cases.

Application 4.4. Next we consider a system of quasi-variational inequality including quasi-linear partial differential operators. Let $a_k(u_1,u_2)$, $k = 1,2$, be continuous functions with respect to u_1 and u_2 on $\mathbf{R} \times \mathbf{R}$ such that

$$c_5 \leq a_k(u_1,u_2) \leq c_5', \quad \forall u_1,\ u_2 \in \mathbf{R},$$

and let $\beta_k(u_k)$, $k = 1,\ 2$, be maximal monotone operators from $D(\beta_k) = \mathbf{R}$ into $\mathbf{R}$ and the primitives $\hat{\beta}_k(r) := \int_0^r \beta_k(s)ds$ satisfy that

$$c_6|r|^{p_k} - c_6' \le \hat{\beta}_k(r) \le c_7|r|^{p_k} + c_7', \quad \forall r \in \mathbf{R},\ k = 1,2,$$

where p_k, $k = 1,2$, are constants satisfying $2 \le p_k < \infty$. Our problem is formulated in the space $X := W^{1,p_1}(0,1) \times W^{1,p_2}(0,1)$, and define an operator $\tilde{A}_0 : X \to X^*$ by

$$< \tilde{A}_0(v,u), w > = \int_0^1 a_1(v_1,v_2)|u_{1,x}|^{p_1-2}u_{1,x}w_{1,x}dx + \int_0^1 a_2(v_1,v_2)|u_{2,x}|^{p_2-2}u_{2,x}w_{2,x}dx,$$

$$\forall v := (v_1,v_2), \quad u := (u_1,u_2), \quad w := (w_1,w_2) \in X,$$

as well as an operator $\tilde{A} : X \to X^*$ by $\tilde{A}(v,u) := \hat{A}_0(v,u) + (\beta_1(u_1), \beta_2(u_2))$, namely, the first and second compornent of $\tilde{A}(v,u)$ are respectively written in the form

$$-(a_1(v_1,v_2)|u_{1,x}|^{p_1-2}u_{1,x})_x + \beta_1(u_1)$$
$$\text{and} \quad -(a_2(v_1,v_2)|u_{2,x}|^{p_2-2}u_{2,x})_x + \beta_2(u_2),$$

which are multivalued, in general, because of $\beta_1(u_1)$ and $\beta_2(u_2)$. As was already seen just before Application 4.1, $\tilde{A}$ is bounded and semimonotone from X into X^*. Further let $k_{kc}(u_1,u_2)$, $k = 1,2$, be Lipschitz continuous functions with respect to $u = (u_1,u_2)$ on $\mathbf{R} \times \mathbf{R}$ such that $k_{kc} \le k^*$ on $\mathbf{R} \times \mathbf{R}$ for a constant k^*. Then, given $f_k \in L^{q_k}(0,1)$, $\frac{1}{p_k} + \frac{1}{q_k} = 1$, $k = 1,2$, by virtue of Theorem 2.2 there exist vector functions $u = (u_1,u_2) \in X$ and $u^* = (u_1^*, u_2^*) \in L^{q_1}(0,1) \times L^{q_2}(0,1)$ such that

$$\begin{cases} u_k \ge k_{kc}(u_1,u_2) \text{ on } (0,1), \quad u_k^* \in \beta_k(u_k) \text{ a.e. on } (0,1), \quad k = 1,2, \\ \displaystyle\int_0^1 a_k(u_1,u_2)|u_{k,x}|^{p_k-2}(u_{k,x} - w_{k,x})dx \\ \qquad \displaystyle + \int_0^1 u_k^*(u_k - w_k)dx \le \int_0^1 f_k(u_k - w_k)dx, \\ \qquad \forall w_k \in W^{1,p_k}(0,1) \text{ with } w_k \ge k_{kc}(u_1,u_2) \text{ on } (0,1), \quad k = 1,2. \end{cases}$$

5. Application to problems with non-local constraints

Let Ω be a bounded domain in $\mathbf{R}^N$, $1 \le N < \infty$, with smooth boundary $\Gamma := \partial\Omega$; and let $X := W^{1,p}(\Omega)$ or $W_0^{1,p}(\Omega)$, $1 < p < \infty$. Let $a_i(x,\eta,\xi)$, $i = 1,2,\cdots,N$, and β be as in the previous section; conditions (4.1), (4.2) and

(4.4) are satisfied as well. Furthmore let $k_c(\cdot)$ be a Lipschitz continuous function on $\mathbf{R}$ with bounded Lipschitz continuous derivative $k_c'(\cdot)$ on $\mathbf{R}$; condition (4.11) is satisfied as well.

Given a singlevalued compact mapping $\Lambda : X \to X$, we define constraint sets $K(v)$ in X by

$$K(v) := \{w \in X; w \geq k_c(\Lambda v) \text{ a.e. on } \Omega\}, \quad \forall v \in X. \tag{5.1}$$

Clearly, $K(v)$ is non-empty, closed and convex in X for every $v \in X$.

Lemma 5.1. *The family $\{K(v); v \in X\}$ given by (5.1) and the set $G_0 := \{k^*\}$ satisfy conditions (K'1) and (K2).*

Proof. Assume that $v_n \to v$ weakly in X and let w be any function in $K(v)$, namely $w \geq k_c(\Lambda v)$ a.e. on Ω. We note that $\Lambda v_n \to \Lambda v$ in X and hence $k_c(\Lambda v_n) \to k_c(\Lambda v)$ in X. Putting $w_n = w - k_c(\Lambda v) + k_c(\Lambda v_n)$, we see that $w_n \in K(v_n)$, $w_n \to w$ in X. Thus (K'1) is verified. Next, assume that $w_n \in K(v_n)$, $w_n \to w$ weakly in X and $v_n \to v$ weakly in X. Then $w_n \to w$ and $v_n \to v$ in $L^p(\Omega)$ as well as $k_c(\Lambda v_n) \to k_c(\Lambda v)$ in $L^p(\Omega)$. Hence $w \geq k_c(\Lambda v)$ a.e. on Ω, that is $w \in K(v)$. Thus (K2) is obtained. $\diamondsuit$

For the same mapping $\tilde{A} := \tilde{A}_0 + \beta$ as in the previous section, all the conditions of Theorem 2.2 are verified. Therefore, given a function $f \in L^q(\Omega)$, $\frac{1}{p} + \frac{1}{q} = 1$, there exists a function $u \in X$ and $u^* \in L^q(\Omega)$ such that

$$\begin{cases} u \in X, \ u \geq k_c(\Lambda u) \text{ a.e. on } \Omega, \ u^* \in \beta(u) \ \text{ a.e. on } \Omega, \\ \displaystyle\sum_{i=1}^N \int_\Omega a_i(x, u, \nabla u)(u_{x_i} - v_{x_i})dx + \int_\Omega u^*(u - w)dx \leq \int_\Omega f(u - w)dx, \\ \qquad\qquad \forall w \in X \text{ with } w \geq k_c(\Lambda u) \text{ a.e. on } \Omega. \end{cases} \tag{5.2}$$

Next, consider typical examples of the mapping Λ.

(Example 5.1) Consider the case when $p = 2$ and $X := W^{1,2}(\Omega)$. Let ν be a positive number. Then, for each $v \in X$, the boundary value problem

$$-\nu\Delta v + v = u \ \text{ in } \Omega, \quad \frac{\partial v}{\partial n} = 0 \ \text{ on } \Gamma, \tag{5.3}$$

has a unique solution v in $W^{2,2}(\Omega)$. Now, we define $\Lambda : X \to X$ by $v = \Lambda u$, where v is the solution of (5.3) for $u \in X$. Since Λ is bounded and linear from X into $W^{2,2}(\Omega)$, we see that Λ is compact from X into itself; in fact, $\Lambda = (I - \nu\Delta)^{-1}$. In this case, given $f \in L^2(\Omega)$, the quasi-variational inequality (5.2) with $p = 2$ is a system to find $u \in W^{1,2}(\Omega)$ with $u^* \in L^2(\Omega)$ and $v \in W^{2,2}(\Omega)$ such that

$$\begin{cases} u \geq k_c(v) \text{ a.e. on } \Omega, \quad u^* \in \beta(u) \quad \text{a.e. on } \Omega, \\ \displaystyle\sum_{i=1}^{N} \int_\Omega a_i(x, u, \nabla u)(u_{x_i} - w_{x_i})dx + \int_\Omega u^*(u - w)dx \\ \qquad \leq \displaystyle\int_\Omega f(u - w)dx, \quad \forall w \in X \text{ with } w \geq k_c(v) \text{ a.e. on } \Omega, \\ -\nu \Delta v + v = u \quad \text{a.e. on } \Omega, \\ \frac{\partial v}{\partial n} = 0 \quad \text{a.e. on } \Gamma, \end{cases} \tag{5.4}$$

and by virtue of Theorem 2.2 the above system has at least one solution $\{u_\nu, v_\nu\}$ for each $\nu \in (0,1]$. Also, see [12] for a related non-local quasi-variational inequality.

Proposition 5.1. *Assume that the space dimension is one, $p = 2$ and $\Omega = (0,1)$. Let $\{u_\nu, v_\nu\}_{\nu \in (0,1]}$ be a family of solutions of (5.4). Then $\{u_\nu, v_\nu\}$ is bounded in $W^{1,2}(\Omega) \times W^{1,2}(\Omega)$, and for any weak limit u of $\{u_\nu\}$ as $\nu \downarrow 0$ is a solution of the interior obstacle problem (4.14).*

Proof. By testing k^* of (4.11) in quasi-variational inequality in (5.4) with $u = u_\nu$, it is easy to see that $\{u_\nu\}_{\nu \in (0,1]}$ is bounded in $W^{1,2}(0,1)$, hence is relatively compact in $C([0,1])$. Now, let u be any weak limit of $\{u_\nu\}$ and choose a sequence ν_n with $\nu_n \downarrow 0$ so that $u_n := u_{\nu_n} \to u$ weakly in $W^{1,2}(0,1)$, hence $u_n := u_{\nu_n} \to u$ uniformly on $[0,1]$, as $n \to \infty$. Putting $v_n := v_{\nu_n}$ and multiplying $-\nu_n \Delta v_n + v_n = u_n$ by v_n and $-\Delta v_n$, we obtain

$$\int_0^1 |\nabla v_n|^2 dx + \int_0^1 |v_n|^2 dx \leq \int_0^1 |v_n||u_n| dx,$$

$$\nu_n \int_0^1 |\Delta v_n|^2 dx + \int_0^1 |\nabla v_n|^2 dx \leq \int_0^1 |\nabla v_n||\nabla u_n| dx,$$

whence $\{\nu_n |\Delta v_n|^2_{L^2(0,1)}\}$ is bounded and $\{v_n\}$ is bounded in $W^{1,2}(0,1)$. Since $v_n - u_n = \nu_n \Delta v_n \to 0$ in $L^2(0,1)$ and $u_n \geq k_c(v_n)$, it follows from the above esitimates that

$$v_n \to u \quad \text{weakly in } W^{1,2}(\Omega) \text{ and uniformly on } [0,1] \tag{5.5}$$

as well as

$$k_c(v_n) \to k_c(u) \quad \text{uniformly on } [0,1], \quad u \geq k_c(u) \quad \text{on } (0,1). \tag{5.6}$$

Let w be any function with $w \geq k_c(u)$ on $(0,1)$. Given $\varepsilon > 0$, choose a positive integer n_ε so that

$$\begin{aligned} &w + \varepsilon \geq k_c(u) + \varepsilon \geq k_c(v_n), \\ &u + \varepsilon \geq k_c(u) + \varepsilon \geq k_c(v_n), \qquad \text{on } (0,1), \quad \forall n \geq n_\varepsilon; \end{aligned} \tag{5.7}$$

this is possible on account of (5.5) and (5.6). Now, take $u + \varepsilon$ as w the quasi-variational inequality in (5.4) which u_n satisfies, to get

$$\int_0^1 a_1(x, u_n, u_{n,x})(u_{n,x} - u_x)dx + \int_0^1 u_n^*(u_n - u - \varepsilon)dx \leq \int_0^1 f(u_n - u - \varepsilon)dx,$$

where $u_n^* \in L^2(0,1)$ is a function satisfying $u_n^* \beta(u_n)$ a.e. on $(0,1)$. Then we have

$$H := \limsup_{n\to\infty} \int_0^1 a_1(x, u_n, u_{n,x})(u_{n,x} - u_x)dx + \int_0^1 u_n^*(u_n - u)dx \leq M_0\varepsilon,$$

where M_0 is a positive constant. Since $\varepsilon > 0$ is arbitrary, it holds that $H \leq 0$. This inequality implies by the pseudo-monotonicity property that

$$\lim_{n\to\infty} \int_0^1 a_1(x, u_n, u_{n,x})(u_{n,x} - u_x)dx + \int_0^1 u_n^*(u_n - u)dx = 0 \tag{5.8}$$

and

$$\begin{aligned} &\lim_{n\to\infty} \int_0^1 a_1(x, u_n, u_{n,x})\tilde{w}_x dx + \int_0^1 u_n^* \tilde{w} dx \\ &\qquad = \int_0^1 a_1(x, u, u_x)\tilde{w}_x dx + \int_0^1 u^* \tilde{w} dx, \end{aligned} \tag{5.9}$$

$$\forall \tilde{w} \in W^{1,2}(0,1),$$

where $u^* \in L^2(0,1)$ with $u^* \in \beta(u)$ a.e. on $(0,1)$. Going back to the quasi-variational inequality (5.4) with $\nu = \nu_n$ and $u = u_n$ and $v = v_n$ for any $n \geq n_\varepsilon$, we obtain by (5.7)

$$\int_0^1 a_i(x, u_n, u_{n,x})(u_{n,x} - w_x)dx + \int_0^1 u_n^*(u_n - w - \varepsilon)dx \leq \int_0^1 f(u_n - w - \varepsilon)dx$$

for all $w \in W^{1,2}(0,1)$ with $w \geq k_c(u)$ on $(0,1)$. Letting $n \to \infty$ in this equality and using (5.8) and (5.9), we see that

$$\begin{aligned} H(u, w) := &\int_0^1 a_i(x, u, u_x)(u_x - w_x)dx \\ &+ \int_0^1 u^*(u - w)dx - \int_0^1 f(u - w)dx \leq M_0\varepsilon \end{aligned}$$

for all $w \in W^{1,2}(0,1)$ with $w \geq k_c(u)$ on $(0,1)$. By the arbitrariness of $\varepsilon > 0$ we conclude that $H(u,w) \leq 0$ and u is a solution of quasi-variational inequality (4.14). $\diamondsuit$

(Example 5.2) The second example of Λ is given as an integral operator as follows. Let $\rho(\cdot,\cdot,\cdot)$ be a smooth function on $\mathbf{R}^N \times \mathbf{R}^N \times \mathbf{R}$ such that

$$|\rho(x,y,r)| \leq c_4|r| + c_4', \quad \forall x, \ y \in \mathbf{R}^N, \ \forall r \in \mathbf{R},$$

where c_4 and c_4' are positive constants. Then we define $\Lambda : X := W^{1,p}(\Omega) \to X$ by

$$(\Lambda v)(x) = \int_\Omega \rho(x,y,v(y))dy, \quad \forall x \in \Omega, \quad \forall v \in X.$$

Clearly Λ is compact from X into itself.

Now we give some concrete examples of integral operator Λ.

(Case 1) We consider as Λ the usual convolution operator by means of mollifier ρ_ε with real parameter $0 < \varepsilon < 1$. Let

$$\rho_\varepsilon(x,y) := \frac{1}{\varepsilon^N}\rho_0\left(\frac{x-y}{\varepsilon}\right), \quad \forall x,y \in \mathbf{R}^N,$$

where

$$\rho_0(x) := \begin{cases} c \cdot \exp\left\{-\frac{1}{1-|x|^2}\right\}, & \text{if } |x| < 1, \\ 0, & \text{otherwise,} \end{cases}$$

the constant $c > 0$ being chosen so that $\displaystyle\int_{\mathbf{R}^N} \rho_0 dx = 1$. Then, the mapping $\Lambda = \Lambda_\varepsilon$ is given by

$$(\Lambda_\varepsilon v)(x) = \int_\Omega \rho_\varepsilon(x-y)v(y)dy, \quad \forall v \in X, \tag{5.10}$$

as well as the constraint set $K(v) = K_\varepsilon(v)$ is of the form:

$$K_\varepsilon(v) = \left\{w \in X; w \geq k_c\left(\int_\Omega \rho_\varepsilon(\cdot - y)v(y)dy\right) \ \text{a.e. on } \Omega\right\}.$$

According to the above existence result, for each $\varepsilon > 0$ problem (5.2) with $\Lambda := \Lambda_\varepsilon$ has at least one solution u_ε $(\in K_\varepsilon(u_\varepsilon))$. For the family $\{u_\varepsilon\}$ of solutions we see by taking $w = k^*$ in inequality (5.2) that $\{u_\varepsilon\}_{0<\varepsilon<1}$ is bounded in X. It is a quite interesting question what happens as $\varepsilon \to 0$.

Proposition 5.2. *Assume that the space dimension is one and $\Omega = (0,1)$. Let $\{u_\varepsilon\}_{\varepsilon\in(0,1]}$ be a family of solutions of (5.2) with $\Lambda = \Lambda_\varepsilon$. Then $\{u_\varepsilon\}$ is* *bounded in $W^{1,p}(0,1)$ and any weak limit u of $\{u_\varepsilon\}$ as $\varepsilon \downarrow 0$ is a solution of the interior obstacle problem (4.14).*

Proof. We take $w \equiv k^*$ in (5.2) with $u = u_\varepsilon$ and $\Lambda = \Lambda_\varepsilon$ to get

$$\int_0^1 a_1(x, u_\varepsilon, u_{\varepsilon,x}) u_{\varepsilon,x} dx + \int_0^1 u_\varepsilon^*(u_\varepsilon - k^*) dx \leq \int_0^1 f(u_\varepsilon - k^*) dx.$$

It is easy to see from this inequality that $\{u_\varepsilon\}$ is bounded in $W^{1,p}(0,1)$. Now let u be any weak limit of $\{u_\varepsilon\}$ in $W^{1,p}(0,1)$ and choose a sequence $\varepsilon_n \downarrow 0$ so that $u_n := u_{\varepsilon_n} \to u$ weakly in $W^{1,p}(0,1)$ and uniformly on $[0,1]$. In this case we have that $\Lambda_{\varepsilon_n} u_n \to u$ uniformly on $[0,1]$ and $u \geq k_c(u)$ on $(0,1)$. Therefore, in a way similar to that of Proposition 5.1, we can prove that u is a solution of the quasi-variational inequality (4.14). ◇

(Case 2) Consider $\Omega = (0,1)$ and define $\rho(x,y)$ by

$$\rho(x,y) := \begin{cases} s(x-y), & \text{if } y \in [x,1), \\ 0, & \text{if } y \in (0,x), \end{cases} \tag{5.11}$$

where $s(\cdot)$ is a non-negative smooth function on $\mathbf{R}$ such that $s(x) = 0$ for $x \leq 0$. Then, we define $\Lambda : X := W^{1,p}(0,1) \to X$ by

$$(\Lambda v)(x) = \int_0^1 s(x-y) v(y) dy, \quad \forall v \in X,$$

where $k_c(\cdot)$ is as before. In this case, by the definition of quasi-variational inequality any solution u of (5.2) has to satisfy that

$$u(x) \geq k_c\left(\int_0^1 s(x-y) u(y) dy\right), \quad \text{a.e. } x \in (0,1). \tag{5.12}$$

As is easily understood, the behaviour of u satisfying (5.12) is controlled by that of $s(\cdot)$, and some suitable classes of functions $s(\cdot)$ are expected to play a role of an effective control space, for instance, in optimal design problems. In this sense, it is interesting to derive necessary conditions of u in order to satisfy (5.12). For instance, it is easily checked from (5.12) that $u(1) \geq k_c(0)$, because, by (5.11), $s(1-y) = 0$ for all $y \in (0,1)$. This means that it may be possible to control the solution u of problem (5.2) by appropriate choice of convolution kernel $s(\cdot)$.

(Case 3) In case $\rho(x,y)$ is constant $|\Omega|$ on $\mathbf{R}^N \times \mathbf{R}^N$, where $|\Omega|$ is the volume of Ω, we define

$$(\Lambda v)(x) = \frac{1}{|\Omega|} \int_\Omega |v(y)| dy, \quad \forall x \in \Omega, \quad \forall v \in X := W^{1,p}(\Omega).$$

In this case, we see that Λ is a continuous mapping from the weak topology of X into $\mathbf{R}$, which is regarded as a compact mapping from X into itself.

References

(1) C.Baiocchi and A.Capelo, Variational and Quasivariational Inequalities, John Wiley and Sons, Chichester-New York-Brisbane-Toronto-Singapore, 1984.

(2) A. Bensoussan and J.L. Lions, Nouvelle formulation de problémes de controle impulsionnel et applications, C. R. Acad. Sci. Paris Sér. A, 276(1973), 1189-1192.

(3) A. Bensoussan, M. Goursat, J.L. Lions, Contrôle impulsionnel et inéquations quasi-variationnelles stationnaires, C. R. Acad. Sci. Paris Ser. A, vol.276(1973), 1279-1284.

(4) H.Brézis, Équations et inéaires dans les espaces vectoriels en dualité, Ann. Inst. Fourier, Grenoble, 18 (1968), 115-175.

(5) F.E. Browder, Nonlinear monotone operators and convex sets in Banach spaces, Bull. Amer. Math. Soc., 71(1965), 780-785.

(6) J.L. Joly and U. Mosco, Sur les inéquations quasi-variationnelles, C. R. Acad. Sci. Paris Ser. A, 279(1974), 499-502.

(7) J.L. Joly and U. Mosco, A propos de l'existence et de la régularité des solutions de certaines inéquations quasi-variationnelles, J. Functional Analysis, 34(1979), 107-137.

(8) N.Kenmochi, Nonlinear operators of monotone type in reflexive Banach spaces and nonlinear perturbations, Hiroshima Math. J., 4(1974), 229-263.

(9) T. Laetsch, A uniqueness theorem for elliptic quasi-variational inequalities, J. Functional Analysis, 18(1975), 286–287.

(10) J.L. Lions and G. Stampacchia, Variational inequalities, Comm. Pure Appl. Math., 20(1967), 493-519.

(11) U. Mosco, Convergence of convex sets and of variational inequalities, Advances Math., 3(1969), 510-585.

(12) N. Nassif and K. Malla, Étude de l'existence de la solution d'une inégalité quasi variationnelle apparaissant dans la théorie des semiconducteurs, C. R. Acad. Sci. Paris Serie 1, 294(1982), 119-122.

(13) L. Tartar, Inéquations quasi variationnelles abstraites, C. R. Acad. Sci. Paris Ser. A, 278(1974), 1193-1196.

ASYMPTOTIC BEHAVIOR OF SOLUTIONS OF SOME SEMILINEAR HEAT EQUATIONS IN $\mathbf{R}^N$

KAZUHIRO ISHIGE* AND TATSUKI KAWAKAMI†

Mathematical Institute, Tohoku University, Aoba, Sendai 980-8578, Japan

Abstract: **We consider the Cauchy problem of the semilinear heat equation,**

$$\begin{cases} \partial_t u = \Delta u + f(u) & \text{in} \quad \mathbf{R}^N \times (0,\infty), \\ u(x,0) = \phi(x) \geq 0 & \text{in} \quad \mathbf{R}^N, \end{cases}$$

where $N \geq 1$, $f \in C^1([0,\infty))$, and $\phi \in L^1(\mathbf{R}^N) \cap L^\infty(\mathbf{R}^N)$ satisfying some suitable conditions. We study the asymptotic behavior of the solutions in the L^q space for $q \in [1,\infty]$ by using the relative entropy methods .

Key Words and Phrases. asymptotic behavior, semilinear heat equation, Cauchy problem, relative entropy methods.

2000 Mathematics Subject Classification Numbers. 35K55, 35B40.

1. Introduction

We consider the Cauchy problem of the semilinear heat equation,

$$\begin{cases} \partial_t u = \Delta u + f(u) & \text{in} \quad \mathbf{R}^N \times (0,\infty), \\ u(x,0) = \phi(x) & \text{in} \quad \mathbf{R}^N, \end{cases} \tag{1.1}$$

where $N \geq 1$, $f \in C^1([0,\infty))$, $\partial_t u = \partial u/\partial t$, and

$$\phi \in L^1(\mathbf{R}^N) \cap L^\infty(\mathbf{R}^N), \qquad \phi \geq 0 \quad \text{in} \quad \mathbf{R}^N. \tag{1.2}$$

Throughout this paper we assume that there exists a continuous function h on $[0,\infty)$, satisfying

(H1) h is increasing on $[0,\infty)$ such that $\left|\dfrac{f(\tau)}{\tau}\right| \leq h(\tau)$ for all $\tau > 0$;

(H2) $\displaystyle\int_1^\infty h(\tau^{-N/2})d\tau < \infty.$

*ishige@math.tohoku.ac.jp

†sa4m10@math.tohoku.ac.jp

The typical examples of $f(u)$ which we treat in this paper are

$$\pm u^p \quad (p > 1 + 2/N), \qquad \pm u^{1+2/N}(1 + (\log u)^2)^\alpha, \quad (\alpha < -1/2).$$

In this paper we study the asymptotic behavior of the solutions for the semilinear heat equations (1.1).

The asymptotic behavior of the solutions of semilinear heat equations has been studied by many mathematicians (see [6], [7], [8], [9], [10], [12], [13], [16], [17], [18], and the references therein). Among others, Kawanago [13] studied the asymptotic behavior of the solutions of the semilinear heat equation with $f(u) = u^p$. In particular, for the case $p > 1 + 2/N$ and $p < (N+2)/(N-2)$, he proved that, for any nonzero, nonnegative initial data $\phi \in L^1(\mathbf{R}^N) \cap L^\infty(\mathbf{R}^N)$ satisfying $|x|\phi \in L^1(\mathbf{R^N})$, if the solution u exists time globally and $\sup_{t>0} \|u(t)\|_{L^1(\mathbf{R}^N)} < \infty$, the solution u satisfies

$$\lim_{t\to\infty} t^{(1-1/q)N/2}\|u(t) - c_* G(t)\|_{L^q(\mathbf{R}^N)} = 0, \qquad q \in [1, \infty], \tag{1.3}$$

where

$$c_* = \|\phi\|_{L^1(\mathbf{R}^N)} + \int_0^\infty \int_{\mathbf{R}^N} f(u)dxdt, \quad G(x,t) = (4\pi t)^{-N/2} \exp\left(-\frac{|x|^2}{4t}\right).$$

(See also Remark 1.1.) Gmira and Veron [9] studied the asymptotic behavior of the solution u of (1.1) with $f(u) = -u^p$. In particular, for the case $p > 1 + 2/N$, they proved that, for any nonzero, nonnegative initial date $\phi \in L^1(\mathbf{R}^N)$,

$$\lim_{t\to\infty} t^{N/2} \max_{x\in P_a(t)} |u(x,t) - c_* G(x,t)| = 0, \qquad a > 0,$$

where $P_a(t) = \{x \in \mathbf{R}^N : |x| \le at^{1/2}\}$.

On the other hand, Toscani [19] studied the asymptotic behavior of the solutions of the linear heat equation by using the relative entropy methods, and proved

$$\|u(t) - c_* G(t)\|_{L^1(\mathbf{R}^N)} \le C(1+t)^{-1/2}, \quad t \ge 0,$$

for some positive constant C (see also [1]). Furthermore Carrillo and Toscani [2] extended the arguments in [1] and [19], and studied the asymptotic behavior of the solutions of the porous medium equation $u_t = \Delta u^m$ $(m > 1)$. Recently, Dolbeault and Karch [4] applied the arguments in [19] to the solution of (1.1), and obtained the asymptotic behavior of the solutions under the nonnegativity condition of the function f.

In this paper, for the Cauchy problem (1.1), we give a sufficient condition for (1.3) to hold true. Furthermore, by modifying the arguments in [2], we study the decay rate of $t^{\frac{N}{2}(1-\frac{1}{q})}\|u(t) - c_* G(t)\|_{L^q(\mathbf{R}^N)}$ as $t \to \infty$.

Now we state the main results of this paper. The first result gives the asymptotic behavior of the solutions of (1.1) in the L^q spaces $(1 \le q \le \infty)$.

Theorem 1.1. *Assume* (H1) *and* (H2). *Let* u *be the solution of* (1.1) *under the condition* (1.2) *such that*

$$\sup_{t>0} \|u(t)\|_{L^1(\mathbf{R}^N)} < +\infty, \quad \lim_{t\to\infty} \|u(t)\|_{L^\infty(\mathbf{R}^N)} = 0. \tag{1.4}$$

Then, for any $q \in [1,\infty]$, *there holds*

$$\lim_{t\to\infty} t^{\frac{N}{2}(1-\frac{1}{q})} \|u(t) - c_* G(t)\|_{L^q(\mathbf{R}^N)} = 0, \tag{1.5}$$

where

$$c_* = \|\phi\|_{L^1(\mathbf{R}^N)} + \int_0^\infty \int_{\mathbf{R}^N} f(u)\,dxdt > 0.$$

Remark 1.1. Assume (H1) and (H2). Let u be the solution of (1.1) under the condition (1.2).
(1) Let $f(u) = -u^p$ with $p > 1 + 2/N$. Then, by the comparison principle, the solution u satisfies the condition (1.4). So, if $f(u) = -u^p$ with $p > 1 + 2/N$, Theorem 1.1 holds without the condition (1.4).
(2) Let $f(u) = u^p$ with $p > 1 + 2/N$ and $(N-2)p < N+2$. Assume $\phi \in L^2(\mathbf{R}^N, e^{|x|^2/4}dx)$. Then, by [13], the solution u satisfies either

(i) the solution u blows-up in a finite time,

(ii) the solution u exists time globally and $\|u(t)\|_{L^\infty(\mathbf{R}^N)} \asymp t^{-\frac{N}{2}}$ as $t \to \infty$,

or

(iii) the solution u exists time globally and $\|u(t)\|_{L^\infty(\mathbf{R}^N)} \asymp t^{-\frac{1}{p-1}}$ as $t \to \infty$.

Furthermore, for the time global solution u of (1.1), if u satisfies (ii), then $\lim_{t\to\infty} \|u(t)\|_{L^1(\mathbf{R}^N)} < \infty$, and if u satisfies (iii), then $\lim_{t\to\infty} \|u(t)\|_{L^1(\mathbf{R}^N)} = \infty$. In particular, we see that, if $\lim_{t\to\infty} \|u(t)\|_{L^1(\mathbf{R}^N)} = \infty$, Theorem 1.1 does not necessarily holds.

The second theorem is the main result of this paper. We study the decay rate of $\|u(t) - c_* G(t)\|_{L^q(\mathbf{R}^N)}$ as $t \to \infty$, without the nonnegativity of f. (Compare with [4].)

Theorem 1.2. *Assume the same conditions as in Theorem* 1.1. *Furthermore assume the conditions*

(H3) $$\limsup_{\tau\to 0} \tau^{-\frac{2}{N}} h(\tau) < \infty$$

and

$$\int_{\mathbf{R}^N} (1+|x|^2)\phi(x)dx < \infty. \tag{1.6}$$

Then, for any $q \in [1,\infty]$ and $T > 0$, there exist positive constants C and δ such that

$$t^{\frac{N}{2}(1-\frac{1}{q})}\|u(t) - c_* G(t)\|_{L^q(\mathbf{R}^N)} \le Ct^{-\frac{1}{2}} + C\left[\int_{\delta t}^{\infty} h(\tau^{-\frac{N}{2}})d\tau\right]^{\frac{1}{2}} \tag{1.7}$$

for all $t \ge T$.

Remark 1.2. For the case $f(s) \equiv 0$ on $[0,\infty)$, the inequality (1.7) gives an optimal decay rate of $\|u(t) - G(t)\|_{L^q(\mathbf{R}^N)}$ as $t \to \infty$ (see [11]).

The rest of this paper is organized as follows: In Section 2 we study the L^∞ estimate of the solutions of (1.1) by the comparison principle and the explicit representation of the solutions for the heat equation, and prove Theorem 1.1. In Section 3, we study the asymptotic behavior of the entropy functional as $s \to \infty$. Furthermore, by using the Csiszar-Kullback type inequality, we study the asymptotic behavior of u as $s \to \infty$ in $L^1(\mathbf{R}^N)$, and prove Theorem 1.2.

2. Proof of Theorem 1.1

In this section we give an upper estimate of the solution u of (1.1), and prove Theorem 1.1.

In order to prove Theorem 1.1, we prove the following lemma. This lemma is the key lemma of proving Theorem 1.1.

Lemma 2.1. *Assume the same conditions as in Theorem* 1.1. *Then there exists a constant C_1 such that*

$$\|u(t)\|_{L^\infty(\mathbf{R}^N)} \le C_1(1+t)^{-\frac{N}{2}}, \qquad t > 0. \tag{2.1}$$

Furthermore there exists a positive constant C_2 such that

$$\int_t^\infty \int_{\mathbf{R}^N} |f(u)|dxd\tau \le C_2 \int_t^\infty h(C_1(1+\tau)^{-\frac{N}{2}})d\tau < \infty, \quad t \ge 0. \tag{2.2}$$

Proof. In order to prove (2.1), we recall the L^p-L^q estimate of the heat equation. In fact, Let $1 \le q \le p \le \infty$. Then there exists a constant $c_{p,q}$ such that

$$\|e^{t\Delta}\phi\|_{L^p(\mathbf{R}^N)} \le c_{p,q} t^{-\frac{N}{2}(\frac{1}{q}-\frac{1}{p})}\|\phi\|_{L^q(\mathbf{R}^N)}, \quad t > 0. \tag{2.3}$$

By (1.4), we may put $\lambda = \sup_{t>0}\|u(t)\|_{L^1(\mathbf{R}^N)}$. By (H2) and (2.3), there exists a positive constant t_0 such that

$$\int_{t_0}^{\infty} h(2c_{\infty,1}\lambda\tau^{-\frac{N}{2}})d\tau < \frac{1}{2}\log 2. \tag{2.4}$$

By (H1) and (H2), we have $h(0) = 0$, and may take a positive constant ϵ so that

$$t_0 h(\epsilon) < \frac{1}{2}\log 2. \tag{2.5}$$

Then, by (1.4), there exists a positive constant T such that

$$\sup_{t\geq T}||u(t)||_{L^\infty(\mathbf{R}^N)} \leq \epsilon. \tag{2.6}$$

Let z be a solution of

$$\begin{cases} z_t = \Delta z + h\left(\|u(t+T)\|_{L^\infty(\mathbf{R}^N)}\right) z \text{ in } & \mathbf{R}^N \times (0,\infty), \\ z(x,0) = u(x,T) & \text{in } \mathbf{R}^N. \end{cases} \tag{2.7}$$

By (H1), we have

$$\left|\frac{f(u)}{u}\right| \leq h(u) \leq h\left(\|u(t)\|_{L^\infty(\mathbf{R}^N)}\right), \qquad (x,t) \in \mathbf{R}^N \times (0,\infty). \tag{2.8}$$

Then, by the comparison principle and (2.8), we have

$$0 \leq u(x,t+T) \leq z(x,t), \quad (x,t) \in \mathbf{R}^N \times (0,\infty). \tag{2.9}$$

By (2.6), we may put

$$T_* = \sup\left\{s > 0\,;\, \|z(t)\|_{L^\infty(\mathbf{R}^N)} < 2c_{\infty,1}\lambda t^{-\frac{N}{2}} \text{ for all } 0 < t < s\right\} > 0, \tag{2.10}$$

where $c_{\infty,1}$ is the constant given in (2.3). Assume that $T_* < \infty$. Then we have

$$\|z(T_*)\|_{L^\infty(\mathbf{R}^N)} = 2c_{\infty,1}\lambda T_*^{-\frac{N}{2}}. \tag{2.11}$$

On the other hand, by (2.7), the function

$$w(x,t) = z(x,t)\exp\left(-\int_0^t h\left(\|u(\tau+T)\|_{L^\infty(\mathbf{R}^N)}\right)d\tau\right) \tag{2.12}$$

satisfies the heat equation with $w(0) = u(T)$. Then, by (2.3), we have

$$\|w(t)\|_{L^\infty(\mathbf{R}^N)} \leq c_{\infty,1}t^{-\frac{N}{2}}\|u(T)\|_{L^1\mathbf{R}^N} \leq c_{\infty,1}\lambda t^{-\frac{N}{2}}, \qquad t > 0. \tag{2.13}$$

Furthermore, by (H1) and (2.4)–(2.6), (2.9), and (2.10), we have

$$\begin{aligned}
&\int_0^{T_*} h(\|u(\tau+T)\|_{L^\infty(\mathbf{R}^N)})d\tau \\
&\le t_0 h(\epsilon)d\tau + \int_{t_0}^{\max\{t_0,T_*\}} h(\|z(\tau)\|_{L^\infty(\mathbf{R}^N)})d\tau \\
&\le t_0 h(\epsilon)d\tau + \int_{t_0}^{\infty} h(2c_{\infty,1}\lambda\tau^{-\frac{N}{2}})d\tau < \log 2.
\end{aligned} \tag{2.14}$$

By (2.12)–(2.14), we have

$$\begin{aligned}
\|z(T_*)\|_{L^\infty(\mathbf{R}^N)} &= \|w(T_*)\|_{L^\infty(\mathbf{R}^N)} \exp\left(\int_0^{T_*} h(\|u(\tau+T)\|_{L^\infty(\mathbf{R}^N)})d\tau\right) \\
&< 2c_{\infty,1}\lambda T_*^{-\frac{N}{2}}.
\end{aligned}$$

This contradicts (2.11), and we see $T_* = \infty$. Then, by (2.9), we have

$$\|u(t+T)\|_{L^\infty(\mathbf{R}^N)} \le 2c_{\infty,1}\lambda t^{-\frac{N}{2}}, \qquad t > 0.$$

This inequality together with the boundedness of the solution u implies (2.1). Furthermore, by (H2), (2.1), (2.8), and (2.9), there exists a positive constant C such that we have

$$\begin{aligned}
\int_t^\infty \int_{\mathbf{R}^N} |f(u)|dxd\tau &\le \int_t^\infty h\left(\|u(\tau)\|_{L^\infty(\mathbf{R}^N)}\right)\left(\int_{\mathbf{R}^N} |u|dx\right)d\tau \\
&\le C\int_t^\infty h\left(C(1+\tau)^{-\frac{N}{2}}\right)d\tau < \infty, \quad t \ge 0.
\end{aligned}$$

Therefore we obtain (2.2), and the proof of Lemma 2.1 is complete. □

By using Lemma 2.1 and (2.3), we obtain Theorem 1.1 (see [11]).

3. Proof of Theorem 1.2

In this section, by using the relative entropy methods, we prove Theorem 1.2. In order to prove Theorem 1.2, we put

$$v(y,s) = (1+2t)^{\frac{N}{2}}u(x,t), \quad y = (1+2t)^{-\frac{1}{2}}x, \quad s = \frac{1}{2}\log(1+2t).$$

Then the function v satisfies

$$\begin{cases} \partial_s v = \operatorname{div}(yv + \nabla v) + e^{(N+2)s}f(e^{-Ns}v) & \text{in } \mathbf{R}^N \times (0,\infty), \\ v(y,0) = \phi(y) \ge 0 & \text{in } \mathbf{R}^N. \end{cases} \tag{3.1}$$

Let $g(y) = (2\pi)^{-N/2}\exp(-|y|^2/2)$. Then the function g is a stationary solution of the linear Fokker-Planck equation $\partial_s v = \operatorname{div}(yv+\nabla v)$. Furthermore, by Theorem 1.1, we have

$$\lim_{s\to\infty} ||v(s) - c_* g||_{L^p(\mathbf{R}^N)} = 0, \qquad 1 \le p \le \infty.$$

Let

$$H[v(s)] = \int_{\mathbf{R}^N} (|y|^2 + 2\log v) v dy,$$

which is a Lyapunov functional of the equation $\partial_s v = \operatorname{div}(yv + \nabla v)$. Let $S > 0$. Then, by following the strategy in [2], we study the asymptotic behavior of entropy functional$H[v(s)]$. First, by formal calculation, we obtain the exponential decay of the relative entropy.

Lemma 3.1. *Assume the same conditions as in Theorem* 1.2. *Then there exists a positive constants* C_1 *such that*

$$H[v(s_1)] - H[v(s_2)] \le C_1 e^{-2s_1} + C_1 \int_{s_1}^{s_2} e^{2\tau} h(Le^{-N\tau}) d\tau \tag{3.2}$$

for all $0 < s_1 < s_2$. *Furthermore there holds that*

$$\lim_{s\to\infty} H[v(s)] = H(c_* g)$$

Next we obtain the convergence of the mass of the solution of (3.1).

Lemma 3.2. *Assume the same conditions as in Theorem* 1.2. *Then there exists a constant* C_1 *and* $S > 0$ *such that*

$$|H[c_* g] - H[g(s)]| \le C_1 \int_s^\infty e^{2\tau} h(Le^{-N\tau}) d\tau,$$

where $g(s) = g(y,s) = d(s)g(y)$ *and* $d(s) = \|v(s)\|_{L^1(\mathbf{R}^N)} = c(t)$ *with* $s = 1/2\log(1+2t)$ *for all* $s > S$. *Furthermore we obtain*

$$H[v(s)] - H[g(s)] = 2\int_{\mathbf{R}^N} \frac{v(y,s)}{g(y,s)} \log\frac{v(y,s)}{g(y,s)} g(y,s) dy, \quad s > 0.$$

Finally we obtain a generalized Csiszar-Kullback inequality (see [3] and [14]), which is an estimate for L^1-distance of two functions $v(s)$ and $g(s)$ in term of their relative entropy. Here we remark that

$$\int_{\mathbf{R}^N} v(y,s) dy = \int_{\mathbf{R}^N} g(y,s) dy = d(s).$$

Lemma 3.3. *Assume the same conditions as in Theorem* 1.2. *Then we obtain*

$$\|v(s) - g(s)\|^2_{L^1(\mathbf{R}^N)} \le 4d(s)(H[v(s)] - H[g(s)])$$

for all $s > 0$.

By Lemmas 3.1, 3.2 and 3.3, there exists a positive constant C such that

$$\begin{aligned}\|v(s) - c_* g\|_{L^1(\mathbf{R}^N)} &\le \|v(s) - d(s)g\|_{L^1(\mathbf{R}^N)} + \|c_* g - d(s)g\|_{L^1(\mathbf{R}^N)}\\ &\le Ce^{-s} + C\left[\int_s^\infty e^{2\tau} h(Le^{-N\tau})d\tau\right]^{1/2}\end{aligned}$$

for all $s > S$. This inequality implies the inequality (1.7) with $q = 1$. Furthermore, by using the explicit representation of the solutions for the heat equation and the inequality (1.7) with $q = 1$, we obtain the inequality (1.7) with $q = \infty$. Finally, by the inequality (1.7) with $q = 1$ and $q = \infty$, we obtain the inequality (1.7) with $1 < q < \infty$, and we obtain Theorem 1.2 formally.

The main difficulty of proving Theorem 1.2 rigorously is to prove the inequality (3.2). It seems difficult to prove the inequality (3.2) by using the arguments in [1], [2] and [4] directly. In order to prove the inequality (3.2), we construct the approximate solutions v_n $(n = 1, 2, \dots)$ to (3.1),

$$\begin{cases} \partial_s v_n = \operatorname{div}(y v_n + \epsilon_n y + \nabla v_n) \\ \qquad\qquad + e^{2s} \dfrac{f(e^{-Ns}\min\{v_n, L\})}{e^{-Ns}\min\{v_n, L\}} v_n \quad \text{in } B_{M_n} \times (0, \infty), \\ v_n(y, s) = 0 \qquad\qquad \text{on } \partial B_{M_n} \times (0, \infty), \\ v_n(y, 0) = \phi(y)\chi_{B_{M_n}}(y) \quad \text{in } B_{M_n}, \end{cases}$$

where $\chi_{B_{M_n}}$ is the characteristic function of the ball B_{M_n}, $\epsilon_n = n^{-\alpha}$, $M_n = [1 + 2\log n]^{1/2}$, and α is a sufficiently large constant. Furthermore we consider the functional,

$$H_n(s) = \int_{B_{M_n}} (|y|^2 + 2\log(v_n + \epsilon_n))(v_n + \epsilon_n)dy, \qquad s > 0,$$

instead of $H[v(s)]$. By the term $\epsilon_n y$, we may obtain some estimates of v_n, and have several estimates of the time derivative of $H_n(s)$. Then we see that there exist positive constants C_1 and C_2 such that

$$H_n(s_1) - H_n(s_2) \le C_1 e^{-2s_1} + C_1 \int_{s_1}^{s_2} e^{2\tau} h(Le^{-N\tau})d\tau + C_2 n^{-1}$$

for all $S < s_1 < s_2$ and all $n = 1, 2, \ldots$. Furthermore, taking the limit of $H_n(s)$ as $n \to \infty$ and $s \to \infty$, we obtain the inequality (3.2) rigorously, and complete the proof of Theorem 1.2.

References

[1] J. A. Carrillo and G. Toscani, Exponetial convergence toward equilibrium for homogeneous Fokker-Planck type equations, Math. Meth. Appl. Sci., 21 (1998), 1269-1286.

[2] J. A. Carrillo and G. Toscani, Asymptotic L^1-decay of solutions of the porous medium equation to self-similarity, Indiana Univ. Math. J., 49 (2000), 113-142.

[3] I. Csiszar, Information-type measures of difference of probability distributions and indirect observations, Stud. Sci. Math. Hung., 2 (1967), 299-318.

[4] J. Dolbeault and G. Karch, Large time behavior of solutions to nonhomogeneous diffusion equations, Banach Center Publ., 74, (2006), 133-147.

[5] M. Escobedo and E. Zuazua, Large time behavior for convection-diffusion equations in $\mathbf{R}^N$, J. Func. Anal., 100 (1991), 119-161.

[6] M. Fila, M. Winkler and E. Yanagida, Convergence rate for a parabolic equation with supercritical nonlinearity, J. Dyna. Diff. Eq., 17 (2005), 249-269.

[7] C. Gui, W.-M. Ni and X. Wang, On the stability and instability of positive steady states of a semilinear heat equation in R^N, Comm. Pure Appl. Math., 45 (1992), 1153-1181.

[8] C. Gui, W.-M. Ni and X. Wang, Further study on a nonlinear heat equation, J. Diff. Eq., 169 (2001), 588-613.

[9] A. Gmira and L. Veron, Large time behavior of the solutions of a semilinear parabolic equation in $\mathbf{R}^N$, J. Diff. Eq., 53 (1984), 258-276.

[10] L. A. Herraiz, Asymptotic behaviour of solutions of some semilinear parabolic problems. Ann. Inst. H. Poincaré Anal. Non Linéaire., 16 (1999), 49-105.

[11] K. Ishige and T. Kawakami, Asymptotic behavior of solutions for some semilinear heat equations in $\mathbf{R}^N$, preprint.

[12] S. Kamin and L. A. Peletier, Large time behavior of solutions of the heat equation with absorption, Ann. Scu. Norm. Sup. Pisa., 12 (1985), 393-408.

[13] T. Kawanago, Asymptotic behavior of solutions of a semilinear heat equation with subcritical nonlinearity, Ann. Inst. H. Poincaré Anal. Non Linéaire., 13 (1996), 1-15.

[14] S. Kullback, A lower bound for dicrimination information in terms of variation, IEEE Trans. Info. Theory, 4 (1967), 126-127.

[15] O. A. Ladyzenskaja, V. A. Solonnikov, N. N. Ural'ceva, Linear and Quasilinear Equations of Parabolic Type, American Mathematical Society, Providence, RI, 1968.

[16] P. Poláčik and E. Yanagida, On bounded and unbounded global solutions of a supercritical semilinear heat equation, Math. Annal., 327 (2003), 745-771.

[17] P. Poláčik and E. Yanagida, A Liouville property and quasiconvergence for a semilinear heat equation, J. Diff. Eq., 208 (2005), 194-214.

[18] P. Quittner, The decay of global solutions of a semilinear heat equation, preprint.

[19] G. Toscani, Kenetic approch to the asymptotic behavior of the solution to diffusion equations, Rend. Mate. Serie VII, 16 (1996), 329-346.

WELL-POSEDNESS AND PERIODIC STABILITY FOR QUASILINEAR PARABOLIC VARIATIONAL INEQUALITIES WITH TIME-DEPENDENT CONSTRAINTS

KEN SHIRAKAWA

Department of Applied Mathematics, Faculty of Engineering, Kobe University, 1-1 Rokkodai, Nada, Kobe 657-8501, Japan

MASAHIRO KUBO

Department of Mathematics, Nagoya Institute of Technology, Gokiso-cho, Showa-ku, Nagoya-city, Aichi, 466-8555, Japan

NORIAKI YAMAZAKI

Department of Mathematical Science, Common Subject Division, Muroran Institute of Technology, 27-1 Mizumoto-cho, Muroran, 050-8585, Japan

Abstract: In this paper, we deal with Cauchy problems for some quasilinear parabolic variational inequalities subject to time-dependent constraints. With regard to this type of problem, the authors of [8] recently developed a general theory concerned with the existence and uniqueness of solutions. The main objective of this paper is to study the key-properties of dynamical systems generated by our Cauchy problems. To this end, we first prove the well-posedness (including continuous dependence) of solutions on the basis of the earlier work [8]. Eventually, we focus on the periodic situation of dynamical systems, to validate the results concerned with the existence of periodic solutions, and the asymptotic stability from the viewpoint of attractors.

Introduction

Let $\Omega \subset \mathbb{R}^N$ ($N \in \mathbb{N}$) be a bounded domain with a regular boundary $\Gamma := \partial\Omega$. This paper is devoted to the observation of the asymptotic stability of a dynamical system generated by the following Cauchy problem of quasilinear parabolic variational inequality:

$$\begin{cases} \displaystyle\int_\Omega u_t(t)(u(t)-z)\,dx + \int_\Omega a(x,u(t),\nabla u(t))\cdot\nabla(u(t)-z)\,dx \\ \displaystyle + \int_\Omega (b(x,u(t)) - f(t))(u(t)-z)\,dx \le 0, \ \forall z \in K(t), \text{ a.e. } t>0, \\ u(x,0) = u_0(x), \ \text{ a.e. } x \in \Omega; \end{cases} \tag{0.1}$$

subject to a time-dependent constraint:

$$u(t) \in K(t), \text{ a.e. } t > 0; \tag{0.2}$$

by convex subsets $K(t)$ ($t \geq 0$) in appropriate functional space (e.g. $H^1(\Omega)$). In the context, $a = (a_1, \cdots, a_N) : \overline{\Omega} \times \mathbb{R} \times \mathbb{R}^N \longrightarrow \mathbb{R}^N$ is a given vectorial function with a potential $A : \overline{\Omega} \times \mathbb{R} \times \mathbb{R}^N \longrightarrow \mathbb{R}$ in the sense that $a(x, r, y) = \nabla_y A(x, r, y)$ for $(x, r, y) \in \overline{\Omega} \times \mathbb{R} \times \mathbb{R}^N$. Also, $b : \overline{\Omega} \times \mathbb{R} \longrightarrow \mathbb{R}$ is a given function, $u_0 = u_0(x)$ is a given initial value, and $f = f(t, x)$ is a given forcing term.

Parabolic variational inequalities subject to time-dependent constraints have been studied by numerous mathematicians from various viewpoints (cf. [2, 3, 5, 6, 7, 8, 10, 12]). We will focus on [8] as being representative of like-minded research. In [8], the Hilbert space $L^2(\Omega)$ is taken as the central range of solutions, and the constraint $\{K(t)\}$ is settled as a family of closed and convex subsets in $H^1(\Omega)$. Also, it is assumed that the vector field $a = a(x, u, y)$ is uniformly elliptic with respect to y and $b = b(x, r)$ is Lipschitz continuous with respect to r with an x-independent Lipschitz constant. Using these settings, the authors of [8] have proved theorems concerned with the existence and the comparison principle (including uniqueness) of solutions, and they have further exemplified some applications including the following two situations.

(AP1) Time-dependent double obstacle.
(AP2) Time-dependent boundary constraints.

The results in [8] certainly provide a theory of the solvability in quite general terms. But, their theory is actually not sufficient to guarantee the continuous dependence of solutions in the topology of the central range $L^2(\Omega)$, because it is made only for solutions starting from the initial range $K(0) \subset H^1(\Omega)$, and on top of that, the demonstration method of uniqueness is based on the L^1-technique devised by Bénilan [1].

Thus, the study in this paper will focus on a continuation of the research in [8], with the main objective being to construct a general theory that can cover advanced topics, such as continuous dependence and attractors. To this end, the assumptions made in [8] will be partially restricted by considering applications to (AP1)-(AP2). Above all, it should be noted that the constraint $\{K(t)\}$ and the forcing term $f = f(t, x)$ generally cause the dynamical system to be non-autonomous; yet the notion of attractors was originally defined for autonomous systems generated by appropriate semigroups. Hence, in the observation by attractors, we will need some special setting to indicate governing semigroups for the dynamical systems under review. As candidates for this setting, we can quote the ideas found in [4](setting of time-independent constraints), [12](asymptotically autonomous setting), [5](periodic setting) and [6](setting from pull-back viewpoints) amongst others. From all these ideas, we have adopted the periodic setting as in [5](Section 3), on the grounds of the variety of applications.

Consequently, the assertions in the main results will be:

(a1) the existence and uniqueness of the solutions that start from the L^2-closure of the initial range $K(0)$;
(a2) continuous dependence of solutions;
(a3) existence of periodic solutions under periodic settings of dynamical systems;
(a4) compactness of flows in periodic dynamical systems, based on the global boundedness of solutions;
(a5) periodic stability, namely attractors for periodic dynamical systems.

Notations. These are the notations that are used throughout this paper. Let $Q := (0, +\infty) \times \Omega$ be the time-space coordinate system consisting of the time-interval $(0, +\infty)$ and the spatial domain $\Omega \subset \mathbb{R}^N$.

For an abstract Banach space X, the norm of X is denoted by $|\cdot|_X$. In particular, when X has the Hilbert structure, the inner product in X is denoted by $(\cdot,\cdot)_X$. For all $x_0 \in X$ and $R > 0$, we denote by $B_X(x_0; R)$, the closed ball in X with the center x_0 and the radius R. Additionally, we denote by $\mathrm{dist}_X(\cdot,\cdot)$ the Hausdorff semi-distance, defined as: $\sup\limits_{y \in Y} \inf\limits_{z \in Z} |y - z|_X$ for all $Y, Z \subset X$.

For simplicity, the functional spaces $L^2(\Omega)$ and $H^1(\Omega)$ are denoted by H and V, respectively. In particular, we simply denote by $(\cdot,\cdot)$ the usual inner product in H.

1. Statement of main results

First, we give an exact definition of our Cauchy problem. In the Cauchy problem {(0.1),(0.2)}, the vector field $a : \overline{\Omega} \times \mathbb{R} \times \mathbb{R}^N \longrightarrow \mathbb{R}^N$ and the function $b : \overline{\Omega} \times \mathbb{R} \longrightarrow \mathbb{R}$, are assumed to satisfy the following conditions.

(A) There exists a function $A : \overline{\Omega} \times \mathbb{R} \times \mathbb{R}^N \longrightarrow \mathbb{R}$ such that:

$$a(x, r, y) = \nabla_y A(x, r, y), \ \forall (x, r, y) \in \overline{\Omega} \times \mathbb{R} \times \mathbb{R}^N;$$

and A and $\nabla_y A$ ($= a$) satisfy the Carathéodory condition. Additionally, there exist positive constants μ and C_1 such that

$$\big(a(x,r,y) - a(x,r,\hat{y})\big) \cdot (y - \hat{y}) \geq \mu|y-\hat{y}|^2, \ \forall x \in \overline{\Omega}, \ \forall r \in \mathbb{R}, \ \forall y, \forall \hat{y} \in \mathbb{R}^N,$$

$$\begin{cases} |a(x,r,y)|^2 + |A(x,r,y)| + |\partial_r A(x,r,y)|^2 \leq C_1(1+|y|^2), \\ |a(x,r,y) - a(x,\hat{r},y)| \leq C_1|r-\hat{r}|, \ \forall x \in \overline{\Omega}, \ \forall r, \forall \hat{r} \in \mathbb{R}, \ \forall y \in \mathbb{R}^N. \end{cases} \tag{1.1}$$

(B1) b is continuous on $\overline{\Omega} \times \mathbb{R}$, and for any $x \in \overline{\Omega}$, $b(x,\cdot) : \mathbb{R} \longrightarrow \mathbb{R}$ is Lipschitz continuous with an x-independent Lipschitz constant L_b; more precisely $|b(x,r) - b(x,\hat{r})| \leq L_b|r-\hat{r}|, \ \forall x \in \overline{\Omega}, \ \forall r, \forall \hat{r} \in \mathbb{R}$.

Also, the constraint $\{K(t) \,|\, t \geq 0\}$ is prescribed as a family of closed and convex subsets in V, the initial value u_0 is a given function in H and the forcing term f is a given function in $L^2_{\mathrm{loc}}([0,+\infty); H)$.

Let us denote by (CP; $\{K(t)\}, u_0, f$) the problem $\{(0.1),(0.2)\}$ when the data of $\{K(t)\}$, u_0 and f are specified. Then, the solutions of this problem are rigorously defined as follows.

Definition 1.1. (Definition of solutions) Let us fix the data of the constraint $\{K(t)\}$, the initial value $u_0 \in H$ and the forcing term $f \in L^2_{\mathrm{loc}}([0,+\infty); H)$. For any fixed $T > 0$, a function $u : [0,T] \longrightarrow H$ is called a solution to the Cauchy problem (CP; $\{K(t)\}, u_0, f$) on the compact interval $[0,T]$, iff $u \in C([0,T]; H) \cap W^{1,2}_{\mathrm{loc}}((0,T]; H) \cap L^2(0,T;V)$, and u fulfills (0.1) and (0.2). Also, a function $u : [0,+\infty) \longrightarrow H$ is called a solution to the Cauchy problem (CP; $\{K(t)\}, u_0, f$) on $[0,+\infty)$, iff u forms the solution on any compact interval $[0,\tilde{T}]$ with any $\tilde{T} > 0$.

Recently, a study of the Cauchy problem (CP; $\{K(t)\}, u_0, f$) was reported in [8], including some results, concerned with the existence, uniqueness and comparison principle of solutions, under more general settings of a and b. According to this study, the following condition is the key-assumption in proving the solvability of our problem.

(K) (Assumption for existence) There exists a function $\alpha \in W^{1,2}_{\mathrm{loc}}[0,+\infty)$, satisfying the following property $(*)$:

$(*)$ $\sup_{t \geq 0} |\alpha'|_{L^2(t,t+1)} < +\infty$, and $\forall s, \forall t \geq 0$, $\forall z \in K(s)$, $\exists \tilde{z} \in K(t)$, s.t.

$$\begin{cases} |\tilde{z} - z|_H \leq |\alpha(t) - \alpha(s)|(1 + |\nabla z|_{(H)^N}), \\ \displaystyle\int_\Omega A(x,w,\nabla\tilde{z})\,dx - \int_\Omega A(x,w,\nabla z)\,dx \\ \quad \leq |\alpha(t) - \alpha(s)|(1 + |\nabla z|^2_{(H)^N} + |w|_V |\nabla z|_{(H)^N}), \ \forall w \in V. \end{cases}$$

Remark 1.1. The assumptions listed above are actually restricted versions of those given in [8]. We adopt the restricted situations, especially after considering the following.

(I) Instead of the inequalities (1.1) as in (A), the authors of [8] have assumed more general ones. The adoption of the restricted inequalities (1.1) is essential to derive key-inequalities (2.2) and (3.2) which will be seen in the proofs of uniqueness and boundedness of solutions, respectively.

(II) In (∗) of (K), the authors of [8] originally chose the two values s, t of time such that $s < t$. This modification is quite important to show the absolute continuity of energies (potentials), mentioned in the next section.

Remark 1.2. Other than the modifications reported in Remark 1.1, the authors of [8] have made an additional assumption for uniqueness. Furthermore, it should be noted that the solutions (on $[0, +\infty)$) treated in [8] actually show the regularity in stronger topology of $W^{1,2}_{\mathrm{loc}}([0, +\infty); H) \cap L^\infty_{\mathrm{loc}}([0, +\infty); V)$ than that given in Definition 1.1. This strong regularity essentially comes from the point that the initial values are taken from more regular class $K(0) \subset V$ than the central range H. Consequently, we see that the existence of solutions starting from $K(0)$ is a direct consequence of the general theory obtained in [8](Theorem 2.1).

The objective of this paper is to challenge some advanced theme on the basis of [8], with the main focus being to characterize the asymptotic stability of the dynamical system, generated by (CP; $\{K(t)\}, u_0, f$), with use of attractors. For the sake of the existence of attractors, some compactness property is required of the dynamical systems under consideration. Hence, the immediate task is to verify whether our Cauchy problems absolutely generate dynamical systems with compactness. To this end, we will first prove the following two theorems.

Theorem 1.1. *(Solvability for general initial values) Let us assume conditions (A), (B1) and (K). Then, for any* $u_0 \in \overline{K(0)}^H$ *and any* $f \in L^2_{\mathrm{loc}}([0, +\infty); H)$*, the Cauchy problem (CP;* $\{K(t)\}, u_0, f$*) admits a unique solution* u *on* $[0, +\infty)$*.*

Theorem 1.2. *(Continuous dependence of solutions) Under conditions (A), (B1) and (K), let us fix any* $T > 0$*, any* $s_0 \geq 0$*, any* $u_0 \in \overline{K(s_0)}^H$ *and any* $f \in L^2_{\mathrm{loc}}([0, +\infty); H)$*. For given sequences* $\{s_n\} \subset [0, +\infty)$*,* $\{u_{n,0} \,|\, u_{n,0} \in \overline{K(s_n)}^H \ (n = 1, 2, 3, \cdots)\}$ *and* $\{f_n\} \subset L^2_{\mathrm{loc}}([0, +\infty); H)$*,* *let $\{u_n\}$ be the sequence consisting of solutions u_n $(n = 1, 2, 3, \cdots)$ of Cauchy problems $(CP; \{K(t + s_n)\}, u_{n,0}, f_n)$ $(n = 1, 2, 3, \cdots)$ on the compact interval $[0, T]$. If $u_{n,0} \to u_0$ in H and $f_n \to f$ in $L^2(0, T; H)$ as $n \to +\infty$, then u_n converges to the solution u of the Cauchy problem $(CP; \{K(t + s_0)\}, u_0, f)$ in the sense that:*

$$u_n \to u \quad \text{in } C([0, T]; H) \text{ as } n \to +\infty.$$

On account of the above two theorems, the Cauchy problem (CP; $\{K(t)\}, u_0, f$) will generate dynamical systems $\{U(t + s, s) \,|\, t \geq 0\}$ $(s \geq 0)$ consisting of solution operators $U(t+s, s) : \overline{K(s)}^H \longrightarrow \overline{K(t+s)}^H$ $(t, s \geq 0)$ that map initial values $u_0 \in \overline{K(s)}^H$ onto the values $u(t)$ of the solutions of Cauchy problems (CP; $\{K(t + s)\}, u_0, f(\cdot + s)$) at time t. Then, it is easily verified that all dynamical systems $\{U(t + s, s) \,|\, t \geq 0\}$ $(s \geq 0)$ form continuous evolution dynamical systems.

As mentioned in the introduction, the attractors for dynamical systems $\{U(t + s, s)\}$ are founded under the following periodic setting on the time-dependence of the constraint and the forcing term.

(P) (Periodic condition) There exists a constant (period) $T_0 > 0$, such that $K(t + T_0) = K(t)$ and $f(t + T_0) = f(t)$ in H, for all $t \geq 0$.

We further assume the following condition for the perturbation b.

(B2) (Growth condition for b) There exist positive constants C_2 and C_3, independent of x and r, such that $b(x, r)\, r \geq C_2 r^2 - C_3$, $\forall x \in \overline{\Omega}$, $\forall r \in \mathbb{R}$.

These conditions will be key to guarantee the compactness of flows.

Including the above two conditions, we see the existence of the periodic solutions, stated as follows.

Theorem 1.3. *(Periodic solutions) Under the assumptions (A), (B1)-(B2), (K) and (P), there exists at least one initial value $u_{p,0} \in \overline{K(0)}^H$ such that the solution u_p of the Cauchy problem $(CP; \{K(t)\}, u_{p,0}, f)$ on $[0, +\infty)$ belongs to $W^{1,2}_{\text{loc}}([0, +\infty); H) \cap L^\infty(0, +\infty; V)$, and forms a periodic function in time with the period T_0; more precisely $u_p(t + T_0) = u_p(t)$ in H, for all $t \geq 0$.*

Next, we prove the following theorem, concerned with the global estimates of solutions.

Theorem 1.4. *(Global estimates of solutions) Under assumptions (A), (B1)-(B2), (K) and (P), let us fix the data of $\{K(t)\}$, $u_0 \in \overline{K(0)}^H$ and $f \in$*

$L^2_{\rm loc}([0,+\infty);H)$, *to take the solution* u *of the problem* $(CP;\{K(t)\},u_0,f)$ *on* $[0,+\infty)$. *Then, there exists a positive constant* R_0, *independent of* u_0 *and* f, *such that*

$$\sup_{t\geq 0}|u(t)|^2_H+\sup_{t\geq 0}\int_t^{t+T_0}|\nabla u(\tau)|^2_{(H)^N}\,d\tau \leq R_0(1+|u_0|^2_H+|f|^2_{L^2(0,T_0;H)}). \quad (1.2)$$

Moreover, for any $\delta>0$, *there exists a positive constant* R_δ, *depending only on* δ, *such that*

$$\sup_{t\geq \delta}|\nabla u(t)|^2_{(H)^N}+\sup_{t\geq \delta}\int_t^{t+T_0}|u_t(\tau)|^2_H\,d\tau \leq R_\delta(1+|u_0|^2_H+|f|^2_{L^2(0,T_0;H)}). \quad (1.3)$$

Now, in the light of embedding $V\hookrightarrow H$, we ascertain that the estimates (1.2)-(1.3) imply the compactness of the flow after a certain large length of time. Moreover, under the periodic situation of (P), it is possible to construct a discrete autonomous dynamical system $\{U^k_\tau\,|\,k=0,1,2,\cdots\}$ $(\tau\geq 0)$ using k-iterations U^k_τ $(k=0,1,2,\cdots)$ of operators $U_\tau:=U(\tau+T_0,\tau):\overline{K(\tau)}^H\longrightarrow\overline{K(\tau)}^H(=\overline{K(\tau+T_0)}^H)$ for $\tau\geq 0$. Then, by virtue of Theorems 1.1-1.2, we immediately see that all discrete dynamical systems $\{U^k_\tau\,|\,k=0,1,2,\cdots\}$ $(\tau\geq 0)$ form semigroups on $\overline{K(\tau)}^H$ $(\tau\geq 0)$.

With help from the above discrete semigroups $\{U^k_\tau\}$, the asymptotic stability for the continuous non-autonomous systems $\{U(t+s,s)\}$ can be characterized as follows.

Theorem 1.5. *Let us assume (A), (B1)-(B2), (K) and (P). Then, the following two statements hold.*

(I) *(Attractors for discrete autonomous systems) For any* $\tau\geq 0$, *there exists a nonempty, connected and compact set* $\mathscr{A}_\tau\subset K(\tau)$, *called the global attractor, such that:*

(i) *(invariance)* $U_\tau\mathscr{A}_\tau=\mathscr{A}_\tau$ *in* H;

(ii) *(attractiveness) for any* $\tau\geq 0$ *and any bounded subset* $B\subset\overline{K(\tau)}^H$, $\mathrm{dist}_H(U^k_\tau B,\mathscr{A}_\tau)\to 0$ *as* $k\to+\infty$.

(II) *(Attractors in continuous non-autonomous systems) Let us take the global attractors* $\mathscr{A}_\tau$ $(\tau\geq 0)$ *as in assertion (I), to set* $\mathscr{A}:=\bigcup_{\tau\in[0,T_0]}\mathscr{A}_\tau$. *Then:*

(iii) *(continuous dependence)* $\mathscr{A}_{t+s}=U(t+s,s)\mathscr{A}_s$ *for all* $t,s\geq 0$;

(iv) *(attractiveness) for any bounded subset* $B\subset H$,

$$\sup_{\tau\in[0,T_0]}\mathrm{dist}_H(U(t+\tau,\tau)(B\cap\overline{K(\tau)}^H),\mathscr{A})\to 0 \quad \text{as } t\to+\infty.$$

2. Well-posedness of solutions

Throughout this section, we assume conditions (A), (B1) and (K), to prove Theorems 1.1 and 1.2 concerned with the well-posedness of solutions. We start with the proof of uniqueness, summarized in the following lemma.

Lemma 2.1. *(Proof of uniqueness) For arbitrary data of two initial values $u_{i,0} \in \overline{K(0)}^H$ $(i = 1, 2)$ and two forcing terms $f_i \in L^2_{\mathrm{loc}}([0, +\infty); H)$ $(i = 1, 2)$, let u_i $(i = 1, 2)$ be two solutions to Cauchy problems (CP; $\{K(t)\}, u_{i,0}, f_i$) $(i = 1, 2)$ on $[0, +\infty)$, respectively. Then, there exists a positive constant C_4, independent of $u_{i,0}$ and f_i $(i = 1, 2)$, such that:*

$$|u_1 - u_2|^2_{C([0,T];H)} \le e^{C_4 T}(|u_{1,0} - u_{2,0}|^2_H + |f_1 - f_2|^2_{L^2(0,T;H)}), \ \forall T > 0. \quad (2.1)$$

Proof. Let us fix $T > 0$, and for each solution u_i of (CP; $\{K(t)\}, u_{i,0}, f_i$) $(i = 1, 2)$, let us assign the other solution to the test function z in (0.1) to add both sides of the resultant inequalities. Then, using conditions (A), (B1) and Schwarz inequality, we calculate that:

$$\frac{d}{dt}|(u_1 - u_2)(t)|^2_H \le C_4|(u_1 - u_2)(t)|^2_H + |(f_1 - f_2)(t)|^2_H; \quad (2.2)$$

a.e. $t \in (0, T)$, by putting $C_4 := (C_1^2/\mu + 2L_b + 1)$. The above Gronwall type inequality implies the asserted inequality (2.1). □

Next, for the proofs of the other assertions, we define additional notations as follows.

For any $t \ge 0$, any $C > 0$ and any $w \in H$, let $\varphi^t_C(w; \cdot) : H \longrightarrow [0, +\infty]$ be a functional, defined as:

$$\varphi^t_C(w; z) := \begin{cases} \displaystyle\int_\Omega A(x, w, \nabla z)\, dx + C(1 + |w|^2_H), & \text{if } z \in K(t), \\ +\infty, & \text{otherwise.} \end{cases}$$

As shown in [8](Lemma 3.1), each of $\varphi^t_C(w; \cdot)$ $(t \ge 0,\ C > 0,\ w \in H)$ forms a proper l.s.c. and convex function on H with the effective domain $K(t)$, and there exist positive constants C_* and C^*, independent of t, w and z, such that:

$$\varphi^t_C(w; z) \ge C_*|\nabla z|^2_{(H)^N} + 1, \ \forall t \ge 0, \forall C \ge C^*, \forall w \in H, \forall z \in K(t). \quad (2.3)$$

Additionally, for any interval $I \subset [0, +\infty)$, any $t \in I$ and any $\omega \in C(I; H)$, we denote by $\Phi^t_{\omega(t)}$ the convex function $\varphi^t_{C^*}(\omega(t); \cdot)$ on H.

Remark 2.1. The above notations are to show the variational structure of our Cauchy problems more precisely. In fact, using the subdifferentials

$\partial\Phi^t_{\omega(t)}$ of convex functions $\Phi^t_{\omega(t)}$ ($t \geq 0$, $\omega \in C([0,+\infty); H)$), the variational inequality as in (0.1) can be reformulated into the following form of evolution equation:

$$u_t(t) + \partial\Phi^t_{\omega(t)}(u(t)) \ni f(t) - b(\cdot, \omega(t)) \quad \text{in } H,\, t > 0; \tag{2.4}$$

subject to $u(t) = \omega(t)$ in H, $t \geq 0$.

Evolution equations kindred to (2.4) have been studied comprehensively by Kenmochi [7], and hence the results from [7] will be referred to frequently in the rest of this section. Then, the following fact is essential to the solvability of Cauchy problems associated with (2.4).

(Fact1) Let us fix any compact interval $J \subset [0,+\infty)$, and let us assume $\omega \in W^{1,2}(J;H) \cap L^\infty(J;V)$, to take any constant $R_\omega(J)$ such that $R_\omega(J) \geq |\omega|_{L^\infty(J;V)} + 1$. Then, for all $s, t \in J$ and $z \in K(s)$, there exist $\tilde{z} \in K(t)$ and a positive constant C_5, independent of any of the above data, such that:

$$\begin{cases} |\tilde{z} - z|_H \leq C_5 R_\omega(J)^4 |\alpha(t) - \alpha(s)|(1 + \Phi^s_{\omega(s)}(z)^{1/2}), \\ \Phi^t_{\omega(t)}(\tilde{z}) - \Phi^s_{\omega(s)}(z) \leq C_5 R_\omega(J)^4 \Big\{ |\alpha(t) - \alpha(s)|(1 + \Phi^s_{\omega(s)}(z)) \\ \qquad + (|\omega(t) - \omega(s)|_H + R_\omega(J)|\alpha(t) - \alpha(s)|)\, \Phi^s_{\omega(s)}(z)^{1/2} \Big\}. \end{cases}$$

(Fact1) above is verified in a similar way to [9](Lemma 4.1). Also, according to [7](Theorem 2.7.1), the following fact will play quite an important role in the argument of continuous dependence.

(Fact2) Let $J \subset [0,+\infty)$ be any compact interval. If $\{s_n\} \subset [0,+\infty)$, $s_0 \geq 0$, $\{\omega_n\} \subset C(J;H)$, $\omega \in C(J;H) \cap L^\infty(J;V)$, and $s_n \to s_0$, $\omega_n \to \omega$ in $C(J;H)$ as $n \to +\infty$, then for any $t \in J$, $\Phi^{t+s_n}_{\omega_n(t)}$ converges to $\Phi^{t+s_0}_{\omega(t)}$ on H, in the sense of Mosco [11], as $n \to +\infty$.

(Fact2) above is obtained by making use of conditions (A) and (K).

Let us fix $T > 0$. Then, in the light of Remark 2.1, it would be expected that convex functions $\Phi^t_{u(t)}$ ($t \in [0,T]$) consisting of the solution u would correspond to some kind of energy (potential) inherent in the generated dynamical system. From this viewpoint, the following lemma is concerned with inequalities to characterize the movement of the energies (energy-inequalities).

Lemma 2.2. *(Energy inequalities) Let us fix the data of the constraint* $\{K(t)\}$ *with the function* $\alpha \in W^{1,2}_{\mathrm{loc}}[0,+\infty)$ *as in (K), the initial value* $u_0 \in K(0)$ *and the forcing term* $f \in L^2_{\mathrm{loc}}([0,+\infty);H)$*, to take the solution* $u \in W^{1,2}_{\mathrm{loc}}([0,+\infty);H) \cap L^\infty_{\mathrm{loc}}([0,+\infty);V)$ *of the Cauchy problem*

(CP; $\{K(t)\}, u_0, f$). Also, let us put

$$G_f(t) := 1 + |\alpha'(t)|^2 + |f(t)|_H^2 \text{ for } t \geq 0,$$

and let us fix any compact interval $J \subset [0, +\infty)$ to take any constant $R_u^\circ(J)$ such that $R_u^\circ(J) \geq |u|_{C(J;H)}+1$. Then, the function $t \mapsto \Phi_{u(t)}^t(u(t))$ is absolutely continuous on J, and there exists a positive constant C_6, independent of u_0 and f, such that:

$$\begin{aligned} &\Phi_{u(t)}^t(u(t)) - \Phi_{u(s)}^s(u(s)) + \frac{1}{8}\int_s^t |u_t(\tau)|_H^2 \, d\tau \\ &\qquad \leq C_6\, R_u^\circ(J)^{16} \int_s^t G_f(\tau) \left(1 + \Phi_{u(\tau)}^\tau(u(\tau))\right) d\tau, \quad \textit{and} \end{aligned} \tag{2.5}$$

$$\begin{aligned} &(t-s)\Phi_{u(t)}^t(u(t)) + \frac{1}{8}\int_s^t (\tau - s)|u_t(\tau)|_H^2 \, d\tau \\ &\leq e^{\int_s^t C_6 R_u^\circ(J)^{16} G_f(\tau) d\tau} \int_s^t \left(C_6 R_u^\circ(J)^{16}|J| G_f(\tau) + \Phi_{u(\tau)}^\tau(u(\tau))\right) d\tau, \end{aligned} \tag{2.6}$$

for all $s, t \in J$ where $s \leq t$, and $|J|$ denotes the length of J.

Proof. From the arguments in [7](Section 1) and [8](Section 4), we immediately check that the function $t \mapsto \Phi_{u(t)}^t(u(t))$ forms a function of bounded variation on J. However, in the restricted situation described in (II) of Remark 1.1, we can refer to the general theory found in [7](Section 2) to see the absolute continuity of this function on J.

On the other hand, the inequality (2.5) is a direct consequence of the line of argument in [8](Sections 3-4), and this also implies that:

$$\frac{d}{d\tau}\Phi_{u(\tau)}^\tau(u(\tau)) + \frac{1}{8}|u_t(\tau)|_H^2 \leq C_6\, R_u^\circ(J)^{16} G_f(\tau)\left(1 + \Phi_{u(\tau)}^\tau(u(\tau))\right), \tag{2.7}$$

a.e. $\tau \in J$. Thus, for every $s, t \in J$ satisfying $s \leq t$, another energy inequality (2.6) is obtained by multiplying both sides of (2.7) by $(\tau - s)\, e^{\int_\tau^t C_6 R_u^\circ(J)^{16} G_f(\sigma) d\sigma}$ and then integrating both sides over $[s, t]$. □

Now, let us prepare two subsections, to complete the proofs of Theorems 1.1 and 1.2.

2.1. Proof of the existence of solutions. Let us fix $T > 0$, and any $u_0 \in \overline{K(0)}^H$ to take a sequence $\{u_{n,0}\} \subset K(0)$ such that $u_{n,0} \to u_0$ in H as $n \to +\infty$. Then, on account of Remark 1.2, we can take solutions $u_n \in W^{1,2}(0,T;H) \cap L^\infty(0,T;V)$ to (CP; $\{K(t)\}, u_{n,0}, f$) ($n = 1, 2, 3, \cdots$) starting from the regular class $K(0) \subset V$. Here, by Lemma 2.1, we immediately have

$$|u_n - u_m|_{C([0,T];H)}^2 \leq e^{C_4 T}|u_{n,0} - u_{m,0}|_H^2, \ \forall n, \forall m \in \mathbb{N}. \tag{2.8}$$

Since the above inequality implies that $\{u_n\}$ is a Cauchy sequence in the Banach space $C([0,T];H)$, we find a certain limit $\bar{u} \in C([0,T];H)$ of this sequence in the topology of $C([0,T];H)$. Hence, the solutions u_n ($n = 1,2,3,\cdots$) fulfill the energy-inequalities (2.5)-(2.6) under a uniform setting of the constants $R^{\circ}_{u_n}([0,T])$ with respect to n.

Now, putting $u = u_n$ ($n \geq 2$) and $z = u_1(t)$ in (0.1), and then integrating both sides of the resultant inequalities over $[0,T]$, (2.8) leads to the boundedness of $\{\Phi^{(\cdot)}_{u_n}(u_n)\}$ in $L^1(0,T)$. So, by virtue of (2.3) and (2.8), $\{u_n\}$ is bounded in $L^2(0,T;V)$. Furthermore, from the inequality (2.6), it is seen that $\{u_n\}$ is bounded in $L^\infty(\delta,T;V)$ and $\{(u_n)_t\}$ is bounded in $L^2(\delta,T;H)$ for every $\delta \in (0,T)$. Consequently, by a standard diagonal argument, we find a subsequence $\{u_{n_k}\} \subset \{u_n\}$ such that:

$$u_{n_k} \to \bar{u} \text{ strongly in } C([0,T];H), \text{ weakly in } L^2(0,T;V),$$

$$\begin{cases} u_{n_k} \to \bar{u} \ \text{ weakly-}* \text{ in } L^\infty(\delta,T;V), \\ (u_{n_k})_t \to \bar{u}_t \ \text{ weakly in } L^2(\delta,T;H), \end{cases} \forall \delta \in (0,T), \text{ as } k \to +\infty.$$

Next, let us fix any $\delta \in (0,T)$ to apply (Fact1)-(Fact2) by putting $J = [\delta, T]$, $\omega = \bar{u}$ and $\omega_n = u_n$ ($n = 1,2,3,\cdots$). Then, in the light of (B1), Remark 2.1 and [7](Theorem 2.7.1), we infer that $u_{n_k}(\cdot + \delta)$ converges to the solution $u^{(\delta)}$ of the following Cauchy problem:

$$\begin{cases} u^{(\delta)}_t(t) + \partial\Phi^{t+\delta}_{\bar{u}(t+\delta)}(u^{(\delta)}(t)) \ni (f - b(\cdot,\bar{u}))(t+\delta) \text{ in } H, \ t \in (0, T-\delta], \\ u^{(\delta)}(0) = \bar{u}(\delta) \ \text{ in } H; \end{cases}$$

in the topology of $C([0,T-\delta];H)$ as $k \to +\infty$. By the uniqueness of limits, $u^{(\delta)}(t)$ should coincide with $\bar{u}(t+\delta)$ for all $\delta \in (0,T)$ and $t \in [0,T-\delta]$. This implies that the function $\bar{u} \in C([0,T];H) \cap W^{1,2}_{\rm loc}((0,T];H) \cap L^2(0,T;V) \cap L^\infty_{\rm loc}((0,T];V)$ is the solution to the Cauchy problem (CP; $\{K(t)\}, u_0, f$).□

2.2. Proof of Theorem 1.2. We divide the proof into two steps.

(Step 1) The case that $\{\Phi^{s_n}_{u_n(0)}(u_{n,0})\}$ is bounded.

In this case, $u_0 \in K(s_0)$ is immediately seen from (Fact2). For the energy-inequalities (2.5) for solutions u_n ($n = 1,2,3,\cdots$), applying Gronwall's lemma yields the boundedness of the sequence $\{\Phi^{(\cdot+s_n)}_{u_n}(u_n)\}$ in the topology of $L^\infty(0,T)$. Also, on account of the boundedness of sequences $\{|u_{n,0}|_H\}$ and $\{\Phi^{s_n}_{u_n(0)}(u_{n,0})\}$, the inequality (2.5) further leads to the boundedness of the sequence $\{u_n\}$ in $W^{1,2}(0,T;H) \cap L^\infty(0,T;V)$. Therefore, taking a subsequence if necessary, Ascoli's theorem and weak/weak-$*$ compactness of the sequence enable us to see the existence of a limit u° of

$\{u_n\}$ in the sense that:

$$u_n \to u^\circ \text{ strongly in } C([0,T];H), \text{ weakly in } W^{1,2}(0,T;H),$$
$$\text{weakly-}* \text{ in } L^\infty(0,T;V), \quad \text{as } n \to +\infty.$$

After this, we can show that u° forms, anyway, a solution to the Cauchy problem (CP; $\{K(t+s_0)\}, u_0, f$), just as in the previous subsection. Thus, by the uniqueness of solutions, we conclude the assertion of this case. □

(Step2) The general case of the sequence $\{\Phi^{s_n}_{u_n(0)}(u_{n,0})\}$.

Based on the results from Theorem 1.1 and (Step1), the assertion of this case will be proved after noting a similar demonstration technique used in [7](Theorem 2.7.1). □

Remark 2.2. Referring to the proof of [7](Theorem 2.7.1), we can further obtain that

$$\int_s^t \Phi^{\tau+s_n}_{u_n(\tau)}(u_n(\tau))\,d\tau \to \int_s^t \Phi^{\tau+s_0}_{u(\tau)}(u(\tau))\,d\tau \quad \text{as } n \to +\infty, \tag{2.9}$$

for all $0 \le s < t \le T$, Therefore, even for solutions starting from $\overline{K(0)}^H$, the energy-inequality (2.6) will be verified through the limiting observation of the inequalities (2.6) for approximating solutions in $W^{1,2}(0,T;H) \cap L^\infty(0,T;V)$.

3. Periodic stability

In this section, we assume all the conditions (A), (B1)-(B2), (K) and (P), to prove Theorems 1.3-1.5. Under these conditions, we can take at least one T_0-periodic function $h_p \in W^{1,2}_{\mathrm{loc}}([0,+\infty);H) \cap L^\infty(0,+\infty;V)$, such that

$$h_p(t+T_0) = h_p(t) \in K(t), \ \forall t \ge 0.$$

In fact, by the general theory given in [5](Theorem 2.1), such function is, for example, found as the T_0-periodic solution to the evolution equation:

$$(h_p)_t(t) + \partial\Phi^t_{\omega(t)}(h_p(t)) + b(\cdot, h_p(t)) \ni f(t) \text{ in } H, \, t \ge 0;$$

when $\omega \equiv 0$. By means of this T_0-periodic function, we can see the dissipativity of the dynamical system, stated in the following lemma.

Lemma 3.1. *(Dissipativity of the dynamical system) Let us fix the data of the constraint $\{K(t)\}$, to take any solution u of the Cauchy problem (CP; $\{K(t)\}, u_0, f$) with any initial value $u_0 \in \overline{K(0)}^H$ and any forcing term*

$f \in L^2_{\text{loc}}([0,+\infty); H)$. *Then, there exist positive constants* C_7 *and* C_8*, independent of* u_0 *and* f*, such that:*

$$\begin{aligned}&\frac{d}{dt}|(u-h_p)(t)|_H^2 + C_7|(u-h_p)(t)|_V^2 \\ &\leq C_8(1+|h_p(t)|_V^2 + |(h_p)_t(t)|_H^2 + |f(t)|_H^2), \quad a.e.\ t \geq 0.\end{aligned} \tag{3.1}$$

Proof. Let us assign $h_p(t)$ to the test function z of the variational inequality in (0.1). Then, from conditions (A) and (B2), we see that:

$$\begin{aligned}&\frac{1}{2}\frac{d}{dt}|(u-h_p)(t)|_H^2 + \mu|\nabla(u-h_p)(t)|_{(H)^N}^2 + \int_\Omega (C_2|u(t)|^2 - C_3)\,dx \\ &\leq (f(t),(u-h_p)(t)) - ((h_p)_t(t),(u-h_p)(t)) + (b(\cdot,u(t)),h_p(t)) \\ &\quad + |C_1^{1/2}(1+|\nabla h_p(t)|)|_H|\nabla(u-h_p)(t)|_{(H)^N}, \text{ a.e. } t\geq 0.\end{aligned} \tag{3.2}$$

Thus, the constants C_7 and C_8 to realize (3.1) will be found by fundamental calculations using condition (B1). □

Now, on account of the above lemma, we can prove Theorems 1.3-1.5, by referring to the demonstration methods found in [5, 12].

Proof of Theorem 1.3. The proof will be a modified version of that in [5](Theorem 2.1). In fact, by virtue of Theorems 1.1-1.2 and Lemma 3.1, we find a sufficiently large positive constant R_p, such that the restriction $U_0|_{B_H(0;R_p)}$ of the solution operator $U_0 = U(T_0,0) : \overline{K(0)}^H \longrightarrow \overline{K(T_0)}^H$ forms a compact map from the closed and convex domain $\overline{K(0)}^H \cap B_H(0;R_p)$ onto itself. Thus, this theorem will be concluded by applying Schauder's fixed point argument for the iteration of $U_0|_{B_H(0;R_p)}$. □

Proof of Theorem 1.4. This theorem will be proved after similar analytical techniques employed in [12](Section 4). Roughly summarized, the estimate (1.2) will be derived from the key-inequality (3.1) as in Lemma 3.1. Also, another estimate (1.3) is obtained by taking account of (1.2), (2.6) and Remark 2.2. □

Proof of Theorem 1.5. For this theorem, we refer to [5](Theorems 3.1-3.3), and omit the detailed proofs. Indeed, due to Theorems 1.1-1.4, the dynamical systems $\{U(t+s,s)\}$ $(s \geq 0)$ and $\{U_\tau^k\}$ $(\tau \geq 0)$, prescribed in Section 1, can be regarded as essentially the same as those in [5](Section 3). Incidentally, according to [5](Theorem 3.1), each of the attractors $\mathscr{A}_\tau$ $(\tau \geq 0)$ is given in the form of the ω-limit set $\mathscr{A}_\tau := \bigcap_{n=0}^{+\infty} \overline{\bigcup_{k\geq n} U_\tau^k A_\tau}$ of a compact set $A_\tau := U_\tau(B_H(0,R_*) \cap \overline{K(\tau)}^H)$ (called the absorbing set), where R_* is a sufficiently large radius obtained using Lemma 3.1. □

4. Applications

In this section, two problems associated with the situations (AP1)-(AP2), described in the introduction, will be stated as the applications.

4.1. Application to the situation (AP1). Let $\Lambda = [\lambda_{ij}(\cdot)] \in C^1(\overline{\Omega})^{N\times N}$ be a given function such that $\Lambda(x)$ is symmetric and positive definite for all $x \in \overline{\Omega}$, and let there be an x-independent constant $\mu_0 > 0$ satisfying $(\Lambda(x)y)\cdot y \geq \mu_0|y|^2$ $(\forall x \in \overline{\Omega}, \forall y \in \mathbb{R}^N)$. Let $\kappa \in W^{1,\infty}(\mathbb{R})^N$ be a given function, and let $g_i \in L^\infty(0,+\infty; V \cap L^\infty(\Omega)) \cap W^{1,2}_{\mathrm{loc}}([0,+\infty); V \cap L^\infty(\Omega))$ $(i = 1,2)$ be prescribed obstacle functions such that $\log(g_2 - g_1) \in L^\infty(Q)$ and $\sup_{\tau\geq 0}\sum_{i=1}^{2}(|(g_i)_t|_{L^2(\tau,\tau+1;V)} + |(g_i)_t|_{L^2(\tau,\tau+1;L^\infty(\Omega))}) < +\infty$. Let $b \in C(\overline{\Omega} \times \mathbb{R})$ be any function satisfying the conditions (B1)-(B2). With the above notations, let us set:

$$a(x,r,y) := \Lambda(x)[y + \kappa(r)],\ \forall(x,r,y) \in \overline{\Omega} \times \mathbb{R} \times \mathbb{R}^N, \text{ and} \tag{4.1}$$

$$K(t) := \{\, z \in V \,|\, g_1(t,\cdot) \leq z \leq g_2(t,\cdot), \ \text{a.e. in } \Omega \,\}. \tag{4.2}$$

Then, referring to [8](Section 5.1), it will be checked that the function $a = a(x,r,y)$ and the constraint $\{K(t)\}$, given in (4.1) and (4.2), fulfill conditions (A) and (K), respectively. Hence, we can apply the general theory, obtained in Theorems 1.1-1.2, to the following double obstacle problem.

Problem 4.1 (Time-dependent double obstacle problem)

$$\begin{cases} (u_t(t), u(t) - z) + (\Lambda[\nabla u + \kappa(u)](t), \nabla(u(t) - z))_{(H)^N} \\ \qquad +(b(\cdot, u(t)) - f(t), u(t) - z) \leq 0, \ \text{a.e. } t > 0, \ \forall z \in K(t); \\ \text{subject to } \ g_1 \leq u \leq g_2 \text{ a.e. in } Q, \ \text{ and } \ u(0,\cdot) = u_0 \in \overline{K(0)}^H. \end{cases}$$

Moreover, the T_0-periodic condition (P) will be verified whenever f and g_i $(i = 1,2)$ are T_0-periodic in time. Therefore, under this periodicity, the dynamical system generated by Problem 4.1 will show the key-properties as in Theorems 1.3-1.5.

4.2. Application to the situation (AP2). In addition to the notations $\Lambda \in C^1(\overline{\Omega})^{N\times N}$, $\kappa \in W^{1,\infty}(\mathbb{R})^N$, $b \in C(\overline{\Omega} \times \mathbb{R})$ and $f \in L^2_{\mathrm{loc}}([0,+\infty); H)$ as in the previous subsection, let us take a function $g \in W^{1,2}_{\mathrm{loc}}([0,+\infty); V) \cap L^\infty(0,+\infty; H^2(\Omega))$, to consider the following time-dependent boundary constraints problem.

Problem 4.2 (Time-dependent boundary constraints problem)

$$\begin{cases} u_t - \nabla \cdot (\Lambda[\nabla u + \kappa(u)]) + b(\cdot, u) = f \ \text{ in } Q, \\ u(t) = g(t) \text{ in } \Gamma_D(t), \ \ \nu_\Gamma \cdot (\Lambda[\nabla u + \kappa(u)])(t) = 0 \text{ on } \Gamma_N(t), \\ \left. \begin{array}{l} u(t) \le g(t), \ \nu_\Gamma \cdot (\Lambda[\nabla u + \kappa(u)])(t) \le 0, \\ (u-g)(t)\left(\nu_\Gamma \cdot (\Lambda[\nabla u + \kappa(u)])(t)\right) = 0, \end{array} \right\} \text{ on } \Gamma_S(t), \\ u(0, \cdot) = u_0(x) \ \text{ in } \Omega. \end{cases}$$

In the context, ν_Γ denotes the unit outer normal vector on Γ, and $\Gamma_D(t), \Gamma_N(t), \Gamma_S(t)$ $(t \ge 0)$ denote pairwise disjoint subsets to form a time-dependent decomposition $\Gamma := \Gamma_D(t) \cup \Gamma_N(t) \cup \Gamma_S(t)$ $(t \ge 0)$ governed by given diffeomorphisms $\Theta(t) = (\theta_1(t,\cdot), \cdots, \theta_N(t,\cdot)) \in C^2(\overline{\Omega})^N$ $(t \ge 0)$ such that $\Theta \in C^2(\overline{Q})^N$, $\Theta(0,x) = x$ $(\forall x \in \overline{\Omega})$, and $\Theta(t, \Gamma_j(0)) = \Gamma_j(t)$ $(t \ge 0$, $j = D, N, S)$. Now, let $a = a(x,r,y)$ as in (4.1), let us set:

$$K(t) := \left\{ z \in V \middle| z \le g(t) \text{ on } \Gamma_S(t) \text{ and } z = g(t) \text{ on } \Gamma_D(t) \right\}, \ t \ge 0;$$

and let us assume that $\sup_{\tau \ge 0} |g_t|_{L^2(\tau,\tau+1;V)} < +\infty$. Then, referring to [8](Section 5.3) (or [9](Section 5.2)), we will see that the general theory stated in Theorems 1.1-1.5 is applicable to Problem 4.2, under the required assumptions by all/part of the conditions (A), (B1)-(B2), (K) and the time-periodicity of f, g and Θ to realize (P).

References

[1] Ph. Bénilan, *Equations d'évolution dans un espace de Banach quelconque et application,* Université de Paris-Sud, Publication Mathématique d'Orsay, 1972.

[2] H. Attouch, Ph. Bénilan, A. Damlamian and C. Picard, *Equations d'évolution avec condition unilatérale,* C. R. Acad. Sci. Paris **279** (1974), 607–609.

[3] M. Biroli, *Sur la solution faible du problème de Cauchy pour des inéquations d'évolution avec convexe dépendant du temps,* C. R. Acad. Sci. Paris **280** (1975), 1209–1212.

[4] V. V. Chepyzhov and M. I. Vishik, *Attractors for Equations of Mathematical Physics,* Colloquium Publications **49**, Amer. Math. Soc., Providence, R. I., 2002.

[5] A. Ito, N. Kenmochi and N. Yamazaki, *Attractors of periodic systems generated by time-dependent subdifferentials,* Nonlinear Anal. **37** (1999), 97–124.

[6] A. Ito, N. Kenmochi and N. Yamazaki, *Time-dependent attractors of bounded dynamical systems generated by subdifferentials,* Commun. Appl. Anal. **5** (2001), no. 3, 403–419.

[7] N. Kenmochi, *Solvability of nonlinear evolution equations with time-dependent constraints and applications,* Bull. Fac. Education, Chiba Univ. **30** (1981), 1–87.

[8] M. Kubo and N. Yamazaki, *Quasilinear parabolic variational inequalities with time-dependent constraints,* Adv. Math. Sci. Appl. 15 (2005), no. 1, 335–354.

[9] M. Kubo and N. Yamazaki, *Elliptic-parabolic variational inequalities with time-dependent constraints,* Discrete Contin. Dyn. Syst., 19 (2007), 335–359.

[10] F. Mignot and J. P. Puel, *Inéquations d'évolution paraboliques avec convexes dépendant du temps: applications aux inéquations quasi-variationelles d'évolution,* Arc. Rational Mech. Anal. **64** (1977), 59–91.

[11] U. Mosco, *Convergence of convex sets and of solutions of variational inequalities,* Advances in Math. **3** (1969), 510–585.

[12] K. Shirakawa, A. Ito, N. Yamazaki and N. Kenmochi, *Asymptotic stability for evolution equations governed by subdifferentials,* Recent Developments in Domain Decomposition Methods and Flow Problems, 287–310, GAKUTO Intern. Ser. Math. Sci. Appl. **11**, Gakkōtosho, Tokyo, 1998.

GLOBAL BIFURCATION FOR SEMILINEAR ELLIPTIC PROBLEMS

MARCELLO LUCIA

Universität zu Köln, Mathematisches Institut
50931 Köln, Germany, mlucia@math.uni-koeln.de

MYTHILY RAMASWAMY

TIFR Center, IISc. Campus, Post Box No. 1234
Bangalore 560012, India, mythily@math.tifrbng.res.in

Abstract: We study the existence of a global branch of solutions for the semilinear elliptic problem

$$-\Delta u = \lambda\big(a(x)u + b(x)r(u)\big), \quad u \in \mathcal{D}_0^{1,2}(\Omega).$$

We work in a general domain Ω of $\mathbb{R}^n$, with indefinite weights a, b belonging to some Lorentz spaces, and the function r is either asymptotically linear or superlinear at infinity. To derive our result we first prove existence, uniqueness and simplicity of a principal eigenvalue for linear problems with weight in Lorentz spaces.

1. Introduction

The present paper deals with the semilinear elliptic problem

$$-\Delta u = \lambda\big(a(x)u + b(x)r(u)\big), \qquad u \in \mathcal{D}_0^{1,2}(\Omega), \tag{1.1}$$

where λ is a real parameter, r a known nonlinearity, a, b are given weights in a domain (connected open set) Ω of $\mathbb{R}^N$ and the space $\mathcal{D}_0^{1,2}(\Omega)$ is defined as the closure of $C_0^\infty(\Omega)$, set of smooth functions having compact support, with respect to the norm

$$\|u\| := \left(\int_\Omega |\nabla u|^2\right)^{1/2}.$$

When $N \geq 3$, or when the domain is bounded, the functional space $\mathcal{D}_0^{1,2}(\Omega)$ can be identified with a space of functions (see [10]). Due to the particularity of the dimensions $N = 1, 2$, for the sake of simplicity we will henceforth assume that $N \geq 3$.

When the function r is such that $r(0) = 0$, then $u \equiv 0$ solves (1.1) and we are interested in finding non-trivial solutions. In such a case the question of existence can be handled by using tools provided by the bifurcation theory like for example the Rabinowitz Theorem [17]. In order to apply such results one needs first to understand the linearized problem which, under suitable assumptions, will be a weighted eigenvalue problem of the type

$$-\Delta u = \lambda w(x)u, \qquad u \in \mathcal{D}_0^{1,2}(\Omega). \tag{1.2}$$

Both problems (1.1) and (1.2) have attracted a lot of attention because they naturally arise in mathematical physics or biology (combustion, population dynamics ...). Since in atomic physics the relevant weights are attractive-repulsive potentials with singularities of the type $\frac{1}{|x|}$, it is desirable to study the above problems with coefficients allowed to be singular and sign changing. Furthermore it is known that Hardy's weight $w(x) = \frac{1}{|x|^2}$ arise as "critical" cases when dealing with existence of eigenfunction for linear problems. Therefore it is of interest to understand how far "small perturbation" of such weights are allowed in Problems (1.1) and (1.2).

An important feature of the linear Problem (1.2) is the existence and simplicity of a principal eigenvalue. These are the value λ for which (1.2) admits a non-negative associated eigenfunction, and by simplicity one means that the associated eigenspace is of dimension one. When $w \equiv 1$ in a bounded domain of $\mathbb{R}^N$, a proof of these properties can be found in [8]. In [16], it has been shown that the smallest positive eigenvalue for (1.2) exists and is simple whenever $w^+ \not\equiv 0$ and $w \in L^r(\Omega)$ with $r > \frac{N}{2}$ in a bounded domain. Sufficient conditions for the existence of principal eigenvalues for the weighted eigenvalue problem in $\mathbb{R}^N$ have been given by many authors. Brown, Cosner, Fleckinger introduced in [6], a sufficient condition for the existence of a positive principal eigenvalue, namely the weight function w be negative and bounded away from 0 at infinity. They also used another sufficient condition for dimensions $N \geq 3$ requiring that w has a positive integral and it decays at infinity faster than $\frac{1}{|x|^2}$. Brown and Tertikas (see [7]) relaxed the first condition in [6], by asking the positive part w^+ to be of compact support. Allegretto proved in [1] the existence of a principal eigenvalue and also infinitely many eigenvalues when w^+ lies in $L^{\frac{N}{2}}(\Omega)$. Szulkin and Willem in [19] studied the problem with w in $L^{\frac{N}{2}}(\Omega)$ or having faster decay than $\frac{1}{|x|^2}$ at infinity or at any finite point. An important aspect of these previous works concerning the simplicity of the principal eigenvalue is that they rely on Harnack's inequality in order to ensure the associated eigenfunctions to be continuous and strictly positive (or negative) in the domain.

In the present paper we extend all the previous results by considering weights w belonging to some Lorentz space $L(p,q)$ which are described in Section 2. It allows us to include some classes of functions not satisfying the earlier conditions described above. Furthermore, by going to Lorentz spaces, the importance of the comparison of decay with $\frac{1}{|x|^2}$ becomes clearer. More precisely the first main result of our paper is the following:

Proposition 1.1. *Let $w \in L(\frac{N}{2}, q_0)$ with $q_0 \in (1,\infty)$ and such that $w^+ \not\equiv 0$. Then*

$$\lambda_1^+(w) := \inf\left\{\frac{\int_\Omega |\nabla u|^2}{\int_\Omega wu^2} : u \in \mathcal{D}_0^{1,2}(\Omega),\ \int_\Omega wu^2 > 0\right\}, \tag{1.3}$$

is a principal eigenvalue and this is the unique positive principal eigenvalue. Furthermore the associated eigenfunction Φ satisfies $\Phi > 0$ a.e. or $\Phi < 0$ a.e. and is unique up to a constant multiple factor.

If $w^- \not\equiv 0$ above proposition applied to $-w$ provides also existence of a negative principal eigenvalue $\lambda_1^-(w)$, which will be simple and the unique negative principal eigenvalue. Let us emphasize that with our assumptions the Harnack's inequality does not apply and we may have $\inf_B |\Phi| = 0$ on some $B \subset\subset \Omega$. Nevertheless thanks to a strong maximum principle as established by Ancona [4] and Brezis, Ponce [5] the set of zeroes of a non-negative eigenfunction has Lebesgue measure zero. This property will be sufficient for proving that the eigenspace associated to $\lambda_1^+(w)$ has dimension one if $w^+ \not\equiv 0$.

Proposition 1.1 opens the possibility of proving existence of global branch of solutions for Problem (1.1) when the coefficients a, b have very less regularity. Positive solution branches have been studied by many authors for nonlinearities which are asymptotically linear (see [3]) and also for superlinear cases as in [11]. In [12] both asymptotically linear and superlinear nonlinearities have been considered with the restriction that the coefficients are bounded. But that is not necessary for the existence of a branch of solutions. Indeed in the present paper we derive existence of at least one global branch under the only following requirements:

(H1) $r \in C^0(\mathbb{R})$, $\lim_{s\to 0} \frac{r(s)}{s} = 0$, $\limsup_{s\to\infty} \frac{|r(s)|}{|s|^\gamma} < \infty$ for some $\gamma \in [1, 2^* - 1)$;

(H2) $a \in L(\frac{N}{2}, q_0)$, $b \in L(\frac{N}{2}, q_0) \cap L(p_0, q_0)$ with $p_0 = \frac{2^*}{2^*-\gamma-1}$ and for some $q_0 \in (1,\infty)$.

As in Proposition 1.1, $L(p,q)$ denotes the Lorentz space. Since $\gamma \in [1, 2^* - 1)$, it follows that $p_0 \in [\frac{N}{2}, \infty)$ and in particular for bounded domain the assumption on b reduces to $b \in L(p_0, q_0)$ with (p_0, q_0) as given in (H2). But for unbounded domain one needs to work on a smaller space. Note also that (H1) implies $r(0) = 0$ and we are in fact interested in existence of non-trivial solutions for Problem (1.1). By defining

$$\mathcal{S} := \{(\lambda, u) \in \mathbb{R} \times \mathcal{D}_0^{1,2}(\Omega) \, : \, (\lambda, u) \text{ solves } (1.1), u \not\equiv 0\},$$

our main result is as follows:

Proposition 1.2. *Assume (H1), (H2) and $a^+ \not\equiv 0$ (resp. $a^- \not\equiv 0$). Then, there exists a set $\mathcal{C}^+$ connected in $\overline{\mathcal{S}}$ bifurcating from $\lambda_1^+(a)$ (resp. $\mathcal{C}^-$ bifurcating from $\lambda_1^-(a)$). Moreover, $\mathcal{C}^+$ (resp. $\mathcal{C}^-$) is*

(i) *either unbounded,*

(ii) *or contains a point $(\lambda, 0)$ with $\lambda \neq \lambda_1^+(a)$ (resp. $\lambda \neq \lambda_1^-(a)$).*

Our results improve considerably the result of [12] since we are working with weights belonging only to some Lorentz space and we work in any domain of $\mathbb{R}^N$. But in the present paper we shall not discuss the positivity of the solution of the branch. This issue will be discussed in more details in a coming work, where we will also see how our assumptions can further be relaxed.

The paper is organized as follows. Section 2 provides the definitions and basic properties of the Lorentz $L(p,q)$ spaces. For weights $w \in L(\frac{N}{2}, q_0)$ with $w^+ \not\equiv 0$, we prove in Section 3 that the positive principal eigenvalue of the linear problem is unique and its associated eigenspace has dimension one. In Section 4, we establish existence of at least one global branch of solutions for the nonlinear problem (1.1) as stated in Proposition 1.2.

2. Prerequisites on Lorentz spaces

The Lorentz $L(p,q)$ spaces, introduced by Lorentz in [14], are generalization of the Lebesgue L^p spaces. We collect here their main properties and refer to [13], [18], [20] for more detailed discussions. To introduce their definitions, we start by recalling that the distribution function and nonincreasing rearrangement of a measurable function $f : \Omega \to \mathbb{R}$ are respectively defined as:

$$\alpha_f(s) := \big|\{x \in \Omega \, : \, |f(x)| > s\}\big|, \quad f^*(t) := \inf\{s > 0 \, : \, \alpha_f(s) \leq t\}. \quad (2.1)$$

Then $\alpha_f : \mathbb{R} \to \mathbb{R}$ is nonnegative, nonincreasing, continuous from the right and we easily verify that

(i) $f^* \geq 0$, nonincreasing and continuous from the right;
(ii) $f^*(\alpha_f(s)) \leq s$, $\alpha_f(f^*(t)) \leq t$;
(iii) if α_f is continuous and strictly decreasing then $f^* = \alpha_f^{-1}$;
(iii) $\alpha_{f^*} = \alpha_f$ and therefore $\int_\Omega |f|^p dx = \int_0^\infty |f^*(t)|^p\, dt$.

Observe that for $f \in L^p(\Omega)$,

$$||f||_p = \left(\int_0^\infty \left[t^{\frac{1}{p}} f^*(t) \right]^p \frac{dt}{t} \right)^{\frac{1}{p}}. \tag{2.2}$$

This can be used to motivate the definition of Lorentz spaces:

Definition 2.1. We define

$$||f||^*_{p,q} = \begin{cases} \left(\dfrac{q}{p} \displaystyle\int_0^\infty \left[t^{\frac{1}{p}} f^*(t) \right]^q \frac{dt}{t} \right)^{\frac{1}{q}}, & \text{if } 1 \leq p,\, q < \infty, \\ \sup\limits_{t>0} \left\{ t^{\frac{1}{p}} f^*(t) \right\}, & \text{if } 1 \leq p \leq \infty,\, q = \infty, \end{cases}$$

$$L(p,q) = \{ f\colon \Omega \to \mathbb{R} \;:\; f \text{ measurable},\; ||f||^*_{p,q} < \infty \}.$$

Notice that with this definition we have

$$L(p,p) = L^p(\Omega), \quad 1 \leq p \leq \infty.$$

The space $L(p,\infty)$ is known as the "weak L^p space" and coincides with the Marcinkiewicz space M^p, defined as follows:

$$M^p := \{ f\colon \Omega \to \mathbb{R} \;:\; f \text{ measurable}, \int_\omega |f| \leq C|\omega|^{1-\frac{1}{p}} \quad \forall\, \omega \subset \Omega \}.$$

When $1 \leq q_1 \leq q_2 \leq \infty$, one can prove (see [[13], Section 1]):

$$||f||^*_{p,q_2} \leq ||f||^*_{p,q_1} \quad \text{and} \quad L(p,q_1) \subset L(p,q_2). \tag{2.3}$$

In general $||\cdot||^*_{p,q}$ need not be a norm as Minkowski inequality may fail. In spite of this, it can be used to define a norm on $L(p,q)$. Indeed let us set

$$f^{**}(t) := \frac{1}{t} \int_0^t f^*(r) dr, \qquad ||f||_{p,q} := ||f^{**}||^*_{p,q}\,.$$

We easily see that

$$||f||_{p,q} = \begin{cases} \left(\displaystyle\int_0^\infty \left[t^{\frac{1}{p}} f^{**}(t) \right]^q \frac{dt}{t} \right)^{\frac{1}{q}}, & \text{if } 1 \leq p, q < \infty, \\ \sup\limits_{t>0} \{ t^{\frac{1}{p}} f^{**}(t) \}, & \text{if } 1 \leq p \leq \infty,\, q = \infty. \end{cases}$$

This is a norm which is "equivalent" to $||\cdot||^*_{pq}$, (see Chapter V, Theorem 3.2, [18]):

$$||f||^*_{p,q} \leq ||f||_{p,q} \leq \frac{p}{p-1}||f||^*_{p,q}\,, \tag{2.4}$$

and enjoys the following monotonicity property:

$$f, g \in L(p,q) \text{ with } |f| \leq |g| \implies ||f||_{p,q} \leq ||g||_{p,q}. \tag{2.5}$$

Under this norm, $L(p,q)$ is a Banach space and the following Hölder's inequality holds (see [13]):

$$||fg||_{p,q} \leq C||f||_{p_1,q_1}\ ||g||_{p_2,q_2}, \qquad \forall (f,g) \in L(p_1,q_1) \times L(p_2,q_2), \tag{2.6}$$

whenever

$$\frac{1}{p} = \frac{1}{p_1} + \frac{1}{p_2}, \quad \frac{1}{q} = \frac{1}{q_1} + \frac{1}{q_2} \quad p_i, q_i \in [1,\infty].$$

This inequality applied with $p = q = 1$ helps to identify the topological dual of the Lorentz spaces:

Proposition 2.1.

(a) *Let $(p,q) \in (1,\infty) \times [1,\infty)$ or $p = q = 1$. Then the dual space of $L(p,q)$ is isomorphic to $L(p',q')$ where $1/p + 1/p' = 1$ and $1/q + 1/q' = 1$.*

(b) *The spaces $L(p,q)$ are reflexive when $1 < p, q < \infty$.*

A main feature of the Lorentz space is that they allow to improve the usual Sobolev embedding $\mathcal{D}^{1,2}_0(\Omega) \hookrightarrow L^{2^*}(\Omega) \simeq L(2^*, 2^*)$ as follows:

Proposition 2.2. ***(Sobolev-Lorentz embedding).***

$$\mathcal{D}^{1,2}_0(\Omega) \hookrightarrow L(2^*, 2).$$

Observe that $2 < 2^*$ for $N \geq 3$ (see for example, appendix in [2]).

3. Existence and uniqueness of principal eigenvalue

For the existence of principal eigenvalue for the problem (1.2), a sufficient condition given by Allegretto in [1] is that the weight function is in $L^{\frac{N}{2}}(\Omega)$. Szulkin and Willem relaxed this in [19] by introducing the following conditions :

$$
\textbf{(H)} \quad \begin{cases} V \in L^1_{\mathrm{loc}}(\Omega), \;\; V^+ = V_1 + V_2 \not\equiv 0, \;\; V_1 \in L^{\frac{N}{2}}(\Omega), \\ \lim\limits_{|x|\to\infty \;\; x\in\Omega} |x|^2 V_2(x) = 0, \qquad \lim\limits_{x\to x_0 \;\; x\in\Omega} |x-x_0|^2 V_2(x) = 0 \;\; \forall x_0 \in \bar{\Omega}. \end{cases}
$$

Under this condition, the existence of a positive principal eigenvalue and a sequence of eigenvalues were proved in [19]. Further, they give examples of weight functions

$$
W_1(x) \; = \; \frac{1}{1+|x|^2}, \qquad W_2(x) \; = \; \frac{1}{|x|^2(1+|x|^2)},
$$

which do not satisfy (H) and the eigenvalue problems do not possess an eigenvalue (see also [21]). But the modified versions of these functions,

$$
\widetilde{W_1}(x) = \frac{W_1(x)}{\big(\log(2+|x|^2)\big)^{\frac{2}{N}}}, \qquad \widetilde{W_2}(x) \; = \; \frac{W_2(x)}{\big(\log(2+\frac{1}{|x|^2})\big)^{\frac{2}{N}}},
$$

satisfy (H) and the eigenvalue problems possess infinitely many eigenvalues. Observe that none of these functions lie in $L^{\frac{N}{2}}(\Omega)$.

By using Lorentz spaces, we will try to relax further the condition (H). In this section we are doing a first step in this direction by showing that the classical results concerning the principal eigenvalue for linear weighted eigenvalue problem holds when

$$
w \in L\left(\frac{N}{2}, q_0\right) \;\text{ for some } 1 < q_0 < \infty. \tag{3.1}
$$

First, let us see some examples to compare different conditions.

Example 1: The function W_1 does not satisfy the condition (3.1). Denoting by ω_N the volume of the unit sphere in $\mathbb{R}^N$ we have for $w = W_1$,

$$
\begin{aligned}
\alpha_w(s) \; &= \; \big|\{x \,:\, w(x) > s\}\big| \; = \; \omega_N |x_0|^N, \qquad s = \frac{1}{1+|x_0|^2}, \;\; 0 < s < 1, \\
& \qquad\qquad w^*(t) = \frac{1}{1+(\frac{t}{\omega_N})^{\frac{2}{N}}}, \qquad t \in [0,\infty).
\end{aligned}
$$

Now it is easy to check that $w \notin L\left(\frac{N}{2}, q\right)$ for any $q \in (1,\infty)$:

$$
(||w||^*_{\frac{N}{2},q})^q \; = \; \int_0^\infty \left\{ t^{\frac{2}{N}} \frac{1}{1+(\frac{t}{\omega_N})^{\frac{2}{N}}} \right\}^q \frac{dt}{t} = \infty \qquad \text{for } q \geq 1.
$$

But $w \in L\left(\frac{N}{2}, \infty\right)$ since

$$||w||^*_{\frac{N}{2},\infty} = \sup_{t\in[0,\infty)} \left\{ t^{\frac{2}{N}} \frac{1}{1+(\frac{t}{\omega_N})^{\frac{2}{N}}} \right\} = \omega_N^{\frac{2}{N}}.$$

Thus condition (3.1) is not satisfied by the function W_1 and similarly for W_2.

Example 2: The functions $\widetilde{W_1}$ and $\widetilde{W_2}$ satisfy condition (3.1). We do the calculation only for $w = \widetilde{W_1}$. We have

$$\alpha_w(s) = \big|\{x : w(x) > s\}\big| = \omega_N |x_0|^N,$$

where

$$s = \frac{1}{(1+|x_0|^2)\left(\log(2+|x_0|)\right)^{\frac{2}{N}}}, \quad 0 < s < \frac{1}{(\log 2)^{\frac{2}{N}}},$$

$$w^*(t) = \frac{1}{\left(1+(\frac{t}{\omega_N})^{\frac{2}{N}}\right)\left(\log(2+(\frac{t}{\omega_N})^{\frac{2}{N}})\right)^{\frac{2}{N}}}, \quad t \in (0,\infty).$$

Now we check if $w \in L\left(\frac{N}{2}, q_0\right)$ for some q_0. A direct calculation shows

$$(||w||^*_{\frac{N}{2},q})^q = \int_0^\infty \left\{ t^{\frac{2}{N}} \frac{1}{\left(1+(\frac{t}{\omega_N})^{\frac{2}{N}}\right)\left(\log(2+(\frac{t}{\omega_N})^{\frac{2}{N}})\right)^{\frac{2}{N}}} \right\}^q \frac{dt}{t},$$

which is finite if $q > \frac{N}{2}$. Thus $\widetilde{W_1}$ satisfies condition (3.1) for any $q_0 \in (\frac{N}{2}, \infty)$.

The following example shows that there are functions failing the condition (H) but satisfying condition (3.1).

Example 3: Consider the following function which is singular all along the x_2-axis in a square $\Omega = \{(x_1, x_2) : |x_i| < R\}$ in $\mathbb{R}^2$ with $R < 1$,

$$W_3(x_1, x_2) = \frac{1}{|x_1 \log(|x_1|)|} \quad \text{in } \Omega, \quad x_1 \neq 0.$$

As in Example 1, we have for $w = W_3$,

$$\alpha_w(s) = \big|\{x : w(x) > s\}\big| = R\,|x_1^0|$$

where

$$s = \frac{1}{|x_1^0 \log(|x_1^0|)|}, \quad 0 < s < \frac{1}{R \log R},$$

$$w^*(t) = \frac{1}{(\frac{t}{R})|\log(\frac{t}{R})|}, \quad t \in (0, R^2).$$

We now check that this function satisfies the condition (3.1) (with $N = 2$):

$$(||w||^*_{\frac{N}{2},q})^q = \int_0^{R^2} \left\{ t \frac{1}{(\frac{t}{R})|\log(\frac{t}{R})|} \right\}^q \frac{dt}{t} = C \int_0^{R^2} \frac{dt}{t|\log(\frac{t}{R})|^q}$$
$$= C \int_{|\log R|}^{\infty} \frac{dy}{y^q} < \infty \quad \text{for } q > 1.$$

But W_3 does not satisfy (H). Indeed the limit of $|x|^2 W_3(x)$, as x tends to 0 along the curve $x_2 = \sqrt{x_1}$, tends to infinity, while along the x_1 axis it tends to 0. Thus the limit does not exist. The same holds for the function in a cube $\Omega = \{(x_1, \cdots, x_N) : |x_i| < R < 1\}$ in $\mathbb{R}^N$,

$$W_3(x) = \frac{1}{|x_1|^{\frac{2}{N}} |\log(|x_1|)|^{\frac{2}{N}}} \qquad \text{in } \Omega.$$

Thus W_3 does not satisfy the condition (H) but satisfies condition (3.1).

There are also functions failing condition (3.1) but satisfying condition (H). Thus the two conditions are independent. Work is in progress towards a unifying condition, which will be a relaxation of both these conditions.

Example 4: Consider the function

$$W_4(x) = \frac{W_1(x)}{\log\log(2 + |x|^2)}.$$

For the function $w = W_4$, we check that

$$w^*(t) = \frac{1}{(1 + (\frac{t}{\omega_N})^{\frac{2}{N}}) \log\log\left(2 + (\frac{t}{\omega_N})^{\frac{2}{N}}\right)}, \quad t \in (0, \infty).$$

Now we check if $w \in L\left(\frac{N}{2}, q_0\right)$ for some q_0. A direct calculation shows

$$(||w||^*_{\frac{N}{2},q})^q = \int_0^{\infty} \left\{ t^{\frac{2}{N}} \frac{1}{(1 + (\frac{t}{\omega_N})^{\frac{2}{N}}) \log\log\left(2 + (\frac{t}{\omega_N})^{\frac{2}{N}}\right)} \right\}^q \frac{dt}{t}$$
$$\leq C + \int_M^{\infty} \frac{dt}{t\left(\log\log(2 + (\frac{t}{\omega_N})^{\frac{2}{N}})\right)^q}$$
$$\leq C + \int_{\log M}^{\infty} \frac{dy}{(\log y)^q}.$$

Since $\log y < y^{\alpha}$ for large y and any $\alpha > 0$, we see that the last integral is divergent. for any q. But this function does satisfy (H) as $|x|$ tends to infinity.

In our setting the existence of a principal eigenvalue will mainly be a consequence of the following result:

Proposition 3.1. *Let $w \in L(\frac{N}{2}, q_0)$ with $1 < q_0 < \infty$. Then the mapping*

$$\mathcal{D}_0^{1,2}(\Omega) \to \mathbb{R}, \qquad u \mapsto \int_\Omega w(x)\, u^2(x)\, dx \tag{3.2}$$

is weakly continuous.

Proof. Note first that for $2 \le p, q < \infty$ the mapping

$$L(p,q) \to L\left(\frac{p}{2}, \frac{q}{2}\right), \quad v \mapsto v^2, \tag{3.3}$$

is well-defined and continuous. Indeed, the inequality $||v^2||_{(\frac{p}{2},\frac{q}{2})} \le ||v||_{p,q}||v||_{p,q}$ holds for any $v \in L(p,q)$ (see (2.6)) and for any sequence $u_n \to u$ in $L(p,q)$, we have

$$\begin{aligned} ||u_n^2 - u^2||_{\frac{p}{2},\frac{q}{2}} &= ||(u_n - u)(u_n + u)||_{\frac{p}{2},\frac{q}{2}} \\ &\le ||u_n - u||_{p,q}\ ||u_n + u||_{p,q}. \end{aligned} \tag{3.4}$$

Let u_n be a sequence bounded in $\mathcal{D}_0^{1,2}(\Omega)$. Then

(i) u_n is bounded in $L(2^*, 2)$ (by Lorentz-Sobolev embedding);
(ii) u_n^2 is bounded in $L(\frac{2^*}{2}, 1)$ (by continuity of (3.3)) and so u_n^2 is bounded in $L(\frac{2^*}{2}, q_0')$.

Since both spaces $\mathcal{D}_0^{1,2}(\Omega)$ and $L(\frac{2^*}{2}, q_0')$ are reflexive (note that $1 < q_0' < \infty$), we deduce:

$$u_n \rightharpoonup f \text{ weakly in } \mathcal{D}_0^{1,2}(\Omega) \quad \text{and} \quad u_n^2 \rightharpoonup g \text{ weakly in } L(\frac{2^*}{2}, q_0').$$

We claim $g = f^2$. Indeed for each $\varphi \in C_0^\infty$ (fixed) we have

$$\int_\Omega |u_n^2 - f^2||\varphi| \le \underbrace{\int_\Omega |u_n - f||u_n||\varphi|}_{A_n} + \underbrace{\int_\Omega |u_n - f||f||\varphi|}_{B_n}. \tag{3.5}$$

We claim that the right hand-side of (3.5) tends to zero as $n \to \infty$. In order to estimate A_n, denote by Ω_0 the interior of the support of φ and apply Hölder inequality (2.6) to derive

$$A_n := \left\|[u_n - f]u_n\varphi\right\|_{1,1} \le \left\|[u_n - f]\chi_{\Omega_0}\right\|_{2,2} \left\|u_n\right\|_{2^*,2} \left\|\varphi\right\|_{N,\infty}.$$

Since there is a compact embedding of $\mathcal{D}_0^{1,2}(\Omega)$ in $L^2_{loc}(\Omega)$ (because $2 < 2^*$) and $\|u_n\|_{2^*,2}$ is bounded (uniformly in n) we deduce that $A_n \to 0$.

To estimate B_n, we simply observe that $(f,\varphi) \in L(2^*,2) \times L(\frac{N}{2},\infty)$ and therefore $f\varphi \in L([2^*]',2)$. Hence by duality we get $\lim_{n\to\infty} B_n = 0$. So (3.5) tends to zero when the test function $\varphi \in C_0^\infty(\Omega)$. Since on the other hand we have also

$$\int_\Omega |u_n^2 - g||\varphi| \to 0, \quad \forall\varphi \in L([\frac{2^*}{2}]', q_0)),$$

we easily deduce that $g = f^2$.

To summarize, we have $u_n^2 \to u^2$ weakly in $L(\frac{2^*}{2}, q_0')$ and $w \in L(\frac{N}{2}, q_0) = \left[L(\frac{2^*}{2}, q_0')\right]'$. Hence the definition of weak convergence implies

$$\left|\int_\Omega w(u_n^2 - u^2)\right| \to 0.$$

Thus the proposition is proved. □

Based on the above proposition, standard arguments imply that (1.3) is a principal eigenvalue. With the aim of proving that this is the *unique* positive principal eigenvalue and that its associated eigenspace is of *dimension one*, we need the following weak version of the Strong Maximum Principle due to Ancona [4] and Brezis-Ponce [5], that in our setting can be stated as follows:

Theorem 3.1. ***(Strong Maximum Principle).*** *Let $V \in L(\frac{N}{2}, q_0)(\Omega)$ with $V \geq 0$. Assume that the following holds in the sense of distribution :*

$$-\Delta u + V(x)u \geq 0, \quad u \in \mathcal{D}_0^{1,2}(\Omega)\setminus\{0\}, \quad u \geq 0,$$

and consider the "precise representative" $\tilde{u}$ of u. Then, $\{x \in \Omega : \tilde{u}(x) = 0\}$ is a set of H^1-capacity zero and in particular of Lebesgue measure zero.

The result holds in much wider generality and we refer to [5] for a more detailed statement. Note that if in the above theorem we make the stronger assumption $V \in L^p_{loc}(\Omega)$ for some $p > N/2$, the usual Strong Maximum Principle would imply $\inf_B u > 0$ for any $B \subset\subset \Omega$. To avoid any confusion, we make the following definition:

Definition 3.1. We say that $\lambda \in \mathbb{R}$ is a "principal eigenvalue" and $u \in \mathcal{D}_0^{1,2}(\Omega)\setminus\{0\}$ a "principal eigenfunction" for Problem (1.2) if (λ, u) solves (1.2) and $u \geq 0$.

In our setting a principal eigenfunction is not necessarily locally bounded and may vanish. But note that if $\lambda > 0$ is a principal eigenvalue and u an associated principal eigenfunction we have

$$-\Delta u + \lambda w^- u \;=\; \lambda w^+ u \geq 0, \quad u \geq 0, \quad u \not\equiv 0, \tag{3.6}$$

and so Theorem 3.1 implies that the set $u^{-1}(0)$ is of H^1-capacity zero. A similar argument holds if the principal eigenvalue is negative. We can now prove

Proof of Proposition 1.1:

Existence: As remarked in [[19], Remark 2.4(b)], the weak continuity of the mapping (3.2) is enough to ensure the existence of $\Phi \geq 0$ solving the minimizing Problem (1.3). Therefore

$$-\Delta\Phi \;=\; \lambda_1^+(w)w\Phi, \quad \Phi \geq 0, \quad \int_\Omega w\Phi^2 > 0, \tag{3.7}$$

and so $\lambda_1^+(w)$ is a principal eigenvalue.

Uniqueness: To prove that $\lambda_1^+(w)$ is the unique positive principal eigenvalue, let us assume the existence of another pair $(\lambda, \varphi) \in (0, \infty) \times \mathcal{D}_0^{1,2}(\Omega)$ satisfying

$$-\Delta\varphi \;=\; \lambda w\varphi, \quad \varphi \geq 0, \quad \varphi \not\equiv 0. \tag{3.8}$$

From the definition (1.3), we immediately see that

$$\lambda \geq \lambda_1^+(w). \tag{3.9}$$

To show that equality holds we modify slightly some of the arguments in [Prop. 4.1, [9]], by taking into account that our eigenfunctions may be unbounded. Let us define for each $k \geq 0$ the following truncated function:

$$\Phi_k(x) := \begin{cases} k & \text{if } \Phi(x) \geq k, \\ \Phi(x) & \text{if } \Phi(x) \in [0, k). \end{cases}$$

Clearly $\Phi_k \in L^\infty(\Omega)$ and it is well-known that $\Phi_k \in \mathcal{D}_0^{1,2}(\Omega)$. Hence, both Φ_k and $\frac{\Phi_k^2}{\varphi+\epsilon}$ are legitimate trial functions in (3.7), respectively (3.8) and therefore

$$\int_\Omega \nabla\Phi\nabla\Phi_k - \int_\Omega \nabla\varphi\nabla(\frac{\Phi_k^2}{\varphi+\epsilon}) = \int_\Omega \lambda_1^+(w)w\Phi\Phi_k - \int_\Omega \lambda w\varphi\frac{\Phi_k^2}{\varphi+\epsilon},$$

which is equivalent to

$$\int_\Omega \left\{|\nabla\Phi_k|^2 - \nabla\varphi\nabla(\frac{\Phi_k^2}{\varphi+\epsilon})\right\} = \int_\Omega \left\{\lambda_1^+(w)w\Phi\Phi_k - \lambda w\varphi\frac{\Phi_k^2}{\varphi+\epsilon}\right\}. \tag{3.10}$$

But, a direct calculation shows that the following "Picone's identity" holds:

$$|\nabla\Phi_k|^2 - \nabla\varphi\nabla(\frac{\Phi_k^2}{\varphi+\epsilon}) = \left|\nabla\Phi_k - (\frac{\Phi_k}{\varphi+\epsilon})\nabla\varphi\right|^2. \tag{3.11}$$

By plugging (3.11) in (3.10), we get

$$0 \le \int_\Omega \left|\nabla\Phi_k - (\frac{\Phi_k}{\varphi+\epsilon})\nabla\varphi\right|^2 = \int_\Omega \left\{\lambda_1^+(w)w\Phi\Phi_k - \lambda w\varphi\frac{\Phi_k^2}{\varphi+\epsilon}\right\}. \tag{3.12}$$

Since by Theorem 3.1 the set $\{\varphi = 0\}$ is of measure zero, (3.12) is equivalent to

$$\begin{aligned} 0 &\le \int_{\{\varphi>0\}} \left|\nabla\Phi_k - (\frac{\Phi_k}{\varphi+\epsilon})\nabla\varphi\right|^2 \\ &= \int_{\{\varphi>0\}} \left\{\lambda_1^+(w)w\Phi\Phi_k - \lambda w\varphi\frac{\Phi_k^2}{\varphi+\epsilon}\right\}. \end{aligned} \tag{3.13}$$

Now, letting $\epsilon \to 0$ and $k \to \infty$ in (3.13) and applying Lebesgue dominated Theorem to the right handside, we get

$$0 \le (\lambda_1^+(w) - \lambda)\int_\Omega w\Phi^2. \tag{3.14}$$

Using (3.9) and the fact that $\int_\Omega w\Phi^2 > 0$, (3.14) implies $\lambda = \lambda_1^+(w)$.

Simplicity: let $V(\lambda_1^+(w))$ be the eigenspace associated to the principal eigenvalue (1.3). Before proving that V_1 has dimension one, we observe that any $\Phi \in V(\lambda_1^+(w))$ satisfies precisely one of the alternative:

$$\text{(i) } \Phi > 0 \text{ a.e. in } \Omega, \qquad \text{(ii) } \Phi < 0 \text{ a.e. in } \Omega, \qquad \text{(iii) or } \Phi \equiv 0. \tag{3.15}$$

Indeed, since

$$\int_\Omega |\nabla\Phi|^2 = \int_\Omega |\nabla|\Phi||^2 = \lambda_1^+(w) \quad \text{and} \quad \int_\Omega w|\Phi|^2 = \int_\Omega w\Phi^2,$$

we deduce that both Φ and $|\Phi|$ solve the minimization problem (1.3). As a consequence

$$\Phi^+ = \Phi + |\Phi| \in V(\lambda_1^+(w)) \quad \text{and} \quad \Phi^- = \Phi - |\Phi| \in V(\lambda_1^+(w)). \tag{3.16}$$

Now by considering the Euler-Lagrange equation satisfied by Φ^+, Φ^- in the form (3.6) and applying Theorem 3.1 we conclude that $\Phi^+ > 0$ a.e., or $\Phi^- > 0$ a.e. or else $\Phi^+ \equiv \Phi^- \equiv 0$. This proves (3.15).

As in [15], we can now modify the "continuation argument" given in Lemma 7 of [16]. More precisely, since $V(\lambda_1^+(w)) \neq \{0\}$, consider $\Phi_1, \Phi_2 \in$

$V(\lambda_1^+(w)) \setminus \{0\}$. By (3.15) we can assume without loss of generality that $\Phi_1, \Phi_2 > 0$ a.e.. Let us consider the set

$$T := \{t \in \mathbb{R} : \; \Phi_1 + t\Phi_2 > 0 \text{ a.e.}\}.$$

We shall show that $\Phi_1 + t_0\Phi_2 \equiv 0$ when $t_0 = \inf T$.

Claim 1: $T \neq \emptyset$ and $\inf T > -\infty$.
Since $0 \in T$, we see that $T \neq \emptyset$. To prove that this set is bounded from below, let us consider for each $\delta, M > 0$, the set

$$A_{\delta,M} := \{\Phi_1 < M, \; \Phi_2 > \delta\}.$$

Since $\Phi_2 > 0$ a.e., there exists $\widetilde{\delta} > 0$ such that $|\{\Phi_2 > \widetilde{\delta}\}| > 0$. Moreover, since $\cup_{M>0} A_{\widetilde{\delta},M} = \{\Phi_2 > \widetilde{\delta}\}$, we deduce the existence of $\widetilde{M}$ such that $|A_{\widetilde{\delta},\widetilde{M}}| > 0$. Now, for any $x \in A_{\widetilde{\delta},\widetilde{M}}$ and $t < -\frac{\widetilde{M}}{\widetilde{\delta}}$, we get

$$(\Phi_1 + t\Phi_2)(x) < \widetilde{M} + t\widetilde{\delta} < 0.$$

Hence, when $t < -\frac{\widetilde{M}}{\widetilde{\delta}}$, the function $\Phi_1 + t\Phi_2$ is negative on a set of positive measure, which proves $\inf T > -\infty$.

Claim 2: Setting $t_0 := \inf T$, we claim $\Phi_1 + t_0\Phi_2 \equiv 0$.
Since $\Phi_1 + t_0\Phi_2 \in V(\lambda_1^+(w))$ the alternative (3.15) applies and so one of the following case holds:

(a) $\Phi_1 + t_0\Phi_2 > 0$ a.e. in Ω, (b) $\Phi_1 + t_0\Phi_2 < 0$ a.e. in Ω, (c) $\Phi_1 + t_0\Phi_2 \equiv 0$.

Assume (a) holds. By setting

$$E_{\delta,M} := \{\Phi_1 + t_0\Phi_2 > \delta, \; \Phi_2 < M\},$$

and arguing as in claim (1), we prove $|E_{\widetilde{\delta},\widetilde{M}}| > 0$ for some $\widetilde{\delta}, \widetilde{M} > 0$. Thus, for any $x \in E_{\widetilde{\delta},\widetilde{M}}$ and $\varepsilon \in (0, \frac{\widetilde{\delta}}{\widetilde{M}})$, we get

$$\big(\Phi_1 + (t_0 - \varepsilon)\Phi_2\big)(x) > \widetilde{\delta} - \varepsilon\widetilde{M} > 0.$$

Alternative (3.15) implies then $\Phi_1 + (t_0 - \varepsilon)\Phi_2 > 0$ a.e., in contradiction with the definition of t_0. A similar contradiction is reached if we assume (b). Therefore, the alternative (c) holds, which concludes the proof of the proposition. $\square$

Remark 3.1.

(a) In [15], Proposition 1.1 has been stated for weight belonging to $L^{\frac{N}{2}}(\Omega)$. Some of the arguments has been repeated for the sake of completeness.
(b) Existence of a principal eigenvalue when the weight belongs to some Lorentz space has already been obtained by [22]. Our arguments are different [22] and here we have completed the result by emphasizing that uniqueness and simplicity of the positive (or negative) principal eigenvalue holds.
(c) Note that in [19] the simplicity of a principal eigenvalue is proved under stronger conditions.

4. Global bifurcation

Existence of global branches for Problem (1.1) will mainly rely on the following result:

Theorem 4.1. ***(Rabinowitz,*** [17]*) Given a Banach space $(B, \|\cdot\|)$, consider a mapping*

$$G : \mathbb{R} \times B \to B, \qquad (\lambda, u) \mapsto \lambda L(u) + H(\lambda, u),$$

where $L : B \to B$ is a compact linear operator and $H(\lambda, \cdot) : B \to B$ is a continuous compact mapping satisfying $\lim\limits_{\|u\| \to 0} \dfrac{\|H(\lambda, u)\|}{\|u\|} = 0$. Denote by

$$r(L) := \big\{\mu \in \mathbb{R} : \mu^{-1} \text{ is an eigenvalue of } L \text{ with odd multiplicity}\big\},$$

$$\mathcal{S} := \big\{(\lambda, u) \in \mathbb{R} \times B : (\lambda, u) \text{ is solution of } u = G(\lambda, u), u \not\equiv 0\big\}.$$

Then, given $\mu \in r(L)$, $\overline{\mathcal{S}}$ has a connected branch $\mathcal{C}_\mu$ bifurcating from $(\mu, 0)$ and

(i) either $\mathcal{C}_\mu$ is unbounded in $\mathbb{R} \times B$, (ii) or, $\mathcal{C}_\mu \ni (\hat{\mu}, 0)$ with $\mu \neq \hat{\mu} \in r(L)$.

Let us recast Problem (1.1) in the framework of Theorem 4.1. To this aim we first note that as a consequence of Riesz Theorem, the mapping

$$\mathcal{D}_0^{1,2}(\Omega) \to L([2^*]', 2') = L(\frac{2N}{N+2}, 2), \qquad u \mapsto -\triangle u, \tag{4.1}$$

is 1-1 with continuous inverse. Based on our hypotheses we will see that the following mappings are well-defined:

$$L : \mathcal{D}_0^{1,2}(\Omega) \to \mathcal{D}_0^{1,2}(\Omega), \qquad u \mapsto (-\triangle)^{-1}\big(a(x)u\big), \tag{4.2}$$

$$H:\mathbb{R}\times\mathcal{D}_0^{1,2}(\Omega)\to\mathcal{D}_0^{1,2}(\Omega),\qquad(\lambda,u)\mapsto(-\Delta)^{-1}\big(\lambda b(x)r(u)\big),\tag{4.3}$$

and so finding a solution to Problem (1.1) is reduced to solving

$$u=\lambda L(u)+H(\lambda,u),\quad u\in\mathcal{D}_0^{1,2}(\Omega).\tag{4.4}$$

In order to prove the continuity and compactness of the mappings L, H we need the following lemma borrowed from [Lemma 4.2,[22]]:

Lemma 4.1. *Let $V\in L(p,q)$ with $1\le p,q<\infty$. Then for each $\varepsilon>0$ there exists a measurable set $\Omega^\varepsilon\subseteq\Omega$ such that*

$$\Omega^\varepsilon\text{ is bounded},\quad V\chi_{\Omega^\varepsilon}\in L^\infty(\Omega),\quad \|V\chi_{\Omega\setminus\Omega^\varepsilon}\|_{p,q}<\varepsilon.$$

Proposition 4.1. *Assume (H1)-(H2) hold. Then the mappings L and H defined by* (4.2) *and* (4.3) *are continuous, compact and furthermore*

$$\lim_{\|u\|\to0}\frac{\|H(\lambda,u)\|}{\|u\|}=0.\tag{4.5}$$

Proof. Let us introduce the two following mappings

$$\begin{array}{ll}\widetilde{L}:L(2^*,2)\longrightarrow L([2^*]',2) & \widetilde{H}:L(2^*,2)\longrightarrow L([2^*]',2)\\ \qquad u\longmapsto a(x)u & \qquad u\longmapsto b(x)r(u)\end{array}\tag{4.6}$$

In order to justify the continuity and compactness of the mappings H, L we first prove the following general statement

Claim 1: Let $\bar u\in L(2^*,2)$ and $u_n\in L(2^*,2)$ be a sequence satisfying
(i) for some $\alpha>\frac{2^*}{2^*-1}\gamma$ we have $u_n\to\bar u$ in $L^\alpha(\Omega\cap B)$ for each ball $B\subset\mathbb{R}^N$;
(ii) u_n is bounded in $L(2^*,2)$;

then we claim

$$\big\|a[u_n-\bar u]\big\|_{[2^*]',2}\to0\quad\text{ and }\quad\big\|b[r(u_n)-r(\bar u)]\big\|_{[2^*]',2}\to0.\tag{4.7}$$

We only give the arguments for $\big\|b[r(u_n)-r(\bar u)]\big\|_{[2^*]',2}$. Notice first that (H1), and the fact that the sequence u_n is bounded in $L(2^*,2)$ implies the existence of $C>0$ such that

$$\begin{cases}|r(s)|\le C\big(|s|+|s|^\gamma\big),\\[2ex] \|u_n\|_{2^*,2}+\|\bar u\|_{2^*,2}+\|u_n\|_{2^*,2}^\gamma+\|\bar u\|_{2^*,2}^\gamma\le C.\end{cases}\tag{4.8}$$

Using Lemma 4.1, we can choose a measurable set $\Omega^\epsilon\subseteq\Omega$ such that

$$\begin{aligned}&\Omega^\varepsilon\text{ is bounded},\\ &b\chi_{\Omega^\varepsilon}\in L^\infty(\Omega),\quad\|b\chi_{\Omega\setminus\Omega^\varepsilon}\|_{\frac{N}{2},q_0}+\|b\chi_{\Omega\setminus\Omega^\varepsilon}\|_{p_0,q_0}<\frac{\varepsilon}{4C^2}.\end{aligned}\tag{4.9}$$

We have

$$\begin{aligned}&\left\|b\big[r(u_n)-r(\bar{u})\big]\right\|_{[2^*]',2}\\&\quad\le\left\|b\big[r(u_n)-r(\bar{u})\big]\chi_{\Omega^\varepsilon}\right\|_{[2^*]',2}+\left\|b\big[r(u_n)-r(\bar{u})\big]\chi_{\Omega\setminus\Omega^\varepsilon}\right\|_{[2^*]',2}.\end{aligned}\tag{4.10}$$

Let us estimate the first term appearing in the right hand-side of (4.10). By setting

$$\frac{1}{p}:=\frac{1}{[2*]'}-\frac{\gamma}{\alpha}=1-\frac{1}{2^*}-\frac{\gamma}{\alpha},\quad\text{(notice } 0<\frac{1}{p}<1),$$

we get

$$\begin{aligned}&\left\|b\big[r(u_n)-r(\bar{u})\big]\chi_{\Omega^\varepsilon}\right\|_{[2^*]',2}\\&\quad\le\left\|b\chi_{\Omega^\varepsilon}\right\|_{p,2}\left\|\big[r(u_n)-r(\bar{u})\big]\chi_{\Omega^\varepsilon}\right\|_{\frac{\alpha}{\gamma},\infty}\\&\quad\le\left\|b\chi_{\Omega^\varepsilon}\right\|_{L^\infty(\Omega)}\left\|\chi_{\Omega^\varepsilon}\right\|_{p,2}\left\|\big[r(u_n)-r(\bar{u})\big]\chi_{\Omega^\varepsilon}\right\|_{\frac{\alpha}{\gamma},\frac{\alpha}{\gamma}}.\end{aligned}\tag{4.11}$$

Since $u_n\to\bar{u}$ in $L^\alpha(\Omega\cap B)$ for any ball $B\subset\mathbb{R}^N$, by using Vitali' Theorem we check that $\|r(u_n)-r(\bar{u})\|_{L^{\frac{\alpha}{\gamma}}(\Omega\cap B)}\to 0$. Therefore the expression in (4.11) can be made as small as we wish. Namely there exists $n_0\in\mathbb{N}$ such that

$$\left\|b\big[r(u_n)-r(\bar{u})\big]\chi_{\Omega^\varepsilon}\right\|_{[2^*]',2}\le\frac{\varepsilon}{2},\qquad\forall n\ge n_0.\tag{4.12}$$

To estimate the second term in the right hand-side of (4.10) we use (4.8), (4.9). We get using $\frac{1}{p_0}=\frac{2^*-1}{2^*}-\frac{\gamma}{2^*}$ and $\gamma\ge 1$:

$$\begin{aligned}&\left\|b\big[r(u_n)-r(\bar{u})\big]\chi_{\Omega\setminus\Omega^\varepsilon}\right\|_{[2^*]',2}\\&\quad\le C\left\|b\chi_{\Omega\setminus\Omega^\varepsilon}\right\|_{\frac{N}{2},\infty}\big(\|u_n\|_{2^*,2}+\|\bar{u}\|_{2^*,2}\big)\\&\qquad+C\left\|b\chi_{\Omega\setminus\Omega^\varepsilon}\right\|_{p_0,\infty}\big(\||u_n|^\gamma\|_{\frac{2^*}{\gamma},2}+\||\bar{u}|^\gamma\|_{\frac{2^*}{\gamma},2}\big)\\&\quad\le C.M\left(\|u_n\|_{2^*,2}+\|\bar{u}\|_{2^*,2}+\|u_n\|^\gamma_{2^*,2\gamma}+\|\bar{u}\|^\gamma_{2^*,2\gamma}\right)\\&\quad\le C.M\left(\|u_n\|_{2^*,2}+\|\bar{u}\|_{2^*,2}+\|u_n\|^\gamma_{2^*,2}+\|\bar{u}\|^\gamma_{2^*,2}\right)\\&\quad\le\frac{\varepsilon}{2}.\end{aligned}\tag{4.13}$$

(In the above formula we have set $M=(\|b\chi_{\Omega\setminus\Omega^\varepsilon}\|_{\frac{N}{2},q_0}+\|b\chi_{\Omega\setminus\Omega^\varepsilon}\|_{p_0,q_0}))$. Putting together (4.10), (4.12), (4.13) we deduce

$$\left\|b\big[r(u_n)-r(\bar{u})\big]\right\|_{[2^*]',2}\le\varepsilon,\quad\forall n\ge n_0,$$

Claim 2: *Continuity and compactness.*
Due to the Sobolev-Lorentz embedding and the fact that Δ^{-1} in (4.1) is an isomorphism, it is enough to prove the continuity of the mappings (4.6). This follows immediately from the embedding $L(2^*,2) \hookrightarrow L(2^*,2^*)$ and by applying claim 1 with $\alpha = 2^*$.

Concerning compactness. Let u_n be a bounded sequence in $\mathcal{D}_0^{1,2}(\Omega)$. Then there is a subsequence, still denoted u_n, converging weakly in $\mathcal{D}_0^{1,2}(\Omega)$ to $\bar{u}$. Then on each bounded open set $\omega \subset \Omega$ this sequence converges strongly in $L^p(\omega)$ whenever $p \in [1, 2^*)$. Hence choosing $\alpha \in (\frac{2^*}{2^*-1}\gamma, 2^*)$ the sequence u_n satisfies the conditions stated in claim 1. Therefore (4.7) hold, and the continuity of (4.1) allows to conclude.

Claim 3: *Differentiability.*
To prove (4.5), it is enough to prove that $\widetilde{H}$ is differentiable at $u \equiv 0$. Let us fix $\varepsilon > 0$. Thanks to (H1), there exist $s_0, C_0 > 0$ depending only on ε such that

$$\begin{aligned} \left|\frac{r(s)}{s}\right| &\leq \frac{\varepsilon}{3\,\|b\|_{\frac{N}{2},q_0}} \quad \forall 0 < |s| < s_0, \\ \text{and} \quad \frac{|r(s)|}{|s|^\gamma} &\leq C_0 \quad \forall |s| \geq s_0. \end{aligned} \tag{4.14}$$

For each $u \in L(2^*,2)$, let us introduce the sets

$$E := \{x \in \Omega \,:\, |u(x)| < s_0\} \quad \text{and} \quad F := \{x \in \Omega \,:\, |u(x)| \geq s_0\}.$$

Using triangle inequality, Hölder's inequality and (4.14), we obtain

$$\begin{aligned} \|br(u)\|_{[2^*]',2} &\leq \|br(u)\chi_E\|_{[2^*]',2} + \|br(u)\chi_F\|_{[2^*]',2} \\ &\leq \|b\|_{\frac{N}{2},\infty} \|r(u)\chi_F\|_{2^*,2} + \|br(u)\chi_F\|_{[2^*]',2} \\ &\leq \frac{\varepsilon}{3}\|u\|_{2^*,2} + C_0\|b|u|^\gamma\chi_F\|_{[2^*]',2} \end{aligned} \tag{4.15}$$

If $\gamma > 1$ we simply use Hölder's inequality to estimate the second term in (4.15), and get

$$\begin{aligned} \frac{\|br(u)\|_{[2^*]',2}}{\|u\|_{2^*,2}} &\leq \frac{\varepsilon}{3} + C_0\|b\|_{p_0,\infty}\frac{\||u|^\gamma\|_{\frac{2^*}{\gamma},2}}{\|u\|_{2^*,2}} = \frac{\varepsilon}{3} + C_0\|b\|_{p_0,\infty}\frac{\|u\|^\gamma_{2^*,2\gamma}}{\|u\|_{2^*,2}} \\ &\leq \frac{\varepsilon}{3} + C_0\|b\|_{p_0,\infty}\|u\|^{\gamma-1}_{2^*,2}. \end{aligned}$$

Since $\gamma > 1$, the conclusion follows immediately.

When $\gamma = 1$, this latter estimate is too rough. In this case we first use Lemma 4.1 to get a measurable set F^ε such that

$$F^\varepsilon \subseteq F,\ \ F^\varepsilon \text{ is bounded},\ \ b\chi_{F^\varepsilon} \in L^\infty(\Omega),\ \ \|b\chi_{F\setminus F^\varepsilon}\|_{\frac{N}{2},q_0} < \frac{\varepsilon}{3C_0}. \tag{4.16}$$

And secondly by fixing $1 < \widetilde{\gamma} < 2^* - 1$ we note that

$$|s| \le |s| \left(\frac{|s|}{s_0}\right)^{\widetilde{\gamma}-1}, \qquad \forall |s| \ge s_0. \tag{4.17}$$

Then by using (4.16), (4.17) and setting $p := \frac{2^*}{2^*-2-\widetilde{\gamma}}$, the second term in (4.15) (with $\gamma = 1$) is estimated as follows

$$\begin{aligned}
\|bu\chi_F\|_{[2^*]',2} &\le \|b\chi_{F\setminus F^\varepsilon} u\|_{[2^*]',2} + \|bu\chi_{F^\varepsilon}\|_{[2^*]',2} \\
&\le \|b\chi_{F\setminus F^\varepsilon}\|_{\frac{N}{2},\infty} \|u\|_{2^*,2} + s_0^{1-\widetilde{\gamma}} \|b\chi_{F^\varepsilon}\|_{p,\infty} \||u|^{\widetilde{\gamma}}\|_{\frac{2^*}{\widetilde{\gamma}},2} \\
&\le \|b\chi_{F\setminus F^\varepsilon}\|_{\frac{N}{2},q_0} \|u\|_{2^*,2} + s_0^{1-\widetilde{\gamma}} \|b\chi_{F^\varepsilon}\|_{p,\infty} \|u\|^{\widetilde{\gamma}}_{2^*,2\widetilde{\gamma}} \\
&\le \frac{\varepsilon}{3C_0} \|u\|_{2^*,2} + s_0^{1-\widetilde{\gamma}} \|b\chi_{F^\varepsilon}\|_{L^\infty(\Omega)} \|\chi_{F^\varepsilon}\|_{p,\infty} \|u\|^{\widetilde{\gamma}}_{2^*,2}
\end{aligned} \tag{4.18}$$

Inequality (4.18) shows that we can find $\delta > 0$ such that

$$\|bu\chi_F\|_{[2^*]',2} \le 2\frac{\varepsilon}{3C_0} \|u\|_{2^*,2} \quad \forall \|u\|_{2^*,2} < \delta. \tag{4.19}$$

By putting together (4.15) and (4.19), we deduce that $\widetilde{H}$ is differentiable at $u \equiv 0$, which concludes the proof of (4.5). □

We now have all the elements for proving the existence of at least one global branch of solutions for Problem (1.1):

Proof of proposition 1.2: Proposition 1.1 shows that $\lambda_1^+(a)$ is an eigenvalue of multiplicity one. Hence Proposition 4.1 together with Theorem 4.1 allow to conclude. □

Acknowledgements. The first author was supported by an Alexander von Humboldt fellowship.

References

[1] W. Allegretto, *Principal eigenvalues for indefinite-weight elliptic problems on* $\mathbb{R}^N$, Proc. Am. Math. Monthly **116** (1992), 701–706

[2] A. Alvino, P-L. Lions, G. Trombetti, *On optimization problems with prescribed rearrangements*, Nonlinear Anal. **13** (1989), 185–220.

[3] A. AMBROSETTI, P. HESS, *Positive solutions of asymptotically linear elliptic eigenvalue problems*, J. Math. Anal. Appl. **73** (1980), 411–422.

[4] A. ANCONA, *Une propriété d'invariance des ensembles absorbants par perturbation d'un opérateur elliptique*, Comm. PDE **4** (1979), 321–337.

[5] H. BREZIS, A. PONCE, *Remarks on the strong maximum principle*, Differential Integral Equations **16** (2003), 1–12.

[6] K.J. BROWN, C. COSNER, J. FLECKINGER, *Principal eigenvalues for problems with indefinite weight function on* $\mathbb{R}^N$, Proc. Amer. Math. Soc. **109** (1990), 147–155.

[7] K.J. BROWN, A. TERTIKAS, *On the bifurcation of radially symmetric steady-state solutions arising in population genetics*, Siam J. Math. Anal. **22** (1991), 400–413.

[8] R. COURANT, D. HILBERT Methods of mathematical physics. Vol. I. Interscience Publishers, Inc., New York, N.Y., 1953.

[9] M. CUESTA, *Eigenvalue problems for the p-Laplacian with indefinite weights*, Electron. J. Differential Equations **33** (2001), 1–9.

[10] J. DENY, J.L. LIONS, *Les espaces du type de Beppo Levi*, Ann. Inst. Fourier, Grenoble **5** (1954), 305–370.

[11] A. L. EDELSON, M. FURI, *Global solution branches for semilinear equations in* $\mathbb{R}^n$, Nonlinear Anal. **28** (1997), 1521–1532.

[12] J. GIACOMONI, M. LUCIA, M. RAMASWAMY, *Some elliptic semilinear indefinite problems on* $\mathbb{R}^N$, Proc. Roy. Soc. Edinburgh **134** (2004), 333–361.

[13] R. HUNT, *On* $L(p, q)$ *spaces*, Enseignement Math. (2) **12** (1966), 249–276.

[14] G. G. LORENTZ, *Some new functional spaces*, Annals of Math., **51** (1950), 37–55.

[15] M. LUCIA, *On the uniqueness and simplicity of the principal eigenvalue*, Rend. Accad. Naz. Sci. XL Mem. Mat. Appl. **16** (2005), 132–142.

[16] A. MANES, A.M. MICHELETTI, *Un'estensione della teoria variazionale classica degli autovalori per operatori ellitici del secondo ordine*, Bollettino U.M.I. **7** (1973), 285–301.

[17] P.H. RABINOWITZ, *Some global results for nonlinear eigenvalues problems*, J. Func. Anal. **7** (1971), 487–517.

[18] E.M. STEIN, G. WEISS, Introduction to Fourier analysis on Euclidean spaces. Princeton Mathematical Series, No. 32. Princeton University Press, Princeton, N.J., 1971.

[19] A. SZULKIN, M. WILLEM, *Eigenvalue problems with indefinite weight*, Studia Math. **135** (1999), 191–201.

[20] L. TARTAR, *Imbedding theorems of Sobolev spaces into Lorentz spaces*, Boll. Unione Mat. Ital. Sez B Artic. Ric. Mat. (1998), 479-500.

[21] A. TERTIKAS, *Critical phenomena in linear elliptic problems*, J. Funct. Anal. **154** (1998), 42–66.

[22] N. VISCIGLIA, *A note about the generalized Hardy-Sobolev inequality with potential in* $L^{p,d}(\mathbb{R}^n)$, Calc. Var. Partial Differential Equations **24** (2005), 167–184.

GLOBAL EXISTENCE AND ASYMPTOTIC BEHAVIOR OF GOWDY SYMMETRIC SPACETIMES WITH NONLINEAR SCALAR FIELD

MAKOTO NARITA

Department of Mathematics, National Taiwan University, 1, Sec.4 Roosevelt Rd. Taipei, 106, Taiwan

1. Introduction

Our interest is to solve the Einstein-nonlinear scalar equations globally and to analyze asymptotic behavior of solutions to the equations. Since the Einstein-nonlinear scalar equations are highly nonlinear, it is too difficult to solve them without suitable conditions. Therefore, we will treat Gowdy symmetric spacetimes, which describe spatially compact, expanding universe: under this assumption, the Einstein-nonlinear scalar equations become $1+1$-dimensional nonlinear wave equations with constraint equations.

We need also assumptions for nonlinear term of the scalar field (i.e. potential): (1) slow-rolling condition (Hypothesis 2.2), and (2) ending accelerating expansion of the universe (Hypothesis 4.1). These two assumptions come from inflation paradigm [7], which is the most promising scenario of our universe in modern cosmology. In addition, we will assume the timelike convergence condition, which means that kinetic terms dominates than the potential term of the scalar field, to extend the existence time of solutions into the past direction.

Under these assumptions, we obtain a global existence theorem, existence of singular solutions into the past direction and energy decay into the future direction.

1.1. *Action*

Our action is of the form

$$S = \int d^4x\sqrt{-g}\left[-\,{}^4R + 2(\nabla\phi)^2 + V(\phi)\right], \tag{1.1}$$

where $g_{\mu\nu}$ is a Lorentzian metric on a four dimensional manifold, R is the Ricci scalar, ϕ is a scalar field and V is a potential of ϕ.

1.2. *Field equations for Gowdy symmetric spacetimes*

The Gowdy symmetric spacetimes admit a T^2 isometry group with spacelike orbits and the twists associated to the group vanish [5]. The topology of spatial section can be accepted S^3, $S^2 \times S^1$, T^3 or the lens space [3]. In this paper, we assume T^3 spacelike topology.

Now, we will choose a coordinate, which is the *areal time* one. This means that time t is proportional to the geometric area of the orbits of the isometry group. Explicitly,

$$g_G = -e^{2(\eta-U)}\alpha dt^2 + e^{2(\eta-U)}d\theta^2 + e^{2U}(dx + Ady)^2 + e^{-2U}t^2dy^2, \quad (1.2)$$

where $\partial/\partial x$ and $\partial/\partial y$ are Killing vector fields generating the T^2 group action, and η, α, U and A are functions of $t \in (0,\infty)$ and $\theta \in S^1$. It is also assumed that functions describing behavior of matter fields are ones of t and θ.

Let us show the field equations obtained by varying the action (1.1) in the areal coordinate (1.2).

Constraint equations

$$\frac{\dot\eta}{t} = \dot U^2 + \alpha U'^2 + \frac{e^{4U}}{4t^2}(\dot A^2 + \alpha A'^2) + \dot\phi^2 + \alpha\phi'^2 + \frac{1}{2}\alpha e^{2(\eta-U)}V(\phi), \quad (1.3)$$

$$\frac{\eta'}{t} = 2\dot U U' + \frac{e^{4U}}{2t^2}\dot A A' - \frac{\alpha'}{2t\alpha} + 2\dot\phi\phi'. \quad (1.4)$$

Now we define

$$\gamma := \eta + \frac{1}{2}\ln\alpha. \quad (1.5)$$

By using γ, one can rewrite the above constraint equations as follows:

$$\frac{\dot\gamma}{t} = \dot U^2 + \alpha U'^2 + \frac{e^{4U}}{4t^2}(\dot A^2 + \alpha A'^2) + \dot\phi^2 + \alpha\phi'^2 - \frac{1}{2}e^{2(\gamma-U)}V(\phi), \quad (1.6)$$

$$\frac{\gamma'}{t} = 2\dot U U' + \frac{e^{4U}}{2t^2}\dot A A' + 2\dot\phi\phi'. \quad (1.7)$$

The following constraint is a matching condition by causing the potential term.

$$\begin{aligned}\dot{\alpha} &= -2t\alpha^2 e^{2(\eta-U)}V(\phi) \\ &= -2t\alpha e^{2(\gamma-U)}V(\phi).\end{aligned} \tag{1.8}$$

Evolution equations

$$\begin{aligned}\ddot{\eta} - \alpha\eta'' = &\ \frac{\eta'\alpha'}{2} + \frac{\dot{\eta}\dot{\alpha}}{2\alpha} - \frac{\alpha'^2}{4\alpha} + \frac{\alpha''}{2} - \dot{U}^2 + \alpha U'^2 + \frac{e^{4U}}{4t^2}(\dot{A}^2 - \alpha A'^2) \\ &- \dot{\phi}^2 + \alpha\phi'^2 + \frac{1}{2}e^{2(\gamma-U)}V(\phi),\end{aligned} \tag{1.9}$$

$$\begin{aligned}\ddot{U} - \alpha U'' = -\frac{\dot{U}}{t} + \frac{\dot{\alpha}\dot{U}}{2\alpha} + \frac{\alpha' U'}{2} + \frac{e^{4U}}{2t^2}(\dot{A}^2 - \alpha A'^2) \\ + \frac{1}{2}e^{2(\gamma-U)}V(\phi),\end{aligned} \tag{1.10}$$

$$\ddot{A} - \alpha A'' = \frac{\dot{A}}{t} + \frac{\dot{\alpha}\dot{A}}{2\alpha} + \frac{\alpha' A'}{2} - 4(\dot{A}\dot{U} - \alpha A'U'), \tag{1.11}$$

$$\ddot{\phi} - \alpha\phi'' = -\frac{\dot{\phi}}{t} + \frac{\dot{\alpha}\dot{\phi}}{2\alpha} + \frac{\alpha'\phi'}{2} - \frac{1}{4}e^{2(\gamma-U)}\frac{\partial V(\phi)}{\partial\phi}. \tag{1.12}$$

Hereafter, dot and prime denote derivative with respect to t and θ, respectively. We will call this system of partial differential equations (PDEs) *Gowdy-nonlinear scalar system.* Note that these equations are not independent because the wave equation (1.9) for η can be derived from other equations. Indeed, there are only two dynamical degree of freedom (i.e. U and A) in the Gowdy symmetric spacetimes.

Wave map The system of the evolution equations is equivalent with the following system of nonlinear wave and wave map equations:

$$S_{\rm NWM} = \int_{S^1} dt d\theta \sqrt{-g}\left(g^{\alpha\beta}h_{AB}\partial_\alpha u^A \partial_\beta u^B + \frac{1}{2}e^{2(\gamma-u^1)}V(u^3)\right), \tag{1.13}$$

where

$$g = -dt^2 + \frac{1}{\alpha}d\theta^2 + t^2 d\psi^2, \qquad 0 \le \theta, \psi \le 2\pi,$$

and

$$h = dU^2 + \frac{e^{4U}}{4t^2}dA^2 + d\phi^2.$$

Every functions depend on time t and θ. The energy-momentum tensor $T_{\alpha\beta}$ for this system is given of the form:

$$T_{\alpha\beta} = h_{AB}\left(\partial_\alpha u^A \partial_\beta u^B - \frac{1}{2} g_{\alpha\beta} \partial_\lambda u^A \partial^\lambda u^B\right) - \frac{1}{4} g_{\alpha\beta} e^{2(\gamma - u^1)} V(u^3). \tag{1.14}$$

The energy is defined as follows:

$$\begin{aligned} E(t) &= \int_{S^1} T_{tt}\mu \\ &= \frac{1}{2}\int_{S^1}\left[h_{AB}\left(\partial_t u^A \partial_t u^B + \partial_\theta u^A \partial_\theta u^B\right) + \frac{1}{2} e^{2(\gamma-u^1)} V(u^3)\right]\mu, \end{aligned} \tag{1.15}$$

where

$$\mu = \frac{1}{\sqrt{\alpha}} d\theta.$$

2. Global existence

Now, consider global existence problem. We will show the following theorem:

Theorem 2.1. *(See [8]) Let (M, g, ϕ) be the maximal Cauchy development of C^∞ initial data for the Gowdy-nonlinear scalar system. Suppose that Hypotheses 2.1 and 2.2 hold. Then, M can be covered by compact Cauchy surfaces of constant areal time t with each value in the range $[t_0, \infty)$.* □

Hypotheses are as follows:

Hypothesis 2.1. $V(\phi)$ is regular with respect to ϕ, i.e. V is bounded if ϕ is bounded.

Hypothesis 2.2. There are positive constants C_{SR1} and C_{SR2}, such that

$$\sup_{S^1}\left|\frac{1}{V(\phi)}\frac{\partial V(\phi)}{\partial \phi}\right| < C_{SR1}, \quad \sup_{S^1}\left|\frac{1}{V(\phi)}\frac{\partial^2 V(\phi)}{\partial \phi^2}\right| < C_{SR2}. \tag{2.1}$$

Corollary 2.1. *If the following condition holds on $t \in (0, t_0]$:*

$$\dot{\phi}^2 \geq \frac{1}{2}\alpha e^{2(\eta - U)} V, \tag{2.2}$$

then the existence time in Theorem 2.1 can be extended to $(0, \infty)$. □

By a theorem in [4], one can show the inextendibility of our spacetimes.

Theorem 2.2. *The maximal globally hyperbolic development of data for the Gowdy-nonlinear scalar system cannot be extended into the future direction in C^2-category, that is, no future extension of the original spacetime is possible.* □

3. Initial singularity

Assume $V = e^{2\lambda\phi}$, where λ is a coupling constant. Now, we can construct singular solutions near singularity $t = 0$. To do this, expansion of the solutions near $t = 0$ will be assumed as follows:

$$\eta = \left(k(\theta)^2 + \frac{\kappa(\theta)^2}{4}\right)\ln t + \eta_0(\theta) + t^\epsilon \nu(t,\theta), \tag{3.1}$$

$$\alpha = \alpha_0(\theta) + t^\epsilon \beta(t,\theta), \tag{3.2}$$

$$U = k(\theta)\ln t + U_0(\theta) + t^\epsilon W(t,\theta), \tag{3.3}$$

$$A = h(\theta) + t^{2-4k}\left(A_0(\theta) + B(t,\theta)\right), \tag{3.4}$$

$$\phi = \kappa(\theta)\ln t + \phi_0(\theta) + t^\epsilon \Phi(t,\theta), \tag{3.5}$$

where

$$\epsilon > 0, \qquad 0 < k(\theta) < \frac{1}{2}, \qquad \alpha_0 > 0, \tag{3.6}$$

$$-1 < \lambda\kappa(\theta) < 0, \tag{3.7}$$

and

$$\eta_0' - 2kU_0' - e^{4U_0}(1-2k)h'A_0 - \frac{\kappa\phi_0'}{2} + \frac{\alpha_0'}{2\alpha_0} = 0. \tag{3.8}$$

Here, ν, β, W, B and Φ are regular parts of the solutions, whereas other parts are singular, which are solutions to homogenization equations of Gowdy-nonlinear scalar system. Now, we have the following theorem:

Theorem 3.1. *(See [8]) Choose data such that conditions (3.6), (3.7) and (3.8) are satisfied. Suppose that ϵ is a positive constant less than* $\min\{4k, 2-4k, -2\lambda\kappa, 2+2\lambda\kappa, 2K\}$. *For any choice of the analytic singular data $\eta_0(\theta)$, $\alpha_0(\theta)$, $k(\theta)$, $U_0(\theta)$, $h(\theta)$, $A_0(\theta)$, $\kappa(\theta)$, $\phi_0(\theta)$, $\omega(\theta)$ and $\sigma_0(\theta)$, the Gowdy-nonlinear scalar system has a solution of the form (3.1)-(3.5), where ν, β, W, B, Φ and Σ tend to zero as $t \to 0$.* □

To prove the above theorem, we have used the theorem on *Fuchsian algorithm* [6]. From this, we can conclude there exist singular solutions near $t = 0$. In this case, one can check the curvature invariant of our spacetimes blows up as $t \to 0$.

4. Energy decay into the future

In this section, asymptotic behavior of the solutions will be discussed. To do this, we need the following assumption:

Hypothesis 4.1. There is a positive constant T, such that for any $t \geq T$ $\phi(t,\theta) = \psi(t,\theta)$, which satisfies $V(\psi) = 0$.

From this, one can show the energy is decay as t^{-1}:

Theorem 4.1. *(See [9]) Consider a solution to the Gowdy-nonlinear scalar system. Suppose that Hypotheses 2.2 and 4.1 hold. Then, there are constants τ and C such that*

$$E(t) \leq \frac{C}{t}, \qquad \forall t \geq \tau. \tag{4.1}$$

□

To prove this theorem, small data decay estimate is needed at first:

Proposition 4.1. *(decay for small data case) Consider a solution to the Gowdy-nonlinear scalar system. Suppose that Hypotheses 2.2 and 4.1 hold. Then, there exists a constant $\epsilon > 0$ such that if $E(t_0) \leq \epsilon$ for some t_0, there are constants τ and C such that*

$$E(t) \leq \frac{C}{t}, \qquad \forall t \geq \tau. \tag{4.2}$$

□

The method to get the above is *corrected energy estimate* developed by Choquet-Bruhat and Moncrief [1, 2, 10].

References

[1] Choquet-Bruhat, Y., *Future complete* U(1) *symmetric Einsteinian spacetimes, the unpolarized case*, in *The Einstein equations and the large scale behavior of gravitational fields*, 251-298, Birkhaüser, Basel, 2004.

[2] Choquet-Bruhat, Y. and Moncrief, V., *Future global in time Einsteinian spacetimes with* U(1) *isometry group*, Ann. Henri Poincaré **2** (2001), 1007-1064.

[3] Chruściel, P. T., *On space-times with* $U(1) \times U(1)$ *symmetric compact Cauchy surfaces*, Ann. Physics, NY **202**, (1990) 100-150.

[4] Dafermos, M. and Rendall, A. D., *Inextendibility of expanding cosmological models with symmetry*, Classical Quantum Gravity **22**, (2005) L143-L147.

[5] Gowdy, R. H., *Vacuum spacetimes and compact invariant hypersurfaces: Topologies and boundary conditions*, Ann. Physics, NY **83**, (1974) 203-224.

[6] Kichenassamy, S. and Rendall, A. D., *Analytic description of singularities in Gowdy spacetimes*, Class. Quantum Grav. **15**, (1998) 1339-1355.

[7] Lyth, D. H. and Riotto, A., *Particle physics models of inflation and the cosmological density perturbation*, Phys. Rep. **314**, (1999) 1-146.

[8] Narita, M., *On initial conditions and global existence for accelerating cosmologies from string theory*, Ann. Henri Poincaré **6**, (2005), 821-847.

[9] Narita, M., *On Gowdy symmetric spacetimes with nonlinear scalar fields*, in preparation.

[10] Ringström, H., *On a wave map equation arising in general relativity*, Comm. Pure Appl.Math. **57**, (2004) 657-703.

INTERFACES DRIVEN BY REACTION, DIFFUSION AND CONVECTION

J. HÄRTERICH

Insititut für Mathematik I, Freie Universität Berlin, Arnimallee 2-6, 14195 Berlin, Germany

K. SAKAMOTO

Department of Mathematical and Life Sciences, Hiroshima University, 1-3-1 Kagamiyama, Higashi-Hiroshima, 739-8526 Japan

1. Introduction

1.1. *Interfaces driven by reaction-diffusion equations*

Interfacial phenomena have been studied in terms of reaction-diffusion equations. In particular the Allen-Cahn equation

$$\frac{\partial u}{\partial t} = \Delta u + g(u), \quad t > 0,\, x \in \mathbb{R}^N, \tag{A-C}$$

where g is a bi-stable reaction term, has been a widely used model to describe various physical phenonena. Here, bi-stability of g means that it is the minus of the derivative, $g(u) = -W'(u)$, of a double well potential $W(u)$ with non-degenerate wells located at $u = u_\pm$. When $W(u_-) \neq W(u_+)$, a hyperbolic scaling gives rise to

$$\frac{\partial u}{\partial t} = \varepsilon \Delta u + \frac{1}{\varepsilon} g(u), \qquad (\varepsilon > 0 \text{ is a scaling parameter})$$

in which case the interface is driven by the difference of potential value $W(u_-) - W(u_+)$;

$$V = c,$$

where V stands for the normal speed of the interface and c is a constant determined from W, i.e., $c \propto W(u_+) - W(u_-)$.

When the two wells have the same depth, then $c = 0$ and hence the interface equation above does not give any information on the motion of the interface. In this case, we apply a parabolic scaling to (A-C) so that it has the following form;

$$\frac{\partial u}{\partial t} = \Delta u + \frac{1}{\varepsilon^2} g(u).$$

The motion law of the interface for this equaton is the so called "mean curvature flow", namely,

$$V = H,$$

where H stands for the sum of principal curvatures of the interface. Such interface evolutions for reaction-diffusion equations have been established by many authors (see [2]).

The purpose of this paper is to investigate what happens to these interface equations when a convection term is added to (A-C).

1.2. *Reaction-diffusion with convection*

We investigate interfacial phenomena for the following equation;

$$u_t + \text{div } \mathbf{f}(u) = \Delta u + g(u), \quad (x,t) \in \mathbb{R}^N \times (0,\infty). \tag{RDC}$$

This equation is derived as follows.

When a physical quantity u is carried by a flux $\mathbb{J}$ with source term $g(u)$, then the balance equation is expressed as

$$\frac{\partial u}{\partial t} + \text{div } \mathbb{J} = g(u). \tag{BL}$$

If the flux $\mathbb{J}$ is represented by a (vector-valued) function $\mathbf{f} : \mathbb{R} \to \mathbb{R}^N$ and the gradient of u,

$$\mathbb{J} = -\mu \nabla u + \mathbf{f}(u) \qquad (\mu > 0 \text{ viscosity}), \tag{Flux}$$

equation (BL) reduces to (RDC). In diffusive flow fields, the flux $\mathbf{f}$ is supposed to originate from fluid flows, and therefore, should be coupled with Navier-Stokes equations governing the flow field. We are considering here a simplified problem without reference to such flow-field equations. From now on, we set the viscosity equal to 1; $\mu = 1$

To describe the dynamics of (RDC) as $t \to \infty$, we perform a hyperbolic spatio-temporal scaling; $(x,t) \to (x/\varepsilon, t/\varepsilon)$ which reduces (RDC) to

$$u_t + \text{div } \mathbf{f}(u) = \varepsilon \Delta u + \varepsilon^{-1} g(u), \tag{1.1}$$

where $\varepsilon > 0$ is a scaling parameter.

Our objective below is to investigate the dynamics of (1.1) in the singular limit $\varepsilon \to 0$.

To consider the singular limit $\varepsilon \to 0$ means that we are describing the variation of u over large spatial ranges as time $t \to \infty$ in the original system.

We work throughout under the following hypotheses.

(H1): (i) $g \in C^2(\mathbb{R})$, $g(u_*) = 0$ at $u_* = u_-, 0, u_+$,

$$g'(u_-) < 0, \quad g'(0) > 0, \quad g'(u_+) < 0 \text{ (bi-stable reaction term).}$$

(ii) $\mathbf{f} \in C^3(\mathbb{R}, \mathbb{R}^N)$

Well-posedness of initial value problem: Under the hypothesis **(H1)**, the problem (1) with an initial condition

$$u(x,0) = \phi(x) \in \mathrm{BC}^2_{\mathrm{unif}}(\mathbb{R}^N) \tag{1.2}$$

possesses a unique global (in time) solution living in $\mathrm{BC}^2_{\mathrm{unif}}(\mathbb{R}^N)$. This is proved by a standard way by using abstract theories for evolution equations (see [1], for example).

2. Planar Waves

First thing to do is to study planar traveling wave solutions. The traveling wave solution of (1) in the $\nu \in S^{N-1}$-direction; $u(x,t) = U(\frac{x\cdot\nu - st}{\varepsilon})$ satisfies

$$U''(z) + (s - \mathbf{f}'(U(z)) \cdot \nu)\, U'(z) + g(U(z)) = 0, \tag{2.1}$$

$$z \in \mathbb{R}, \qquad (' = \mathrm{d}/\mathrm{d}z),$$

$$U(\pm\infty) = u_\pm, \quad U(0) = 0,$$

where s is the wave speed to be determined together with the wave profile. The following result is obtained by a phase plane analysis.

Proposition 2.1. ([3] and [5])

(i) *For each direction $\nu \in S^{N-1}$, there uniquely exists a wave speed $s = s(\nu)$ for which the problem (3) has a unique heteroclinic orbit connecting $(u_-, 0)$ (at $z = -\infty$) and $(u_+, 0)$ (at $z = +\infty$).*

(ii) *The wave speed $s(\nu)$ depends on ν as smooth as the nonlinear terms $\mathbf{f}'(u)$ and $g(u)$ do on u.*

(iii) *The wave profile $Q(z;\nu)$ with $Q(0;\nu) = 0$ depends on (z,ν) as smooth as the nonlinear terms $\mathbf{f}'(u)$ and $g(u)$ do on u, and it is a (strictly) monotone increasing function of z.*

(iv) *If* $\mathbf{f}(u)$ *is even and* $g(u)$ *is odd in* u*, then* $Q(z)$ *is an odd function of* z *and the wave speed satisfies* $s(\nu) \equiv 0$.

An important feature is that the wave speed is orientation (direction) dependent. This anisotropy later gives rise to anisotropic mean curvature flows.

It is interesting to note that the wave speed and the wave profile are related as follows.

$$s(\nu) = \frac{(\mathbf{f}(u_+) - \mathbf{f}(u_-)) \cdot \nu}{u_+ - u_-} + \frac{1}{u_+ - u_-} \int_{-\infty}^{\infty} g(Q(z;\nu))\mathrm{d}z$$

and

$$s(\nu) = \frac{G(u_-) - G(u_+)}{\int_{-\infty}^{\infty} Q_z(z;\nu)^2 \mathrm{d}z} + \frac{\int_{-\infty}^{\infty} Q_z(z;\nu)^2 \mathbf{f}'(Q(z;\nu)) \cdot \nu \mathrm{d}z}{\int_{-\infty}^{\infty} Q_z(z;\nu)^2 \mathrm{d}z},$$

where $G(u)$ is an anti-derivative of $g(u)$. The first formula looks like a generalized Rankin-Hugoniot condition for viscous shocks for conservation laws, while the second expression resembles the wave speed characterization for bi-stable reaction-diffusion equations, with modification in terms of the entire wave profile. It is also important to note that $s(\nu)$ depends not only on the asymptotic states (which is the case for reation-diffusion equations), but also on the entire viscous wave profile.

Lemma 2.1. *In the nonlinear eigenvalue problem* (2.1)*, if the nonlinearities are given by*

$$\begin{aligned} g(u) &= -R(u-u_-)u(u-u_+), \\ &\quad \textit{with } u_- < 0 < u_+, \quad R > 0, \\ \mathbf{f}(u) &= \frac{1}{2}u^2\mathbf{a} + u\mathbf{b}, \quad \mathbf{a},\ \mathbf{b} \in \mathbb{R}^N, \end{aligned}$$

then, the wave speed $s(\nu)$ *and wave profile* $Q(z)$ *are explictly represented as follows.*

$$s(\nu) = \frac{u_- + u_+}{4}\left\{\mathbf{a}\cdot\nu - \sqrt{(\mathbf{a}\cdot\nu)^2 + 8R}\right\} + \mathbf{b}\cdot\nu,$$

$$Q(z) = \frac{-u_+u_- + u_+u_-\mathrm{e}^{-D(u_+-u_-)z}}{-u_- + u_+\mathrm{e}^{-D(u_+-u_-)z}},$$

where D *is defined by*

$$D = \frac{\sqrt{(\mathbf{a}\cdot\nu)^2 + 8R} - \mathbf{a}\cdot\nu}{4}.$$

Proof. The proof is the same as the case without convection term (see [6]). One may also substitute these functions into (3) to directly verify the lemma. □

Generically, we expect that the wave speed $s(\nu)$ vanishes on a subset in S^{N-1} which has codimension at least one. However, we have been unable to prove or disprove this expectation in general case. The set where $s(\nu)$ vanishes,

$$\mathcal{P} := \{\nu \in S^{N-1} \mid s(\nu) = 0\}$$

is called a set of pinned directions. When the wave speed is given as in Lemma 2.1, then $\mathcal{P}$ is generically of codimension at least one. Althourgh this is the case for the specific cases, we make the following hypothesis.

(H2): The set of pinned directions $\mathcal{P}$ has codimension at least one in S^{N-1}.

3. Results

We now give some of main results. There are two cases, one in which **(H2)** is valid, and the other where $\mathcal{P} = S^{N-1}$.

3.1. *When codim $\mathcal{P} \leq 1$*

We call this case a hyperbolic scaling case.

Theorem 3.1. *Under the hypotheses (H1) and (H2), we consider the following Cauchy problem*

$$\begin{cases} u_t^\varepsilon = \varepsilon\Delta u^\varepsilon - \mathbf{f}'(u^\varepsilon)\cdot\nabla u^\varepsilon + \varepsilon^{-1}g(u^\varepsilon) \\ u^\varepsilon(x,0) = \phi^\varepsilon(x). \end{cases} \tag{3.1}$$

There exist two functions $\underline{u}_0^\varepsilon(x) < \overline{u}_0^\varepsilon(x)$ and a constant $T > 0$ such that the following statement is true: If the initial function satisfies

$$\underline{u}_0^\varepsilon(x) < \phi^\varepsilon(x) < \overline{u}_0^\varepsilon(x),$$

then the solution $u^\varepsilon(x,t)$ converges to a limit $u^0(x,t) = \lim_{\varepsilon\to 0} u^\varepsilon(x,t)$ for almost all $(x,t) \in \mathbb{R}^N \times [0,T]$. The limit function $u^0(x,t)$ is a piece-wise constan function, assuming only two values u_- and u_+. The bulk regions

$$\Omega^\pm(t) := \{x \in \mathbb{R}^N;\ u^0(x,t) = u_\pm\}$$

are separated by a hypersurface $\Gamma(t)$, *and the hypersurface (interface) evolves according to the motion law*

$$V = s(\nu),$$

where V *represents the normal velocity of the interface* $\Gamma(t)$, *and* ν *is a unit normal vector on* $\Gamma(t)$ *pointing into the interior of the bulk region* $\Omega^+(t)$.

If we define ε*-dependent interface* $\Gamma^\varepsilon(t)$ *by*

$$\Gamma^\varepsilon(t) = \{x \in \mathbb{R}^N \mid u^\varepsilon(x,t) = 0\},$$

then its motion law is governed by

$$\begin{aligned} V^\varepsilon =& s(\nu^\varepsilon) \\ &+ \varepsilon \left\{ H^\varepsilon(y,t) + \sum_{p,q=1}^{N} \mathbf{T}^\varepsilon_{pq} K^{pq}_\varepsilon(y,t) \right\} \\ &+ \mathcal{O}(\varepsilon^2), \end{aligned} \tag{3.2}$$

where ν^ε *is the unit normal vector of* $\Gamma^\varepsilon(t)$ *(pointing into the interior of + region),* $s(\nu^\varepsilon)$ *is the wave speed evaluated at* ν^ε, $H^\varepsilon(y,t)$ *is the sum of principal curvatures (mean curvature, for short) of* $\Gamma^\varepsilon(t)$ *at* $y \in \Gamma^\varepsilon(t)$, $(\mathbf{T}^\varepsilon_{pq})$ *is a symmetric, positive semi-definite* $N \times N$ *matrix depending only on* $(\mathbf{f}, g, \nu^\varepsilon)$, K^{pq}_ε *is a symmetric tensor related to the second fundamental form of* $\Gamma^\varepsilon(t)$.

We note that $T > 0$ in the statement above is determined by the time interval where $V = s(\nu)$ has a smooth soluiton.

Now, let us give explicit forms to the quantities appearing in the theorem. For this purpose, we use the travelling wave profile $Q = Q(z;\nu)$. Let $P = P(z;\nu)$ be defined by

$$P(z) = Q_z(z;\nu)\exp\left(\int_0^z [s(\nu) - \mathbf{f}'(Q(\tau;\nu)\cdot\nu]\,\mathrm{d}\tau\right).$$

We also let $\Gamma(t)$ be represented by γ_0 as follows.

$$\gamma_0 : \mathcal{M} \times [0,T] \ni (y,t) \mapsto \gamma_0(y,t) \in \Gamma(t),$$

where $\mathcal{M}$ is a reference manifold.

Then, representing by $g = (g_{ij})$ and $h = (h_{ij})$ the first and second fundamental forms of $\Gamma(t)$, respectively, with $g^{-1} = (g^{ij})$, we have

$$K^{pq} = \frac{\partial(\gamma_0)^p}{\partial y^j}\frac{\partial(\gamma_0)^q}{\partial y^l} g^{jk} h_{ks} g^{sl},$$

$$\mathbf{T} = M_0^{-1} \int_{-\infty}^{\infty} \frac{1}{PQ_z} L(z) \otimes L(z)\,\mathrm{d}z,$$

$$L(z) = \int_{-\infty}^{z} P(z')Q_z(z')\,[\nabla_\nu s(\nu) - \mathbf{f}'(Q(z'))]\,\mathrm{d}z',$$

$$M_0 = \int_{-\infty}^{\infty} P(z)Q_z(z)\,\mathrm{d}z > 0.$$

It is clear from these formula that $\mathbf{T}$ is positive semi-definite. Generically, we expect that the matrix $\mathbf{T}$ is *positive definite*, not only positive *semi-definite*. To see this, let $\mathbf{a} \in \mathbb{R}^N$, then we have

$$\mathbf{a}^\top \mathbf{T}\mathbf{a} = M_0^{-1} \int_{-\infty}^{\infty} \frac{1}{PQ_z} (L(z)\cdot \mathbf{a})^2\,\mathrm{d}z \geq 0.$$

Therefore, $\mathbf{a}^\top \mathbf{T}\mathbf{a} = 0$ implies $L(z)\cdot \mathbf{a} \equiv 0$, which in turn implies

$$(\nabla_\nu s(\nu) - \mathbf{f}'(Q(z;\nu)))\cdot \mathbf{a} \equiv 0.$$

This is possible for non-zero $\mathbf{a}$ only when the vector $\nabla_\nu s(\nu) - \mathbf{f}'(Q(z;\nu))$ is parallel to a constant vector for all $z \in \mathbb{R}$. Generically, we do not expect that this should happen.

On the other hand, when $\mathbf{f}'(u) = \mathbf{b}$ is a constant vector, then

$$s(\nu) = c + \mathbf{b}\cdot \nu$$

where c is the traveling wave speed of

$$U_{zz} + cU_z + g(U) = 0, \quad z \in \mathbb{R}$$
$$\lim_{z\to\pm\infty} U(z) = u_\pm, \quad U(0) = 0.$$

Therefore, we have $\nabla_\nu s(\nu) - \mathbf{f}'(Q(z;\nu)) \equiv 0$, and hence $\mathbf{T} = 0$.

3.2. *When* $s(\nu) \equiv 0$

We have been unable to give general conditions which imply $s(\nu) \equiv 0$ on S^{N-1}. In this subsection, therefore, we assume the following conditions are fulfilled.

(H3): $\mathbf{f}(u)$ is even and $g(u)$ is odd.

Evidently, Proposition 2.1 says that **(H3)** implies $s(\nu) \equiv 0$.

Theorem 3.2. *Under the hypotheses* ***(H1)*** *and* ***(H3)****, we consider the following problem;*

$$\begin{cases} u_t^\varepsilon = \Delta u^\varepsilon - \varepsilon^{-1}\mathbf{f}'(u^\varepsilon)\cdot\nabla u^\varepsilon + \varepsilon^{-2}g(u^\varepsilon) \\ u^\varepsilon(x,0) = \phi^\varepsilon(x). \end{cases} \tag{3.3}$$

There exist a class of initial functions ϕ^ε *and a constant* $T > 0$ *such that the solution* u^ε *of* (3.3) *converges to a limit for almost all* $(x,t) \in \mathbb{R}^N \times [0,T]$*;*

$$u^0(x,t) := \lim_{\varepsilon\to 0} u^\varepsilon(x,t).$$

The limi function $u^0(x,t)$ *is piecewise constant, taking on two values* u_- *and* u_+.

The interface $\Gamma(t)$ *separating two bulk regions* $\Omega^\pm(t) := \{x \in \mathbb{R}^N;\ u^0(x,t) = u_\pm\}$ *evolves according to the following motion law.*

$$V = H + \sum_{p,q=1}^{N} \mathbf{T}_{pq}K^{pq} = \sum_{p,q=1}^{N} (\delta_{pq} + \mathbf{T}_{pq})\,K^{pq}, \tag{3.4}$$

where

$$\begin{aligned} K^{pq} &= \frac{\partial(\gamma_0)^p}{\partial y^j}\frac{\partial(\gamma_0)^q}{\partial y^l}g^{jk}h_{ks}g^{sl}, \\ \mathbf{T} &= M_0^{-1}\int_{-\infty}^{\infty}\frac{1}{PQ_z}L(z)\otimes L(z)\,\mathrm{d}z, \\ L(z) &= \int_{-\infty}^{z} P(z)Q_z(z)\mathbf{f}'(Q)\,\mathrm{d}z', \\ A(z) &= -\int_0^z \mathbf{f}'(Q(\tau;\nu))\cdot\nu\,\mathrm{d}\tau, \\ P(z) &= \mathrm{e}^{A(z)}Q_z(z). \end{aligned}$$

As before, $\mathbf{T}$ is positive semi-definite, and generically positive definite. It is also of interest to note that $\mathbf{T}$ introduces a kind of Riemannian metric (possibly degenerate) in the ambient space $\mathbb{R}^N$. If $\mathbf{T}$ were the $N\times N$ identity matrix, then $\mathbf{T}K = \mathrm{tr}(hg^{-1})$ would be the sum of principal curvatures of the interface.

Proposition 3.1. *The sum $\mathbf{T}_{pq}K^{pq}$ is a weighted sum of principal curvatures:*

$$\mathbf{T}_{pq}K^{pq} = \sum_{i=1}^{N-1} w^i \kappa_i,$$

where κ_i $(i = 1, 2, \ldots, N-1)$ are principal curvatures of $\Gamma(t)$ and

$$w^i = \sum_{p,q=1}^{N} \sum_{j=1}^{N-1} \mathbf{T}_{pq} \frac{\partial(\gamma_0)^p}{\partial y^i} \frac{\partial(\gamma_0)^q}{\partial y^j} g^{ji}.$$

Therefore, (3.4) *is rewritten as*

$$V = \sum_{i=1}^{N-1} (1 + w^i)\kappa_i, \tag{3.4}$$

namely, the interface $\Gamma(t)$ is driven by an anisotropic mean curvature flow.

By inspecting the proof of Theorem 3.2, we obtain the existence result for equilibrium solutions of (3.3).

Corollary 3.1. *Let Γ^* be* anisotropically minimal *in the sense that*

$$0 = \sum_{i=1}^{N-1} (1 + w^i)\kappa_i \qquad \text{on } \Gamma^*.$$

If Γ^* is non-degenerate in the sense that an elliptic linear operator $\mathcal{L}$ defined on Γ^* does not have 0-eigenvalue, then there exists a family of equilibirum solutions $u^\varepsilon(x)$ of (3.3) *for small $\varepsilon > 0$ so that*

$$\lim_{\varepsilon \to 0} u^\varepsilon(x) = \begin{cases} u_+ & x \in \Omega_+ \\ u_- & x \in \Omega_- \end{cases}.$$

The linear operator $\mathcal{L}$ has the following explicit form.

$$\begin{aligned} \mathcal{L}A =& \Delta^{\Gamma^*} A + \mathbf{T}_{pq} \left[\nabla_{\Gamma^*} (\nabla_{\Gamma^*} A)^p\right]^q \\ &+ \mathbb{F} \cdot \nabla_{\Gamma^*} A + \mathbb{G} \cdot \nabla^{(1)}_{\Gamma^*} A \\ &+ \left[H^{(1)} - \mathbf{T}_{pq} \left(\nabla^{(1)}_{\Gamma^*} \nu^p\right)^q\right] A, \end{aligned}$$

where Δ^{Γ^*} is the Laplace-Beltrami operator on Γ^*,

$$H^{(1)} = \sum_{j=1}^{N-1} (\kappa_j)^2,$$

$$\nabla_{\Gamma^*} A = \sum_{j,k=1}^{N-1} \frac{\partial \gamma^*}{\partial y^j} g^{jk} \frac{\partial A}{\partial y^k},$$

$$\nabla_{\Gamma^*}^{(1)} A = \sum_{j,s,l,k=1}^{N-1} \frac{\partial \gamma^*}{\partial y^j} g^{js} h_{sl} g^{lk} \frac{\partial A}{\partial y^k},$$

and $\mathbb{F}$ and $\mathbb{G}$ are vector fields on Γ^* to which we do not give explicit forms. However, we emphasize that $\mathcal{L}$ is the linearization of the right hand side of (3.4) around Γ^*, relative to normal variations of hypersuface.

4. Some Examples

4.1. *Symmetric nonlinearity*

We deal with the following specific nonlinearity;

$$\mathbf{f}(u) = \frac{1}{2} u^2 \mathbf{a} \in \mathbb{R}^N, \qquad g(u) = -u(u^2 - 1)$$

Then we obtain; wave profileF$Q(z) = \tanh(Dz)$ and wave speed: $s(\nu) \equiv 0$. Note that the ν-dependency of the wave profile is only throught the quantity D defined by

$$D = \frac{1}{4}(\sqrt{(\mathbf{a} \cdot \nu)^2 + 8} - \mathbf{a} \cdot \nu)$$

To describe the interface equation, we need to compute:

$$Q_q := \frac{\partial Q}{\partial \nu^q} = -\frac{D \mathbf{a}_q}{\sqrt{(\mathbf{a} \cdot \nu)^2 + 8}} \frac{z}{\cosh^2(Dz)},$$

$$P = Q_z \mathrm{e}^A = D\,(\cosh(Dz))^{-(\frac{\mathbf{a} \cdot \nu}{D} + 2)}$$

$$M_0 = \int_{-\infty}^{\infty} \mathrm{e}^{A(z)} Q_z^2(z)\,\mathrm{d}z = D^2 \int_{-\infty}^{\infty} (\cosh(Dz))^{-(\frac{\mathbf{a} \cdot \nu}{D} + 4)}\,\mathrm{d}z$$

and

$$
\begin{aligned}
M_0\mathbf{T}_{pq} &= -\int_{-\infty}^{\infty} \mathrm{e}^{A(z)} Q_z Q_q \mathbf{f}'_p(Q)\,dz \\
&= \frac{D^2 \mathbf{a}_p \mathbf{a}_q}{\sqrt{(\mathbf{a}\cdot\nu)^2+8}} \int_{-\infty}^{\infty} z\tanh(Dz)\,(\cosh(Dz))^{-(\frac{\mathbf{a}\cdot\nu}{D}+4)}\ \mathrm{d}z \\
&= \frac{D^2 \mathbf{a}_p \mathbf{a}_q}{\sqrt{(\mathbf{a}\cdot\nu)^2+8}} \left(\frac{1}{\mathbf{a}\cdot\nu+4D}\right) \int_{-\infty}^{\infty} (\cosh(Dz))^{-(\frac{\mathbf{a}\cdot\nu}{D}+4)}\ \mathrm{d}z.
\end{aligned}
$$

Therefore, by using $4D + \mathbf{a}\cdot\nu = \sqrt{(\mathbf{a}\cdot\nu)^2+8}$, we obtain

$$
\mathbf{T}_{pq} = \frac{\mathbf{a}_p \mathbf{a}_q}{(\mathbf{a}\cdot\nu)^2+8}.
$$

On the other hand, a simple computation yields

$$
\mathbf{T}_{pq}K^{pq} = \frac{1}{2D^2+1}\mathbf{a}\cdot\nabla_{\Gamma(t)}D,
$$

and the interface equation is given by

$$
V = H(y,t) + \mathbf{a}\cdot\nabla_{\Gamma(t)}\left(\frac{\operatorname{Arctan}(\sqrt{2}D)}{\sqrt{2}}\right) \tag{4.1}
$$

Here $D > 0$ in $Q(z) = \tanh(Dz)$ means "**steepness**" of the wave profile, and (4.1) says that the tangential variation of the "steepness"

$$
\frac{\operatorname{Arctan}(\sqrt{2}D)}{\sqrt{2}}
$$

of the wave profile is converted to the normal speed of the interface in the singular limit. At a formal level, this kind of observation was first given in [4].

As before, the second term in the right hand side of (4.1) $\mathbf{T}_{pq}K^{pq}$ is rewritten and (4.1) reduces to

$$
V = \sum_{i=1}^{N-1}(1+w^i)\kappa_i, \tag{4.2}
$$

where

$$
w^i = \frac{1}{(\mathbf{a}\cdot\nu)^2+8}\sum_{j=1}^{N-1}\left(\frac{\partial\gamma_0}{\partial y^j}\cdot\mathbf{a}\right)\left(\frac{\partial\gamma_0}{\partial y^i}\cdot\mathbf{a}\right)g^{ji}
$$

In other words, (4.2) is interpreted as a weighted mean curvature flow. Note that w^i in (4.2) is 0 when $\mathbf{a}$ is parallel to ν.

Although the matrix $\mathbf{T}$ originates from the first order differential operator div $\mathbf{f}$, it exhibits a curvature effect (a second order differential operator) in the singular limit.

References

[1] D. Daners and P. Koch Medina, *Abstract evolution equations, periodic problems and application*, Pitman Research Notes **273** (1992).

[2] P. de Mottoni and M. Schatzman, *Geometrical evolution of developed interfaces*, Trans. AMS **347** (1995), 1533–1589.

[3] G. F. D. Duff, *Limit-cycles and rotated vector fields*, Ann. of Math. (2)**57** (1953), 15–31.

[4] H. Fan and S. Jin, *Multi–dimensional viscous conservation laws with stiff source terms driven by mean curvature and variation of front thickness*, Quarterly Appl. Math. LXI (4), (2003), 701–721.

[5] J. Härterich and C. Mascia, *Front Formation and Motion in Quasilinear Parabolic Equations*, preprint, 2002.

[6] J. D. Murray, *Mathematical Biology*, Springer (1989).

LIFE SPAN OF SOLUTIONS FOR A SUPERLINEAR HEAT EQUATION

SHOTA SATO

Mathematical Institute, Tohoku University,
Sendai 980-8578, JAPAN.

Abstract: We discuss the initial-boundary problem

$$\begin{cases} u_t = \Delta u + f(u) & \text{in } \Omega \times (0, \infty), \\ u(x,t) = 0 & \text{on } \partial\Omega \times (0, \infty), \\ u(x,0) = \rho\varphi(x) & \text{in } \Omega, \end{cases}$$

where Ω is a bounded domain in $\mathbb{R}^N$ with smooth boundary $\partial\Omega$, $\rho > 0$ is a parameter, $\varphi(x)$ is a non-negative continuous function on $\overline{\Omega}$, $f(u)$ is a non-negative superlinear continuous function on $[0, \infty)$. It is known that if ρ is large, the solution of this problem blows up in finite time. We show that the life span (or blow up time), denoted by $T(\rho)$, satisfies

$$T(\rho) = \int_{\rho\|\varphi\|_\infty}^{\infty} \frac{du}{f(u)} + h.o.t. \quad \text{as } \rho \to \infty.$$

Moreover, when the maximum of φ is attained at a finite number of points in Ω, we can determine the higher order term of $T(\rho)$ which depends on the minimal value of $|\Delta\varphi|$ at the maximal points. The proof is based on a careful construction of a supersolution and a subsolution.

1. Introduction

We consider the initial-boundary value problem

$$\text{(P)} \qquad \begin{cases} u_t = \Delta u + f(u) & \text{in } \Omega \times (0, \infty), \\ u = 0 & \text{on } \partial\Omega \times (0, \infty), \\ u(x,0) = \rho\varphi(x) & \text{in } \Omega, \end{cases}$$

where Ω is a bounded domain on $\mathbb{R}^N$ with a smooth boundary $\partial\Omega$, $\rho > 0$ is a parameter, and $\varphi(x)$ is a non-negative continuous function on $\overline{\Omega}$, $f(u)$ is a non-negative continuous function on $\overline{\Omega}$.

We denote by $T(\rho)$ the maximal existence time of a classical solution of the problem (P), and we call $T(\rho)$ the life span of u. If $T(\rho)$ is finite, then $T(\rho)$ is equal to the blow-up time.

Since the pioneering work of Fujita [2], the blow-up of solutios of the problem (P) has been studied extensively for the nonlinearity $f(u) = u^p$. Among others, Friedmann and Lacey [1] gave a result on the life span of solutions of (P) in the case of small diffusion. Subsequently, Gui and Wang [3], Lee and Ni [4] obtained a leading term of the expansion of $T(\rho)$ as $\rho \simeq \infty$. They proved that $T(\rho)$ is expanded as

$$T(\rho) = \frac{1}{p-1} M^{1-p} \rho^{1-p} + o(\rho^{1-p})$$

as $\rho \to \infty$. Later, Mizoguchi and Yanagida [5] extended the result and determined the second term of the expansion of $T(\rho)$ as $\rho \simeq \infty$. They proved that when φ attains the maximum at only one point $a \in \Omega$, $T(\rho)$ is expanded as

$$T(\rho) = \frac{1}{p-1} M^{1-p} \rho^{1-p} + \frac{2}{p-1} M^{2(1-p)-1} |\Delta\varphi(a)| \rho^{2(1-p)} + o(\rho^{2(1-p)})$$

as $\rho \to \infty$. Moreover, Mizoguchi and Yanagida [6] obtained the result on the life span of solutions of (P) in the case of small diffusion.

In this paper we extend the above results on the life span to more general nonlinearity f.

Throughout this paper, we assume that the nonlinearity $f(u)$ satisfies the assumption

$$\text{(A)} \qquad \begin{cases} \text{(i)} \ f(u) \in C^2((0,\infty)) \cap C([0,\infty)), \\ \text{(ii)} \ f(u),\ f'(u),\ f''(u) > 0 \text{ for all } u > 0, \\ \text{(iii)} \ \displaystyle\int_1^\infty \frac{du}{f(u)} < \infty. \end{cases}$$

Typical examples of such nonlinearity f are

$$u^p,\ e^u \text{ and } (u+1)\{\log(u+1)\}^p.$$

It is easy to see that the assumption (A) implies that

$$\lim_{u\to\infty} \frac{f(u)}{u} = \infty,$$

and a function $F(u)$ is defined by

$$F(u) := \int_u^\infty \frac{dz}{f(z)}.$$

Let M be the maximum of φ on Ω.

Our first result is concerning the leading term of $T(\rho)$ as $\rho \to \infty$.

Theorem 1.1. *Suppose that φ satisfies $\varphi \in C^2(\Omega) \cap C(\overline{\Omega})$, $\varphi > 0$, $\varphi \not\equiv 0$ and $\varphi(x)|_{\partial\Omega} = 0$. Then the life span $T(\rho)$ of the solution of (P) satisfies*

$$T(\rho) \leq F(\rho M) + o(F(\rho M))$$

as $\rho \to \infty$.

We note that the leading term of $T(\rho)$ as $\rho \to \infty$ is exactly equal to the life span of the solution of the ordinary differential equation

$$\begin{cases} z_t(t) = f(z(t)), \ t > 0, \\ z(0) = \rho M. \end{cases}$$

This suggests that the diffusion term has an effect only on higher order terms of $T(\rho)$ as $\rho \to \infty$.

In order to state a result on the second order term, we define $B(\varphi)$ by the set of maximal points of φ and introduce the following additional condition on the nonlinearity f;

$$\text{(B)} \quad \begin{cases} \text{(i)} \ \lim\limits_{u\to\infty} \dfrac{u^{1+\beta} f'(u)}{f(u)^{1+\beta}} = 0 \text{ for all } \beta > 0, \\ \text{(ii)} \ \dfrac{d}{du}\Big(\dfrac{f(u)}{u \log u^{\alpha}}\Big) \geq 0 \text{ for some } \alpha > 2 \text{ and all sufficiently large } u. \end{cases}$$

Remark 1.1. The functions

$$u^p \ (p > 1), \ \text{and } e^u$$

satisfy (A) and (B). The functions

$$(u+1)\{\log(u+1)\}^p \ (p > 1)$$

satisfy (A), but satisfies (B) if and only if $p > 2$.

Under the assumption (B), we can determine the second order term of $T(\rho)$ as $\rho \to \infty$.

Theorem 1.2. *Assume that the assumption (B) holds. Assume that φ satisfies the same conditions as in Theorem 1.1, and in addition, $B(\varphi)$ consists of a finite number of points and*

$$(D^2\varphi(a)x, x) < 0 \text{ for all } x \in \mathbb{R}^N \setminus \{0\}$$

for any $a \in B(\varphi)$. Then the life span of the solution of (P) satisfies

$$T(\rho) = F(\rho M) + \min_{a \in B(\varphi)} |\Delta\varphi(a)| \rho f(\rho M)^{-1} F(\rho M) + o(\rho f(\rho M)^{-1} F(\rho M))$$

as $\rho \to \infty$.

As in [5], we also consider the case where a peak of φ is flat. In this case the higher order term becomes exponentially small as follows.

Theorem 1.3. *Assume the same condition as in Theorem 1.1. Furthermore assume that there are a point $a \in \Omega$ and a constant $d > 0$ such that*

$$\{x \in \mathbb{R}^N;\ |x-a| \le d\} \subset B(\varphi).$$

Then the life span of the solution of (P) satisfies

$$T(\rho) = F(\rho M) + o(\exp(-CF(\rho M)^{-1}))$$

as $\rho \to \infty$ with some constant $C > 0$.

Since u^p satisfies the assumptions (A) and (B), those theorems generalize the results in [5].

In order to prove Theorems 1.1-1.3, we employ the scaling

$$v(x,s) = \rho^{-1}u(x,t), \ s = F(\rho M)^{-1}t.$$

By using this scaling, v satisfies

$$\begin{cases} v_s = F(\rho M)\Delta v + \rho^{-1}F(\rho M)f(\rho v) & \text{in } \Omega \times (0,\infty), \\ v = 0 & \text{on } \partial\Omega \times (0,\infty), \\ v(x,0) = \varphi(x) & \text{in } \Omega. \end{cases}$$

We consider this problem as the case of small diffusion and employ the idea of [6]. However, due to the lack of scaling invariance, the arguments of [6] do not apply directly. Moreover, in [6], to construct the subsolution, they use the fact that the solution of $u_t = u^p$ with the initial data $u(0) = 0$ is identically equal 0. However, since we do not assume $f(0) = 0$, the solution of $u_t = f(u)$ with the initial data $u(0) = 0$ may be positive for $t > 0$. In order to overcome these difficulties, we derive a positive lower estimate of the solution and modify arguments of [6] considerably to obtain the estimate of $T(\rho)$.

This paper is organized as follows: In Section 2 we give some preliminary estimates of solutions and use those estimates to obtain upper estimates of the life span of the solution. In Section 3 we show the outline of the proofs of Theorems 1.1-1.3.

2. Preliminaries

In this section we give some preliminary results which will be used to obtain upper estimates of the life span.

By putting

$$v(x,s) = \rho^{-1}u(x,t), \ s = F(\rho M)^{-1}t,$$

v satisfies

$$\text{(2.1)} \qquad \begin{cases} v_s = F(\rho M)\Delta v + \rho^{-1}F(\rho M)f(\rho v) & \text{in } \Omega \times (0,\infty), \\ v = 0 & \text{on } \partial\Omega \times (0,\infty), \\ v(x,0) = \varphi(x) & \text{in } \Omega. \end{cases}$$

We denote by $S(\rho)$ the life span of the solution v of (2.1). Let $q(s;\alpha)$ be the solution of the ordinary differential equation

$$\text{(2.2)} \qquad \begin{cases} q_s(s;\alpha) = \rho^{-1}F(\rho M)f(\rho q(s;\alpha)), \ s > 0, \\ q(0;\alpha) = \alpha, \end{cases}$$

and $Q(\alpha)$ be the life span of the solution $q(s;\alpha)$. Then we have

$$q(s;\alpha) = \rho^{-1}F^{-1}(F(\rho\alpha) - F(\rho M)s), \ 0 \le s \le Q(\alpha),$$
$$Q(\alpha) = F(\rho M)^{-1}F(\rho\alpha).$$

Let $G_\rho(x,y;s)$ be the Green function of the Cauchy-Dirichlet problem

$$\text{(2.3)} \qquad \begin{cases} V_s = F(\rho M)\Delta V & \text{in } \Omega \times (0,\infty), \\ V = 0 & \text{on } \partial\Omega \times (0,\infty). \end{cases}$$

Comparing $G_\rho(x,y;s)$ with the fundamental solution of $V_s = F(\rho M)\Delta V$ in $\mathbb{R}^N$, we have

$$\text{(2.4)} \qquad 0 < G_\rho(x,y;s) < \frac{1}{(4\pi F(\rho M)s)^{N/2}} \exp(-\frac{|x-y|^2}{4F(\rho M)s})$$

for all $(x,y,s) \in \Omega \times \Omega \times (0,\infty)$. Now for large ρ, we obtain a lower estimate of $G_\rho(x,y;s)$ as follows.

Lemma 2.1. ([6]) *For any $a \in \Omega, \tau > 0$, there exist constants C, α, $\rho_0 > 0$ such that if $\rho_0 \le \rho$, then the Green function $G_\rho(x,y;s)$ of (2.3) satisfies*

$$G_\rho(x,y;s) \le \{1 - \exp(-CF(\rho M)^{-1})\} \frac{1}{(4\pi F(\rho M)s)^{N/2}} \exp(-\frac{|x-y|^2}{4F(\rho M)s})$$

for all $(x,y,s) \in B_{a,\alpha} \times B_{x,\alpha} \times (0,\tau)$, where $B_{a,\alpha} := \{x \in \mathbb{R}^N;\ |x-a| < \alpha\}$.

By using Lemma 2.1 and the representation formula, we obtain the next lemma on a lower estimate of the solution of (2.1) near the maximal point of φ.

Lemma 2.2. *Assume the same conditions as in Theorem 1.1 and $\varphi(a) = M$. Then for any $\tau > 0$, there exist constants α, β, $\rho_1 > 0$ such that* *if* $\rho \geq \rho_1$, *the solution* v *of* (2.1) *satisfies* $v(x,s) > \beta$ *in* $B_{a,\alpha} \subset \Omega$ *for* $0 < s < \min(\tau, S(\rho))$.

3. Proofs

In this section we show the outline of proof of Theorems 1.1-1.3. Let us consider the life span $S(\rho)$ of the solution v of (2.1) satisfying $S(\rho) = F(\rho M)^{-1}T(\rho)$. To prove the theorems, we construct a supersolution and a subsolution of (2.1) and estimate those life span. By the comparison theorem, $S(\rho)$ is bounded below the life span of a supersolution and above the life span of a subsolution. To construct a supersolution and a subsolution, we use the ordinary differential equationand (2.2) and the Cauchy-Dirichlet problem (2.3).

At first, let us construct a supersolution of (2.1). In the proofs of Theorems 1.1 and 1.3, we take $q(s;M)$ as a supersolution of (2.1). In the proof of Theorem 1.2, for arbitrary s_0 with $0 < s_0 < 1$, we take

$$q(s; \max_{\Omega} v(x, s_0))$$

as a supersolution of (2.1) with the initial time s_0. Let $\bar{S}(\rho)$ be the life span of this. Then it holds

$$\bar{S}(\rho) \leq s_0 + S(\rho)$$

by the comparison principle.

Next, let us construct a subsolution of (2.1). We take any $\tau > 1$ and fix it. By Lemma 2.2, there exist a ball $D \subset \Omega$ centered at the origin and β, $\rho_1 > 0$ such that if $\rho > \rho_1$, then $v(x,s) > \beta$ in D for $0 < s < \min(\tau, S(\rho))$. Let V be a solution of the Cauchy-Dirichlet problem (2.3) with $\Omega = D$ and put

$$\tilde{v}(x,s) := q(s; V(x,s)).$$

We denote by $\tilde{S}(\rho)$ the life span of $\tilde{v}$. The function $\tilde{v}(x,s)$ is bounded at (x_0, s_0) if and only if $F(\rho\tilde{V}(x_0,s_0)) > F(\rho M)s_0$. This implies

$$\tilde{S}(\rho) = \sup\{s > 0;\ F(\rho M)^{-1}F(\rho \parallel \tilde{V}(\cdot,\sigma) \parallel_{L^\infty(D)}) > \sigma \text{ for } \sigma \in (0,s)\}.$$

Since we have

$$q_\alpha(s;\alpha) := \frac{\partial q}{\partial \alpha}(s;\alpha) = f(F^{-1}(F(\rho\alpha) - F(\rho M)s))f(\rho\alpha)^{-1} > 0$$

and

$$\begin{aligned} q_{\alpha\alpha}(s;\alpha) = \rho f(F^{-1}(F(\rho\alpha) - F(\rho M)s))f(\rho\alpha)^{-2} \\ \cdot(f'(F^{-1}(F(\rho\alpha) - F(\rho M)s)) - f'(\rho\alpha)) \geq 0, \end{aligned}$$

we see

$$
\begin{aligned}
&\tilde{v}_s(x,s) - F(\rho M)\Delta\tilde{v}(x,s) - \rho^{-1}F(\rho M)f(\rho\tilde{v}(x,s)) \\
&= q_s(s;\tilde{V}(x,s)) + q_\alpha(s;\tilde{V}(x,s))\tilde{V}_s(x,s) - q_{\alpha\alpha}(s;\tilde{V}(x,s))F(\rho M)|\nabla\tilde{V}(x,s)|^2 \\
&\quad - q_\alpha(s;\tilde{V}(x,s))F(\rho M)\Delta\tilde{V}(x,s) - \rho^{-1}F(\rho M)f(\rho q(s;\tilde{V}(x,s))) \leq 0.
\end{aligned}
$$

Moreover, since we have

$$\tilde{v}(x,s) \to 0 \; as \; \rho \to \infty \text{ uniformly } (x,s) \in \partial D \times (0,\tau),$$

if $\rho > 0$ is sufficiently large, we derive

$$\tilde{v}(x,s) \leq \beta \text{ on } \partial D \times (0,\tau).$$

Hence $\tilde{v}$ is a subsolution of (2.1). By the comparison principle, we obtain $S(\rho) \leq \tilde{S}(\rho)$.

In order to estimate $\bar{S}(\rho)$ and $\tilde{S}(\rho)$, we estimate $\max_{\Omega} v(x,s_0)$ and $\| V(\cdot,\sigma) \|_{L^\infty(D)}$ respectively by using Lemma 2.1 and use the assumptions (A) and (B). This prove the theorems.

References

[1] A. Friedman and A. Lacey, *The blow-up time of solutions of nonlinear heat equations with small diffusion*, SIAM J. Math. Anal. **18** (1987) 711–721.

[2] H. Fujita, *On the blowing up of solutions of the Cauchy problem for* $u_t = \Delta u + u^{1+\alpha}$, J. Fac. Sci. Univ. Tokyo Sect. IA Math. **16** (1966) 105–113

[3] C. Gui and X. Wang, *Life span of solutions of the Cauchy problem for a semilinear heat equation*, J. Differential Equations, **115** (1995) 166–172.

[4] T.-Y. Lee and W.-M. Ni, *Global existence,large time behavior and life span of solutions of a semilinear parabolic Cauchy problem*, Trans. Amer. Math. Soc. **333** (1992) 1434–1446.

[5] N. Mizoguchi and E. Yanagida, *Life span of solutions with large initial data in a semilinear prabolic equation*, Indiana Univ. Math. J. **50** (2001) no. 1, 591–610.

[6] N. Mizoguchi and E. Yanagida, *Life span of solutions for a semilinear parabolic problem with small diffusion*, J. Math. Anal. Appl. **261** (2001) 350–368.

OPTIMAL CONTROL PROBLEMS OF QUASILINEAR ELLIPTIC-PARABOLIC EQUATION

NORIAKI YAMAZAKI

Department of Mathematical Science, Common Subject Division
Muroran Institute of Technology
27-1 Mizumoto-chō, Muroran, 050-8585, Japan

Abstract: In this paper we study optimal control problems of a quasilinear elliptic-parabolic equation with a Signorini-Dirichlet-Neumann type mixed boundary condition. We prove the existence of an optimal control by applying the abstract theory. Also, we consider an approximating problem of our original system. Then, we prove the relationship between the original control problem and its approximating one. Moreover, we give a necessary condition of the approximating control problem under some smoothness assumptions.

1. Introduction

We consider a quasilinear elliptic-parabolic equation with a Signorini-Dirichlet-Neumann type mixed boundary condition as follows:

Problem (P) :

$$
\begin{aligned}
& b(u)_t - \nabla \cdot a(x, b(u), \nabla u) = f(t, x) && \text{in } (0,T) \times \Omega, \quad (1.1)\\
& u \le g(t), \qquad \nu \cdot a(x, b(u), \nabla u) \le 0 && \\
\text{and} \quad & (u - g(t))\nu \cdot a(x, b(u), \nabla u) = 0 && \text{on } (0,T) \times \Gamma_S,\\
& u = g(t) && \text{on } (0,T) \times \Gamma_D,\\
& \nu \cdot a(x, b(u), \nabla u) = 0 && \text{on } (0,T) \times \Gamma_N,\\
& b(u(0,\cdot)) = b(u_0) && \text{in } \Omega.
\end{aligned}
$$

Here, Ω is a bounded domain in $\mathbb{R}^N$ $(N \ge 1)$ with a smooth boundary $\Gamma := \partial\Omega$ having disjoint three parts Γ_j $(j = S, D, N)$, and $\nu = (\nu_1, \cdots, \nu_N)$ is the outward normal vector on the boundary Γ. The given function $b : \mathbb{R} \to \mathbb{R}$ is bounded, nondecreasing and Lipschitz continuous, and the term $a(x, s, p)$ is a quasilinear elliptic vector field satisfying some structure condition, in particular we assume $a(x, s, p) = \partial_p A(x, s, p)$ for a potential function A :

$\Omega \times \mathbb{R} \times \mathbb{R}^N \to \mathbb{R}$. Also, $f(t,x)$ is a given function on $(0,T) \times \Omega$, and $g(t,x)$ is a given function on $(0,T) \times \Gamma$.

Note that the equation (1.1) is called an elliptic-parabolic equation, since it is elliptic in the region $\{b'(u) = 0\}$ and parabolic in $\{b'(u) > 0\}$, respectively.

Problem (P) is the model of flows in partially saturated porous media. In fact, Γ_S, Γ_D and Γ_N refer to the parts of the boundary in contact with the atmosphere, reservoirs and impervious layer, respectively. So, the function g is the pressure in the reservoirs on Γ_D, and is the atmospheric pressure on Γ_S.

Elliptic-parabolic problems have been studied by many mathematicians. For instance, we refer to [1, 2, 3, 6, 7, 8, 10, 11, 12, 13, 14]. In particular, Alt-Luckhaus [1] studied the problem (P) in the case where the Signorini part is empty (i.e. $\Gamma_S = \emptyset$). Note that they [1] introduced a class of weak solutions, without any strong time-derivative of $b(u)$ in $L^1(\Omega)$. Later, Otto [13] proved the uniqueness of the weak solution of [1]. Also, various aspects of weak solutions have been studied, for instance in [2, 3, 7]. For studies of strong solutions, we refer to [6, 10, 11, 12].

The aim of the present paper is to investigate the optimal control problem (OP) for (P) as follows:

Problem (OP) : Find the optimal control $f^* \in F$ such that

$$J(f^*) = \inf_{f \in F} J(f).$$

Here, $F := \{ f \in L^2(0,T; H^1(\Omega)) \ ; \ f_t \in L^2(0,T; L^2(\Omega)) \}$ is a control space, and $J(f)$ is a cost functional defined by

$$J(f) := \frac{1}{2}\int_0^T |b(u) - b_d|^2_{L^2(\Omega)} dt + \frac{1}{2}\int_0^T |f|^2_{H^1(\Omega)} dt + \frac{1}{2}\int_0^T |f_t|^2_{L^2(\Omega)} dt, \tag{1.2}$$

where $f \in F$ is the control, u is a unique solution to the state problem (P), and b_d is a given target profile in $L^2(0,T; L^2(\Omega))$.

The Signorini-Dirichlet-Neumann type mixed boundary condition in (P) is very complicated. So, it is difficult to analyze the problem (P) numerically.

Now, from the view-point of numerical analysis, we consider the approximating problems of (P) and (OP). In fact, we employ an approximation of the boundary condition by the penalty method. Namely, for each $\lambda \in (0, 1]$ we study an approximating problem of (P) as follows:

Problem (P)$_\lambda$:

$$b(u_\lambda)_t - \nabla \cdot a(x, b(u_\lambda), \nabla u_\lambda) = f(t,x) \qquad \text{in } (0,T) \times \Omega,$$

$$\nu \cdot a(x, b(u_\lambda), \nabla u_\lambda) = -\frac{u_\lambda - g(t)}{\lambda} \cdot \chi_D - \frac{[u_\lambda - g(t)]^+}{\lambda} \cdot \chi_S$$

$$\text{on } (0,T) \times \Gamma,$$

$$b(u_\lambda(0,\cdot)) = b(u_0) \qquad \text{in } \Omega,$$

where χ_j is the characteristic function of $(0,T) \times \Gamma_j$ $(j = D, S)$, and $[z]^+$ is the positive part of z.

Next, for each $\lambda \in (0,1]$ let us consider an approximating control problem of (OP) as follows:

Problem (OP)$_\lambda$: Find the optimal control $f_\lambda^* \in F$ such that

$$J_\lambda(f_\lambda^*) = \inf_{f \in F} J_\lambda(f).$$

Here, $J_\lambda(f)$ is a cost functional defined by

$$J_\lambda(f) := \frac{1}{2}\int_0^T |b(u_\lambda) - b_d|^2_{L^2(\Omega)} dt + \frac{1}{2}\int_0^T |f|^2_{H^1(\Omega)} dt + \frac{1}{2}\int_0^T |f_t|^2_{L^2(\Omega)} dt, \tag{1.3}$$

where $f \in F$ is the control, u_λ is a unique solution to the state approximating problem (P)$_\lambda$, and b_d is a given target profile in $L^2(0,T; L^2(\Omega))$.

The main object of this paper is to show the relationship between the original control problem (OP) and its approximating one (OP)$_\lambda$. Moreover, we give a necessary condition of an approximating control problem (OP)$_\lambda$ under some smoothness assumptions.

The plan of this paper is as follows. In the next Section 2, we state the main result (Theorem 2.1) concerning the relationship between (OP) and (OP)$_\lambda$. In Section 3, we consider the state problem (P) and the optimal control problem (OP) by applying the abstract theory obtained in [6, 12]. In Section 4, we study the approximating problems (P)$_\lambda$ and (OP)$_\lambda$, and show the relationship between (P) and (P)$_\lambda$. Also, we prove Theorem 2.1. In Section 5, we consider the special form $a(x,s,p) = p + k(s)$, where $k : \mathbb{R} \to \mathbb{R}^N$ is Lipschitz continuous. Then, we give a necessary condition of (OP)$_\lambda$ under some smoothness assumptions.

Notation: Throughout this paper, we put $H := L^2(\Omega)$ with usual real Hilbert space structure. The inner product and norm in H are denoted by $(\cdot,\cdot)$ and by $|\cdot|_H$, respectively. We also put $V := H^1(\Omega)$ with the usual norm $|u|_V := \left(|u|_H^2 + |\nabla u|_H^2\right)^{\frac{1}{2}}$, and denote by V' the dual space of V. Furthermore, we denote by $\langle \cdot, \cdot \rangle$ the duality pairing between V' and V.

Various L^∞-norms, e.g., norms in $L^\infty(\Omega), L^\infty(\mathbb{R})$, etc., are all denoted by the same symbol $|\cdot|_\infty$.

2. Assumptions and main result

At first we list some assumptions on data as follows:

(A1) $a(x,s,p) = \partial_p A(x,s,p)$ for some potential function $A(x,s,p)$. There exist constants $\mu > 0$, $C_1 > 0$ and $C_2 > 0$ such that

$$[a(x,s,p) - a(x,s,\hat{p})] \cdot (p - \hat{p}) \geq \mu |p - \hat{p}|^2,$$

$$|a(x,s,p)|^2 + |A(x,s,p)| + |\partial_s A(x,s,p)|^2 \leq C_1(1 + |s|^2 + |p|^2),$$

$$|a(x,s,p) - a(x,\hat{s},p)| \leq C_2(1 + |p|)|s - \hat{s}|$$

for all $x \in \Omega$, s, $\hat{s} \in \mathbb{R}$, p, $\hat{p} \in \mathbb{R}^N$. Moreover, $a(\cdot,\cdot,\cdot)$ and $A(\cdot,\cdot,\cdot)$ satisfy the Carathéodory condition.

(A2) $b : \mathbb{R} \to \mathbb{R}$ is bounded, nondecreasing and Lipschitz continuous.

(A3) $g \in W^{1,2}(0,T;V)$.

(A4) The boundary Γ of Ω is smooth, and admits a mutually disjoint decomposition such as $\Gamma = \Gamma_S \cup \Gamma_D \cup \Gamma_N$, where Γ_S, Γ_D and Γ_N are measurable subsets of Γ, and Γ_D has positive surface measure.

Now, we give the weak formulation of (P) in the variational sense. To do so, we define a non-empty, closed and convex subset $K(t)$ of V for all $t \in [0,T]$ by

$$K(t) := \{z \in V \ ; \ z \leq g(t) \text{ on } \Gamma_S \text{ and } z = g(t) \text{ on } \Gamma_D\}. \tag{2.1}$$

Definition 2.1. Let $f \in L^2(0,T;H)$ and $u_0 \in H$. Then, a function $u : [0,T] \to V$ is a solution of (P), or (P;f) when the forcing term f is specified, if the following items (a)–(d) are satisfied.

(a) $u \in L^\infty(0,T;V)$ and $b(u) \in W^{1,2}(0,T;H)$.

(b) $u(t) \in K(t)$ for a.e. $t \in (0,T)$.

(c) For a.e. $t \in (0,T)$ the following inequality holds:

$$(b(u)_t, u - v) + \int_\Omega a(x, b(u), \nabla u) \cdot \nabla(u - v)dx \leq (f, u - v)$$

for all $v \in K(t)$.

(d) $b(u(0)) = b(u_0)$ in H.

Kubo-Yamazaki [12] and Hoffmann-Kubo-Yamazaki [6] established the general theory of the Cauchy problem and the optimal control problem for quasilinear elliptic-parabolic variational inequalities with time-dependent constraints. So, by applying the abstract results obtained in [6, 12], we get some results for (P) and (OP) as follows:

Proposition 2.1. *(cf. [6,12]) Assume (A1)–(A4) and* $u_0 \in K(0)$.
(i) Let $f \in W^{1,2}(0,T;H)$*. Then, there is a unique solution* u *of (P;f).*
(ii) Let b_d *be an element in* $L^2(0,T;H)$*. Then, the problem (OP) has at least one optimal control* $f^* \in F$ *such that* $J(f^*) = \inf_{f\in F} J(f)$.

In Section 3, we give the sketch of the proof of Proposition 2.1.

Now, we state the main theorem in this paper, which is concerned with the relationship between (OP) and $(\mathrm{OP})_\lambda$.

Theorem 2.1. *(Relationship between (OP) and* $(OP)_\lambda$*) Assume (A1)–(A4) are satisfied. Let* $\lambda \in (0,1]$, $u_0 \in K(0)$*, and* b_d *be an element in* $L^2(0,T;H)$*. Then, the approximating control problem* $(OP)_\lambda$ *has at least one optimal control* $f_\lambda^* \in F$ *such that* $J_\lambda(f_\lambda^*) = \inf_{f\in F} J_\lambda(f)$.

Moreover, there is a subsequence $\{\lambda_k\} \subset \{\lambda\}$ *and a function* $f^* \in F$ *such that* f^* *is the optimal control of (OP), and* $\lambda_k \to 0$,

$$f_{\lambda_k}^* \rightharpoonup f^* \qquad \text{weakly in } L^2(0,T;V) \qquad \text{as } k \to \infty, \tag{2.2}$$

$$\frac{d}{dt} f_{\lambda_k}^* \rightharpoonup \frac{d}{dt} f^* \qquad \text{weakly in } L^2(0,T;H) \qquad \text{as } k \to \infty. \tag{2.3}$$

3. Original problems (P) and (OP)

In this section we consider the state problem (P) and the optimal control problem (OP) by applying abstract results in [6, 12]. In fact, the following lemmas are key ones to showing Proposition 2.1.

Lemma 3.1. *((K1) of [6,12]) Assume (A1)–(A4). Put*

$$\alpha(t) = C_3 \int_0^t |g'(\tau)|_V d\tau,$$

where the constant $C_3 > 0$ *is dependent only on* $|\Omega|$, $|b|_\infty$, $|g|_{L^\infty(0,T;V)}$ *and* C_1*. Then, we have the following property* $(*)$*:*
$(*)$*: For any* $0 \le s < t \le T$, $w \in V$ *with* $|w(x)| \le |b|_\infty$ *a.e. in* Ω *and* $z \in K(s)$*, there exists* $\tilde{z} \in K(t)$ *such that*

$$|\tilde{z} - z|_H \le |\alpha(t) - \alpha(s)|,$$

$$\int_\Omega A(x, w, \nabla \tilde{z})dx - \int_\Omega A(x, w, \nabla z)dx \le |\alpha(t) - \alpha(s)| \left(1 + |z|_V\right).$$

Proof. Put $\tilde{z} := z - g(s) + g(t)$. Then, we see that $\tilde{z} \in K(t)$ if $z \in K(s)$. Moreover, we observe from [12, Section 5.1] that $(*)$ holds. □

Also, we have the other properties of $K(t) (\subset V)$ as follows.

Lemma 3.2. *((K2)–(K4) of* [6, 12]*) Assume (A1)–(A4). Then, we have:*
(k1) There is a constant $C_4 > 0$ *such that*

$$|z|_V \le C_4(1 + |\nabla z|_H) \quad \text{for all } z \in K(t) \text{ and } t \in [0, T].$$

(k2) For any $z, \overline{z} \in K(t)$ *and* $w, \overline{w} \in V$ *with* $w \le z$, $\overline{z} \le \overline{w}$, *we have*

$$\max\{w, \overline{z}\} \in K(t), \quad \min\{z, \overline{w}\} \in K(t).$$

(k3) If $z, \overline{z} \in K(t)$ *and* $\nabla[z - \overline{z}]^+ \equiv 0$, *then* $z \le \overline{z}$, *where* $[z]^+ := \max\{z, 0\}$.

Proof. We can prove (k1) by (A4) and Poincaré inequality. Also, we can easily show the assertions (k2) and (k3) by the definition (2.1) of $K(t)$. □

Proof. [**Proof of Proposition 2.1**] Kubo-Yamazaki [12] and Hoffmann-Kubo-Yamazaki [6] considered the Cauchy problem and the optimal control problem of elliptic-parabolic variational inequalities with the more general time-dependent constraints. By Lemmas 3.1 and 3.2, we can apply the results obtained in [6, 12] to (P) and (OP), respectively. Thus, we get Proposition 2.1. For the detailed proof, see [6, 12]. □

4. Approximating problems $(\mathrm{P})_\lambda$ and $(\mathrm{OP})_\lambda$

In this section, for each $\lambda \in (0, 1]$ we consider approximating problems $(\mathrm{P})_\lambda$ and $(\mathrm{OP})_\lambda$ of (P) and (OP), respectively.

We begin by defining the solution of the approximating problem $(\mathrm{P})_\lambda$.

Definition 4.1. Let $f \in L^2(0, T; H)$, $u_0 \in H$ and $\lambda \in (0, 1]$. Then, a function $u_\lambda : [0, T] \to V$ is a solution of $(\mathrm{P})_\lambda$, or $(\mathrm{P};f)_\lambda$ when the forcing term f is specified, if the following items (a)–(c) are satisfied.

(a) $u_\lambda \in L^\infty(0, T; V)$ and $b(u_\lambda) \in W^{1,2}(0, T; H)$.
(b) For a.e. $t \in (0, T)$ the following variational identity holds:

$$(b(u_\lambda)_t, z) + \int_\Omega a(x, b(u_\lambda), \nabla u_\lambda) \cdot \nabla z dx + \int_{\Gamma_D} \frac{u_\lambda - g(t)}{\lambda} z d\Gamma$$
$$+ \int_{\Gamma_S} \frac{[u_\lambda - g(t)]^+}{\lambda} z d\Gamma = (f, z) \qquad \text{for all } z \in V.$$

(c) $b(u_\lambda(0)) = b(u_0)$ in H.

Now, for each $\lambda \in (0,1]$ we consider the problem $(\mathrm{P})_\lambda$. By the similar argument in Proposition 2.1 (cf. [12]), we easily get the existence-uniqueness of solutions to $(\mathrm{P})_\lambda$ as follows.

Proposition 4.1. *(cf. [12]) Suppose (A1)–(A4), $f \in W^{1,2}(0,T;H)$, $\lambda \in (0,1]$ and $u_0 \in K(0)$. Then, there is a unique solution u_λ of $(P;f)_\lambda$ such that the following boundedness holds:*

$$\begin{aligned} &|u_\lambda|^2_{L^\infty(0,T;V)} + |b(u_\lambda)|^2_{L^\infty(0,T;V)} + \int_0^T |b(u_\lambda)_t(t)|^2_H dt \\ &\le N_1 \left(|u_0|^2_V + |f|^2_{W^{1,2}(0,T;H)} + 1\right) \end{aligned} \tag{4.1}$$

for some constant $N_1 > 0$ independent of u_0 and λ.

The following Lemma is the key one to proving Proposition 4.1.

Lemma 4.1. *((K1) of [6, 12]) Suppose (A1)–(A4), and $\lambda \in (0,1]$. Put*

$$\beta(t) = C_5 \int_0^t |g'(\tau)|_V d\tau,$$

where the constant $C_5 > 0$ is dependent only on $|\Omega|$, $|b|_\infty$, $|g|_{L^\infty(0,T;V)}$ and C_1. Then, we have the following property (**):
(**): For any $0 \le s < t \le T$, $w \in V$ with $|w(x)| \le |b|_\infty$ a.e. in Ω and $z \in V$, there is $\tilde{z} \in V$ such that

$$|\tilde{z} - z|_H \le |\beta(t) - \beta(s)|,$$

$$\begin{aligned} &\int_\Omega A(x,w,\nabla\tilde{z})dx + \int_{\Gamma_D} \frac{|\tilde{z} - g(t)|^2}{2\lambda} d\Gamma + \int_{\Gamma_S} \frac{([\tilde{z} - g(t)]^+)^2}{2\lambda} d\Gamma \\ &\quad - \int_\Omega A(x,w,\nabla z)dx - \int_{\Gamma_D} \frac{|z - g(s)|^2}{2\lambda} d\Gamma - \int_{\Gamma_S} \frac{([z - g(s)]^+)^2}{2\lambda} d\Gamma \\ &\le |\beta(t) - \beta(s)| \left(1 + |z|_V\right). \end{aligned}$$

Proof. Put $\tilde{z} := z - g(s) + g(t)$. Then, we have $\tilde{z} \in V$ if $z \in V$. Moreover, from the same calculation of [12, Section 5.1], we see that (**) holds. □

By Lemma 4.1, we can prove Proposition 4.1 by the similar method in [12]. Moreover, by taking account of the boundedness (4.1), we easily show the relationship between (P) and $(\mathrm{P})_\lambda$ as follows.

Proposition 4.2. *Suppose (A1)–(A4), $\lambda \in (0,1]$ and $u_0 \in K(0)$. Also, assume $\{f_n\} \subset W^{1,2}(0,T;H)$, $f \in W^{1,2}(0,T;H)$,*

$$\{f_n\} \text{ is bounded in } W^{1,2}(0,T;H),$$
$$f_n \to f \text{ strongly in } L^2(0,T;H) \quad \text{as } n \to \infty.$$

Let $u_{\lambda,n}$ be a unique solution of $(P)_\lambda$ with data $\{b(u_0), f_n\}$. Then, there are subsequences $\{n_k\} \subset \{n\}$, $\{\lambda_k\} \subset \{\lambda\}$ and a function $u \in L^\infty(0,T;V)$ such that u is the unique solution of (P) with data $\{b(u_0), f\}$, and $n_k \to \infty$, $\lambda_k \to 0$ (as $k \to \infty$),

$$\begin{aligned} u_{\lambda_k,n_k} &\rightharpoonup u \quad \text{weakly-* in } L^\infty(0,T;V) && \text{as } k \to \infty, \\ b(u_{\lambda_k,n_k}) &\to b(u) \quad \text{strongly in } C([0,T];H) && \text{as } k \to \infty. \end{aligned}$$

Now, we prove Theorem 2.1 by using the convergence result as above.

Proof. **[Proof of Theorem 2.1]** Note that by the same argument in (ii) of Proposition 2.1, we easily show the existence of optimal control $f_\lambda^* \in F$ of $(\mathrm{OP})_\lambda$ so that $J_\lambda(f_\lambda^*) = \inf_{f \in F} J_\lambda(f)$, so, omit the detailed proof.

Now, we show (2.2)–(2.3). To do so, for any $f \in F$ let u_λ be the solution of $(\mathrm{P})_\lambda$ with data $\{b(u_0), f\}$. Also, let u be the solution of (P) with data $\{b(u_0), f\}$. Then, by Proposition 4.2 there is a subsequence $\{\lambda_k\} \subset \{\lambda\}$ such that $\lambda_k \to 0$,

$$\begin{aligned} u_{\lambda_k} &\rightharpoonup u \quad \text{weakly-* in } L^\infty(0,T;V) && \text{as } k \to \infty, \\ b(u_{\lambda_k}) &\to b(u) \quad \text{strongly in } C([0,T];H) && \text{as } k \to \infty. \end{aligned} \tag{4.2}$$

Since $f_{\lambda_k}^* \in F$ is the optimal control of $(\mathrm{OP})_{\lambda_k}$, we have

$$J_{\lambda_k}(f_{\lambda_k}^*) \le J_{\lambda_k}(f) \tag{4.3}$$
$$\left(= \frac{1}{2}\int_0^T |b(u_{\lambda_k}) - b_d|_H^2 dt + \frac{1}{2}\int_0^T |f|_V^2 dt + \frac{1}{2}\int_0^T |f_t|_H^2 dt \right).$$

Clearly, we infer from (4.2)–(4.3) that $\{f_{\lambda_k}^* \in F \,;\, \lambda_k \in (0,1]\}$ is bounded in F. Hence, there is a subsequence of $\{\lambda_k\}$, denoted by $\{\lambda_k\}$ again, and a function $f^* \in F$ such that $\lambda_k \to 0$,

$$f_{\lambda_k}^* \rightharpoonup f^* \quad \text{weakly in } L^2(0,T;V) \qquad \text{as } k \to \infty, \tag{4.4}$$

$$\frac{d}{dt} f_{\lambda_k}^* \rightharpoonup \frac{d}{dt} f^* \quad \text{weakly in } L^2(0,T;H) \qquad \text{as } k \to \infty. \tag{4.5}$$

Here, taking account of the Aubin's compactness, we may assume that

$$f_{\lambda_k}^* \longrightarrow f^* \quad \text{strongly in } L^2(0,T;H) \qquad \text{as } k \to \infty. \tag{4.6}$$

Now we denote by $u^*_{\lambda_k}$ the solution of $(\mathrm{P})_{\lambda_k}$ with data $\{b(u_0), f^*_{\lambda_k}\}$. Then, by Proposition 4.2, we can find the solution u^* of (P) with data $\{b(u_0), f^*\}$ such that

$$\begin{aligned} u^*_{\lambda_k} &\rightharpoonup u^* \quad \text{weakly-}* \text{ in } L^\infty(0,T;V) && \text{as } k \to \infty, \\ b(u^*_{\lambda_k}) &\to b(u^*) \quad \text{strongly in } C([0,T];H) && \text{as } k \to \infty. \end{aligned} \tag{4.7}$$

Therefore, we observe from (4.2)–(4.7) that

$$J(f^*) \le \liminf_{k\to\infty} J_{\lambda_k}(f^*_{\lambda_k}) \le J(f). \tag{4.8}$$

Since f is the arbitrary function in F, we infer from (4.8) that $f^* \in F$ is the optimal control of (OP). Thus, the proof has been completed. □

5. Necessary condition of optimal control for $(\mathrm{OP})_\lambda$

In this section, we give a necessary condition of the approximating control problem $(\mathrm{OP})_\lambda$ under some smoothness assumptions. For simplicity, we consider the following state problem:

Problem $(\mathrm{SP})_\lambda$:

$$b(u_\lambda)_t - \nabla \cdot [\nabla u_\lambda + k(b(u_\lambda))] = f(t,x) \qquad \text{in } (0,T)\times\Omega,$$

$$\nu \cdot [\nabla u_\lambda + k(b(u_\lambda))] = -\frac{u_\lambda - g(t)}{\lambda}\cdot \chi_D - \frac{h(u_\lambda - g(t))}{\lambda}\cdot \chi_S \qquad \text{on } (0,T)\times\Gamma,$$

$$b(u_\lambda(0,\cdot)) = b(u_0) \qquad \text{in } \Omega,$$

where $k : \mathbb{R} \to \mathbb{R}^N$ is a Lipschitz continuous function, and h is a non-negative continuous function on $\mathbb{R}$ defined by

$$h(r) := \begin{cases} 0 & \text{if } -\infty < r < 0, \\ r^2 & \text{if } 0 \le r < \frac{1}{2}, \\ r - \frac{1}{4} & \text{if } \frac{1}{2} \le r < +\infty. \end{cases} \tag{5.1}$$

Note that $(\mathrm{SP})_\lambda$ can be considered as the approximating problem of (P) by replacing $a(x,s,p)$ and $[z]^+$ in $(\mathrm{P})_\lambda$ by $p + k(s)$ and $h(z)$, respectively. Thus, by the same arguments in Section 4 (cf. Propositions 4.1 and 4.2), we have the following results to $(\mathrm{SP})_\lambda$.

Proposition 5.1. *Assume (A2)–(A4), $k : \mathbb{R} \to \mathbb{R}^N$ is Lipschitz continuous, and h is the non-negative continuous function on $\mathbb{R}$ defined by (5.1). Let $u_0 \in K(0)$, $f \in W^{1,2}(0,T;H)$ and $\lambda \in (0,1]$. Then, there is a* *unique solution* $u_\lambda \in L^\infty(0,T;V)$ *to* $(SP)_\lambda$ *with data* $\{b(u_0), f\}$ *such that* $b(u_\lambda) \in W^{1,2}(0,T;H)$.

Moreover, let b_d *be an element in* $L^2(0,T;H)$. *Then, there exists at least one optimal solution* $f_\lambda^* \in F$ *such that*

$$J_\lambda(f_\lambda^*) = \inf_{f\in F} J_\lambda(f), \tag{5.2}$$

where $F = \{f \in L^2(0,T;V)\ ;\ f_t \in L^2(0,T;H)\}$ *and* $J_\lambda(\cdot)$ *is the cost functional defined in* (1.3).

Definition 5.1. (i) We denote by $\Lambda_\lambda : F \to L^2(0,T;H)$ a solution operator that assigns to any control $f \in F$ the unique state solution $u = \Lambda_\lambda(f)$ to $(SP)_\lambda$ with data $\{b(u_0), f\}$.
(ii) Let $f_\lambda^* \in F$ be the optimal solution to (5.2). Then, $(u_\lambda^*, f_\lambda^*) = (\Lambda_\lambda(f_\lambda^*), f_\lambda^*)$ is called the optimal pair to the problem (5.2).

By Proposition 5.1, we get the optimal control of (5.2) for $(SP)_\lambda$. But, it is very difficult to show the necessary condition of (5.2) because of b and k. So, we assume additional conditions for b and k as follows:

- (A5) There is a positive constant $C_6 > 0$ such that $b'(r) \geq C_6 > 0$ for any $r \in \mathbb{R}$. Moreover, $b'(r)$ is smooth and Lipschitz continuous in r on $\mathbb{R}$.
- (A6) There is a positive constant $C_7 > 0$ such that $|k_i'(r)| \leq C_7$ for any $r \in \mathbb{R}$ and $i = 1, 2, \cdots, N$, where $k(r) = (k_1(r),\ k_2(r),\ \cdots,\ k_N(r))$. Moreover, $k_i'(r)$ is smooth and Lipschitz continuous in r on $\mathbb{R}$ for $i = 1, 2, \cdots, N$.

By the quite standard method (cf. [8]), we can get the necessary condition of the optimal pair $(u_\lambda^*, f_\lambda^*) = (\Lambda_\lambda(f_\lambda^*), f_\lambda^*)$ as follows:

Proposition 5.2. *Assume (A5)–(A6) and the same conditions in Proposition 5.1. For each* $\lambda \in (0,1]$, *let* $f_\lambda^* \in F$ *be the optimal control for* (5.2). *Let* u_λ^* *be a unique solution of* $(SP)_\lambda$ *with the source term* f_λ^*. *Then, there is a unique solution* $p_\lambda \in W^{1,2}(0,T;H) \cap L^\infty(0,T;V)$ *to the adjoint equation:*

$$-b'(u_\lambda^*)(p_\lambda)_t - \Delta p_\lambda + \left[\frac{\partial k}{\partial u_\lambda^*}(b(u_\lambda^*))\right] \cdot \nabla p_\lambda = (b(u_\lambda^*) - b_d)b'(u_\lambda^*) \quad \textit{in } (0,T)\times\Omega, \tag{5.3}$$

$$\frac{\partial p_\lambda}{\partial \nu} = -\frac{1}{\lambda}\chi_D \cdot p_\lambda - \frac{1}{\lambda}\chi_S \cdot h'(u_\lambda^* - g(t))p_\lambda \quad \textit{on } (0,T)\times\Gamma, \tag{5.4}$$

$$p_\lambda(T,\cdot) = 0 \quad \textit{in } \Omega. \tag{5.5}$$

Moreover, $f_\lambda^* \in F$ satisfies the following variational identity:

$$\int_0^T (p_\lambda + f_\lambda^*, \zeta)dt + \int_0^T (\nabla f_\lambda^*, \nabla \zeta)dt + \int_0^T ((f_\lambda^*)_t, \zeta_t)dt = 0 \tag{5.6}$$

for any $\zeta \in F$.

Remark 5.1. We infer from (5.6) that $f_\lambda^* \in F$ is the weak solution of

$$-(f_\lambda^*)_{tt} - \Delta f_\lambda^* + f_\lambda^* + p_\lambda = 0 \qquad \text{in } (0,T)\times\Omega$$

(in a distribution sense),

$$(f_\lambda^*)_t(T,\cdot) = 0 \qquad \text{in } \Omega.$$

In order to prove Proposition 5.2, we prepare some lemmas.

Let $\xi \in (0,1]$. Then, for any $f \in F$ and $\tilde{f} \in F$, we put $u_\xi := \Lambda_\lambda(f + \xi\tilde{f})$, $u := \Lambda_\lambda(f)$ and $v_\xi := \frac{u_\xi - u}{\xi}$. Note that v_ξ satisfies the following equation:

$$\left(\overline{b}_\xi(t,x)v_\xi\right)_t - \sum_{i=1}^N \frac{\partial}{\partial x_i}\left(\frac{\partial v_\xi}{\partial x_i} + \overline{k}_{\xi,i}(t,x)v_\xi\right) = \tilde{f} \quad \text{in } (0,T)\times\Omega, \tag{5.7}$$

$$\sum_{i=1}^N \nu_i\left(\frac{\partial v_\xi}{\partial x_i} + \overline{k}_{\xi,i}(t,x)v_\xi\right) = -\frac{1}{\lambda}\chi_D \cdot v_\xi - \frac{1}{\lambda}\chi_S \cdot \overline{h}_\xi(t,x)v_\xi \tag{5.8}$$

$$\text{on } (0,T)\times\Gamma, \tag{5.9}$$

$$\overline{b}_\xi(0)v_\xi(0) = 0 \qquad \text{in } \Omega, \tag{5.10}$$

where $\nu = (\nu_1, \cdots, \nu_N)$ is the outward normal vector on Γ and

$$\overline{b}_\xi(t,x) = \int_0^1 b'(u(t,x) + s(u_\xi(t,x) - u(t,x)))ds,$$

$$\overline{k}_{\xi,i}(t,x) = \int_0^1 \frac{\partial}{\partial u}k_i(b(u(t,x) + s(u_\xi(t,x) - u(t,x))))ds, \quad (i = 1,\cdots,N),$$

$$\overline{h}_\xi(t,x) = \int_0^1 h'(u(t,x) - g(t,x) + s(u_\xi(t,x) - u(t,x)))ds.$$

Here, we give the uniform estimate of v_ξ with respect to $\xi > 0$. Such a calculation is standard one, so, we omit the detailed proof.

Lemma 5.1. *Suppose the same conditions in Proposition 5.2. Then, there* *is a positive number $N_2 > 0$ independent of $\xi \in (0,1]$ such that*

$$\begin{aligned}
&\sup_{t\in[0,T]} |v_\xi(t)|_H^2 + \int_0^T |\nabla v_\xi(t)|_H^2 dt + \int_0^T |\nabla(\bar{b}_\xi v_\xi)(t)|_H^2 dt \\
&\quad + \frac{1}{\lambda}\int_0^T |v_\xi(t)|_{L^2(\Gamma_D)}^2 dt + \frac{1}{\lambda}\int_0^T \int_{\Gamma_S} \overline{h}_\xi(t,x)|v_\xi(t,x)|^2 d\Gamma dt \\
&\quad + \int_0^T |(v_\xi)_t(t)|_{V'}^2 dt + \int_0^T |(\bar{b}_\xi v_\xi)_t(t)|_{V'}^2 dt \\
&\le N_2 |\tilde{f}|_{L^2(0,T;H)}^2 \qquad \text{for any } f \in F \text{ and } \tilde{f} \in F. \qquad (5.11)
\end{aligned}$$

Now, we mention the result of the differentiability of Λ_λ and J_λ.

Lemma 5.2. *Assume the same conditions in Proposition 5.2. Then, Λ_λ and J_λ are differentiable at any $f \in F$ and any direction $\tilde{f} \in F$. More precisely, for any $f \in F$ and $\tilde{f} \in F$, there is a function $\eta \in L^2(0,T;H)$ such that*

$$\left.\frac{d}{d\xi}\Lambda_\lambda(f+\xi\tilde{f})\right|_{\xi=0} := \lim_{\xi\to 0}\frac{\Lambda_\lambda(f+\xi\tilde{f}) - \Lambda_\lambda(f)}{\xi} = \eta, \qquad (5.12)$$

$$\begin{aligned}
\left.\frac{d}{d\xi}J_\lambda(f+\xi\tilde{f})\right|_{\xi=0} &:= \lim_{\xi\to 0}\frac{J_\lambda(f+\xi\tilde{f}) - J_\lambda(f)}{\xi} \\
&= \int_0^T (b(u)-b_d, b'(u)\eta)dt + \int_0^T (f,\tilde{f})dt \\
&\quad + \int_0^T (\nabla f, \nabla \tilde{f})dt + \int_0^T (f_t, \tilde{f}_t)dt, \qquad (5.13)
\end{aligned}$$

where $u = \Lambda_\lambda(f)$.

Moreover, the function η fulfills the following:

$$\begin{aligned}
&\langle (b'(u)\eta)_t, z\rangle + \int_\Omega \left(\nabla\eta + \frac{\partial}{\partial u}k(b(u))\eta\right)\cdot\nabla z dx + \frac{1}{\lambda}\int_{\Gamma_D}\eta z d\Gamma \\
&+\frac{1}{\lambda}\int_{\Gamma_S} h'(u-g(t))\eta z \, d\Gamma = (\tilde{f}, z), \quad \forall z \in V, \quad a.e.\ t\in(0,T), \qquad (5.14)
\end{aligned}$$

$$(b'(u)\eta)(0) = 0 \qquad \text{in } \Omega. \qquad (5.15)$$

Proof. At first, we show the differentiability of Λ_λ at any $f \in F$ and any direction $\tilde{f} \in F$. To do so, let $\xi \in (0,1]$. For any $f \in F$ and $\tilde{f} \in F$, we put $u_\xi := \Lambda_\lambda(f+\xi\tilde{f})$, $u := \Lambda_\lambda(f)$ and $v_\xi := \frac{u_\xi - u}{\xi}$. Then, by (5.11) and the Aubin's compactness, there is a subsequence $\{\xi_n\} \subset \{\xi\}$ and functions

$\eta, \tilde{\eta} \in L^2(0,T;V) \cap L^\infty(0,T;H)$ with $\eta_t, \tilde{\eta}_t \in L^2(0,T;V')$ such that $\xi_n \to 0$,

$$
\begin{aligned}
v_{\xi_n} &\rightharpoonup \eta \quad \text{weakly-}* \text{ in } L^\infty(0,T;H), \text{weakly in } L^2(0,T;V), && (5.16)\\
&\qquad\quad \text{strongly in } L^2(0,T;H), \text{ strongly in } C([0,T];V'), && (5.17)\\
(v_{\xi_n})_t &\rightharpoonup \eta_t \quad \text{weakly in } L^2(0,T;V'), && (5.18)\\
\overline{b}_{\xi_n} v_{\xi_n} &\rightharpoonup \tilde{\eta} \quad \text{weakly-}* \text{ in } L^\infty(0,T;H), \text{ weakly in } L^2(0,T;V), && (5.19)\\
&\qquad\quad \text{strongly in } L^2(0,T;H), \text{ strongly in } C([0,T];V'), && (5.20)\\
(\overline{b}_{\xi_n} v_{\xi_n})_t &\rightharpoonup \tilde{\eta}_t \quad \text{weakly in } L^2(0,T;V') && (5.21)
\end{aligned}
$$

as $n \to \infty$.

Now, we show that the limit η of v_{ξ_n} satisfies the relations (5.14)–(5.15). Since $\overline{b}_{\xi_n}$ is bounded in $L^2(0,T,H)$ (cf. (A2)), we observe that

$$\overline{b}_{\xi_n} \rightharpoonup \overline{b} \text{ weakly in } L^2(0,T;H) \quad \text{for some } \overline{b} \in L^2(0,T;H) \tag{5.22}$$

as $n \to \infty$. Note from (5.11) that

$$u_{\xi_n} \longrightarrow u \text{ strongly in } C([0,T];H) \text{ and strongly in } L^2(0,T;V) \tag{5.23}$$

as $n \to \infty$. So, from (A5), (5.23) and the definition of $\overline{b}_{\xi_n}$, we see that

$$\overline{b}_{\xi_n}(t,x) \longrightarrow b'(u)(t,x) \quad \text{a.e. } (t,x) \in (0,T) \times \Omega. \tag{5.24}$$

Thus, by (5.16)–(5.24), we have

$$\tilde{\eta} = \overline{b}\eta = b'(u)\eta \quad \text{a.e. on } (0,T) \times \Omega. \tag{5.25}$$

Similarly, we observe from (A6), (5.1) and (5.23) that

$$\overline{k}_{\xi_n,i}\, v_{\xi_n} \rightharpoonup \frac{\partial k_i(b(u))}{\partial u}\eta \quad \text{weakly in } L^2(0,T;H) \tag{5.26}$$

for $i = 1, 2, \cdots, N$, and

$$\overline{h}_{\xi_n} v_{\xi_n} \rightharpoonup h'(u-g)\eta \quad \text{weakly in } L^2(0,T;L^2(\Gamma_S)) \tag{5.27}$$

as $n \to \infty$.

Since v_{ξ_n} is the solution of (5.7)–(5.10), the following relations hold:

$$
\begin{gathered}
\langle (\overline{b}_{\xi_n}(t,x)v_{\xi_n})_t, z\rangle + \int_\Omega \sum_{i=1}^N \left(\frac{\partial v_{\xi_n}}{\partial x_i} + \overline{k}_{\xi_n,i}(t,x)v_{\xi_n} \right) \frac{\partial z}{\partial x_i} dx \\
+\frac{1}{\lambda}\int_{\Gamma_D} v_{\xi_n} z \, d\Gamma + \frac{1}{\lambda}\int_{\Gamma_S} \overline{h}_{\xi_n} v_{\xi_n} z \, d\Gamma = (\tilde{f}, z), \quad \forall z \in V, \quad \text{a.e. } t \in (0,T), \\
\overline{b}_{\xi_n}(0) v_{\xi_n}(0) = 0 \qquad \text{in } \Omega.
\end{gathered}
$$

By (5.16)–(5.27) and passing to the limit in the above relations as $n \to \infty$, we see that the limit function η of v_{ξ_n} satisfies (5.14)–(5.15). Also, we easily show the uniqueness of solutions of (5.14)–(5.15). Hence, v_ξ converges to η as $\xi \to 0$ in the sense of (5.16)–(5.18). Thus, we conclude that the solution operator Λ_λ is differentiable at any $f \in F$ and any direction $\tilde{f} \in F$.

By the differentiability of Λ_λ, we easily see that the cost functional J_λ is also differentiable, and the equation (5.13) holds for any $f \in F$ and any $\tilde{f} \in F$. Thus, the proof of Lemma 5.2 has been completed. □

By taking account of Lemma 5.2 as above, we can obtain the necessary condition of an optimal pair $(u_\lambda^*, f_\lambda^*) = (\Lambda_\lambda(f_\lambda^*), f_\lambda^*)$.

Proof. **[Proof of Proposition 5.2]** At first, we consider the adjoint equation (5.3)–(5.5). Taking account of (A2), (A5), (A6) and (5.1), we easily get a unique solution $p_\lambda \in W^{1,2}(0,T;H) \cap L^\infty(0,T;V)$ to (5.3)–(5.5) for each $\lambda \in (0,1]$. In fact, we define the time-dependent variable norm $|\cdot|_t$ by

$$|z|_t^2 := \int_\Omega b'(u_\lambda^*)(t,x)|z(x)|^2 dx \quad \text{for any } z \in H \text{ and } t \in [0,T].$$

Then, we observe from (A2) and (A5) that any variable norms $|\cdot|_t$ ($t \in [0,T]$) are uniformly equivalent on H. Moreover, there is a positive constant $C_8 > 0$ such that for any $0 \le s < t \le T$ and $z \in H$,

$$\big|\, |z|_t^2 - |z|_s^2 \,\big| \le C_8 |t-s| |z|_s^2.$$

Thus, by applying the abstract theory established by Damlamian [5], we get a unique solution $p_\lambda \in W^{1,2}(0,T;H) \cap L^\infty(0,T;V)$ to (5.3)–(5.5).

Now, let $(u_\lambda^*, f_\lambda^*) = (\Lambda_\lambda(f_\lambda^*), f_\lambda^*)$ be the optimal pair to the problem (5.2). We put $\eta_\lambda^* := \frac{d}{d\xi}\Lambda_\lambda(f_\lambda^* + \xi\zeta)\Big|_{\xi=0}$ for any $\zeta \in F$. Since f_λ^* is a minimizer for $J_\lambda(\cdot)$, then, we have

$$\begin{aligned}
0 &\le \liminf_{\xi \to +0} \frac{J_\lambda(f_\lambda^* + \xi\zeta) - J_\lambda(f_\lambda^*)}{\xi} \\
&= \int_0^T (b(u_\lambda^*) - b_d, b'(u_\lambda^*)\eta_\lambda^*)dt + \int_0^T (f_\lambda^*, \zeta)dt \\
&\qquad + \int_0^T (\nabla f_\lambda^*, \nabla\zeta)dt + \int_0^T ((f_\lambda^*)_t, \zeta_t)dt \\
&= \int_0^T (-b'(u_\lambda^*)(p_\lambda)_t, \eta_\lambda^*)dt - \int_0^T \langle (b'(u_\lambda^*)\eta_\lambda^*)_t, p_\lambda \rangle\, dt \\
&\qquad + \int_0^T (\zeta, p_\lambda)dt + \int_0^T (f_\lambda^*, \zeta)dt + \int_0^T (\nabla f_\lambda^*, \nabla\zeta)dt + \int_0^T ((f_\lambda^*)_t, \zeta_t)dt
\end{aligned}$$

$$= \int_0^T (p_\lambda + f_\lambda^*, \zeta)dt + \int_0^T (\nabla f_\lambda^*, \nabla \zeta)dt + \int_0^T ((f_\lambda^*)_t, \zeta_t)dt \tag{5.28}$$

for any $\zeta \in F$, here, we use the relations (5.14)–(5.15) and (5.3)–(5.5) for η_λ^* and p_λ, respectively. Since $\zeta \in F$ is arbitrary, we infer from (5.28) that the optimal control f_λ^* satisfies the variational identity (5.6). □

References

[1] H. W. Alt and S. Luckhaus, *Quasilinear elliptic-parabolic differential equations*, Math. Z., **183**(1983), 311–341.

[2] H. W. Alt, S. Luckhaus and A. Visintin, *On nonstationary flow through porous media,* Ann. Mat. Pura. Appl., **136**(1984), 303–316.

[3] J. Carrillo and P. Wittbold, *Uniqueness of renormalized solutions of degenerate elliptic-parabolic problems*, J. Differential Equations, **156**(1999), 93–121.

[4] M. Chipot and A. Lyaghfouri, *The dam problem for non-linear Darcy's laws and non-linear leaky boundary conditions*, Math. Methods Appl. Sci., **20**(1997), 1045–1068.

[5] A. Damlamian, *Non Linear Evalution Equations with Variable Norms,* Thesis, Harvard University, 1974.

[6] K.-H. Hoffmann, M. Kubo and N. Yamazaki, *Optimal control problems for elliptic-parabolic variational inequalities with time-dependent constraints,* Numerical Functional Analysis and Optimization, **27**(2006), 329–356.

[7] A. V. Ivanov and J.-F. Rodrigues, *Weak solutions to the obstacle problem for quasilinear elliptic-parabolic equations,* St. Petersburg Math. J., **11**(2000), 457–484.

[8] A. Kadoya and N. Kenmochi, *Optimal Shape Design in Elliptic-Parabolic Equations*, Bull. Fac. Education, Chiba Univ., **41**(1993), 1–20.

[9] N. Kenmochi, *Solvability of nonlinear evolution equations with time-dependent constraints and applications*, Bull. Fac. Education, Chiba Univ., **30**(1981), 1–87.

[10] N. Kenmochi and M. Kubo, *Periodic stability of flow in partially saturated porous media,* Free Boundary Problems, Int. Series Numer. Math., Vol. 95, Birkhäuser, Basel, 1990, pp. 127–152.

[11] N. Kenmochi and I. Pawłow, *Parabolic-elliptic free boundary problems with time-dependent obstacles,* Japan J. Appl. Math., **5**(1988), 87–121.

[12] M. Kubo and N. Yamazaki, *Elliptic-parabolic variational inequalities with time-dependent constraints,* Discrete Contin. Dyn. Syst., **19**(2007), 335–359.

[13] F. Otto, *L^1-contraction and uniqueness for quasilinear elliptic-parabolic equations,* J. Differential Equations, **131**(1996), 20–38.

[14] I. Pawłow, *Analysis and Control of Evolution Multi-Phase Problems with Free Boundaries,* Prace habilitacyjne, Polska Akademia Nauk, Instytut Badań Systemowych, 1987.

[15] N. Yamazaki, *Doubly nonlinear evolution equation associated with elliptic-parabolic free boundary problems*, Discrete Contin. Dyn. Syst., 2005, suppl., 920–929.

